AF339218

ON THIN ICE

a synthesis of the Canadian Arctic Shelf Exchange Study (CASES)

Chief Editors: Louis Fortier, David Barber and Josée Michaud

Aboriginal Issues Press

On Thin Ice: a synthesis of the Canadian Arctic Shelf Exchange Study (CASES) is a refereed publication. Information and perspectives presented in this book are the sole opinions of the authors and not those of the University of Manitoba or the Aboriginal Issues Press, its employees, editors and volunteers. All profits from the sale of this book are used to support the Aboriginal Issues Press Scholarship at the University of Manitoba.

© 2008 Aboriginal Issues Press

Clayton H. Riddell Faculty of Environment, Earth, and Resources
440 Wallace Building
University of Manitoba
Winnipeg, Manitoba, R3T 2N2

Phone: (204) 474-7252   Fax: (204) 275-3147

Email: Aboriginal_Issues_Press@umanitoba.ca

Cover and text design and layout by Karen Armstrong Graphic Design, Winnipeg.

PRINTED IN CANADA by Friesen Printing, Altona, MB

COVER PHOTOGRAPH by Francoise Proust. As the sea freezes, free-floating circular pieces of ice or pancake ice (background photo) forms when slush ice and ocean spray accumulate at the edge of new floes bumping into each other under wave action. INSET PHOTOGRAPHS (l to r): Keith Levesque, Martin Fortier, James Ford, Martin Fortier, Corinne Pomerleau.

BACK COVER PHOTOGRAPH by Thomas Juul-Pedersen. INSET PHOTOGRAPHS (L to R): Alfonso Mucci, Gérald Darnis, David Barber,Christina Blouw, Martin Fortier.

Library and Archives Canada Cataloguing in Publication

On thin ice: a synthesis of the Canadian Arctic Shelf Exchange Study
(CASES) / chief editors: Louis Fortier, David Barber and Josée Michaud.
Includes bibliographical references.
ISBN 978-0-9738342-6-0

1. Climatic changes--Beaufort Sea. 2. Ecology--Beaufort Sea. 3. Sea
ice--Beaufort Sea. 4. Biogeochemistry--Beaufort Sea. 5. Canada,
Northern--Environmental conditions. 6. Beaufort Sea--Environmental
conditions. 7. Climatic changes--Arctic Ocean. 8. Canada, Northern--Climate.
I. Fortier, Louis, 1953- II. Barber, David G.,
1960- III. Michaud, Josée, 1969-

QC985.5.A7O5   2008   551.46'1327   C2008-903588-7

Table of Contents

Acknowledgements

The CASES Research Network and the operation of the CCGS *Amundsen, Pierre Radisson* and *Sir Wilfrid Laurier* were funded by the Natural Science and Engineering Research Council of Canada (NSERC) and the Canada Foundation for Innovation, with additional financial support from Fisheries and Oceans Canada (Canadian Coast Guard), the US National Science Foundation, the Japanese Ministry of Education, Culture, Sports, Science and Technology and Devon Petroleum. We gratefully acknowledge the in kind support of our numerous partners in Canada and abroad, including, among many others, the Canadian Foundation for Climate and Atmospheric Sciences, the Canadian Ice Service, the Canadian Museum of Nature, Fisheries and Oceans Canada, the Geological Survey of Canada, Indian and Northern Affairs Canada, the Institut de Ciències del Mar-CSIC of Barcelona, the Institute of Oceanology at the Polish Academy of Sciences, the Inuvialuit Game Council, the Inuvialuit Joint Fisheries Management Committee, the Inuvialuit Joint Secretariat, the Japan Marine Science and Technology Centre, the National Institute of Polar Research (Japan), Natural Resources Canada, the US National Aeronautics and Space Agency (NASA), the US National Oceanic and Atmospheric Administration (NOAA), and the US Naval Research Laboratory.

Special thanks to Ms. Leah Braithwaite, former NSERC Officer, for decisive support at critical times in the unfolding of CASES, and to Dr. Martin Fortier and M. Marc Ringuette for their indefatigable efforts in coordinating the Network. Dr. Denis St-Onge of the Geological Survey of Canada aptly chaired the Board of Directors. Ms. Anne Alper, Ms. Barbara Muir and M. Robert Therrien, NSERC Officers, supervised the activities of the Network at different stages of its ontogeny. We gratefully acknowledge the support of Inuvialuit Communities and organizations, with special thanks to Dr. Nellie Cournoyea, Chair of the Inuvialuit Regional Corporation, M. Frank Pokiak, Chair of the Inuvialuit Game Council, Dr. Norm Snow, Executive Director of the Joint Secretariat, M. Larry Carpenter, Chair of the Wildlife Management Advisory Council, and M. Duane Smith, President of the Inuit Circumpolar Council - Canada. Thanks to the numerous Inuit Wildlife Observers who contributed to the science effort and the security of the scientists at sea and on the ice, as well as in making CASES a wonderful inter-cultural experience for Canadian and international participants. The ambitious field program of CASES would not have been possible without the dedication and superb seamanship of the commanding officers, officers, and crew members of the Canadian Coast Guard.

In memory of Grant Ingram, for his contributions to Canadian research in the Arctic.

Foreword

The Canadian Arctic Shelf Exchange Study (CASES) was arguably the first truly inter- disciplinary integrated research program in the Canadian Arctic Ocean. Vilhjalmur Stefansson's Canadian Arctic Expedition, 1913-1918, was largely concerned with describing Arctic biology, ethnology, archaeology, hydrography, geology, and geography. Since that time there have been international ice projects and several research initiatives in the Canadian Beaufort Sea which were mainly driven by the desire to explore for offshore hydrocarbon resources. These latter projects, although intense and often multi disciplinary, did not benefit from a high degree of coordination. The considerable variability in sea-ice thickness, extent and dynamics has been recognised by Inuit and southern scientists alike, but changes in the arctic sea-ice regime observed in recent years, points out the need for assessments of the effects of such variability and human adaptation to this.

There is a need to further our understanding of marine processes in general. This includes the interaction between freshwater and salt-water, air/sea/ice interaction, marine ecosystems and geochemical fluxes. This need is especially acute for the relatively shallow, coastal (continental shelf) areas where there is considerable sea-ice variability. This seasonal ice-zone is highly significant to the Inuvialuit, and indeed all Inuit peoples, in terms of traditional use and cultural pursuits. It is also a zone which may be subjected to considerable development activity.

From 2001 until 2006, CASES has been addressing these needs by way of a large scientific network with international involvement and attention. The ten chapters of this present synopsis cover the physical environment of the area, the biological production from viruses to whales, geochemical cycling and the archiving of scientific results. The Inuvialuit have been partners in the project since its inception and the program as a whole, including the operations of CCGS *Amundsen*, has provided an extraordinary opportunity for Inuvialuit youth to become involved in a broad spectrum of marine science of the highest level.

Since the settlement of their Land Claim in 1984, the Inuvialuit have developed a world-class integrated resource management process via the Inuvialuit Game Council (IGC) and the Co-Management Bodies (CMBs) established pursuant to the Inuvialuit Final Agreement. The CASES results will significantly augment the database which the IGC and the CMBs utilise to arrive at their management decisions.

It is gratifying to know that networking and collaboration which CASES initiated is continuing through ArcticNet and the International Polar Year's Circumpolar Flaw Lead study. We hope that this spirit of meaningful cooperation will also continue into all future arctic research programs.

Norman B. Snow, PhD.
Executive Director, Joint Secretariat
Inuvialuit Final Agreement
Inuvik, NT.
May, 2008

LEFT: Polar bear on ice floe.
Photo: Marc Tawil/ArcticNet.

CHAPTER 1

An Introduction to the Canadian Arctic Shelf Exchange Study

Louis Fortier[1]*, David G. Barber[2]

1.1 Rationale for the study

The planning of the Canadian Arctic Shelf Exchange Study (CASES) in 2000 was motivated by the notion that the warming of Earth's climate anticipated by climatologists would begin and be most intense at arctic latitudes. This notion was not new. As early as 1895, the famous scientist Svende Arrhenius (Nobel prize for chemistry in 1903) had predicted that a 2.5 fold increase in carbon dioxide concentrations would warm the lower atmosphere of arctic regions by 8 to 9 °C. By 2000, sophisticated General Circulation Models (GCM) confirmed Arrhenius' alarming prophecies by anticipating a 3.5°C rise in mean atmospheric temperature north of the Arctic Circle for a doubling of atmospheric CO_2 by 2070 (e.g. Shindell et al, 1999; Flato et al, 2000). Actually, the GCMs were telling us that the Arctic should already have started to warm up significantly in response to the on-going massive emissions of greenhouse gases by human industry. By 2005, the Arctic Climate Impact Assessment was confirming the anticipated symptoms of an arctic amplification of climate warming: glaciers and ice shelves were regressing, the vegetation was changing, precipitation and river runoff were increasing, the melting of the Greenland Inlandsis

was quickening, and salinity in the deep thermohaline circulation was decreasing (ACIA, 2005). Since then, the accelerated warming of the lower atmosphere and the spectacular regression of the sea ice cover in 2007 suggest that the Arctic is warming much faster than expected, and that Arrhenius after all may have been closer to reality than the conservative predictions of the GCMs on which official bodies such as the ACIA or the Intergovernmental Panel on Climate Change (IPCC) have based their analyses to date.

The on-going reduction of the Arctic Ocean sea ice cover, observed since the 1950's, was expected to severely disrupt the marine ecosystem on which the Inuit depended to maintain their economy, traditional way of life and culture. In 2000, both simulations and observations were indicating that by 2080, and perhaps earlier, the Arctic Ocean could be free of ice during summer (Comiso 2002; Johannessen et al., 2004; Stroeve et al., 2005). It was hypothesized that, by increasing photosynthetic fixation of atmospheric carbon, the shrinking of the ice cover could boost biogeochemical fluxes on Arctic shelves, therefore accelerating the export of carbon to the pelagic and benthic food webs, and to the deep basins where it can be sequestered. However,

LEFT: Recovering instrument from the ice. Photo: David Barber.

[1] *Québec-Océan, Département de Biologie, Université Laval, Québec, QC, G1V 0A6, Canada*

[2] *Centre for Earth Observation Science, Clayton H. Riddell Faculty of Environment, Earth and Resources, University of Manitoba, Winnipeg, MB, R3T 2N2, Canada*

* Corresponding author: Louis.Fortier@bio.ulaval.ca

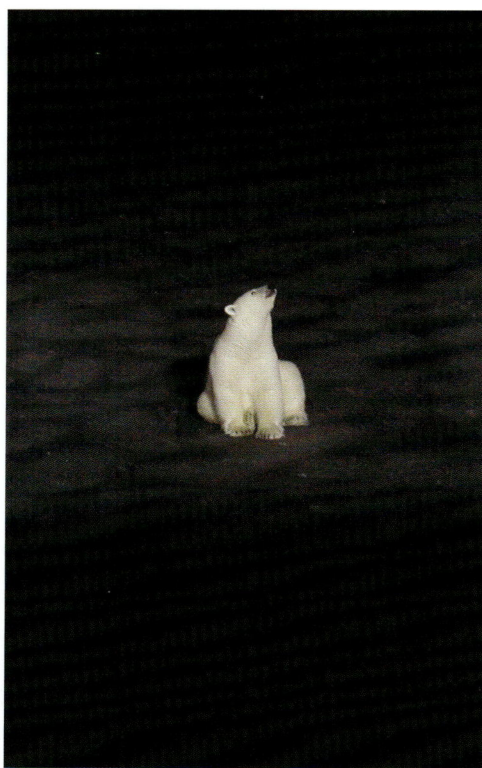

Polar bear in the dark. Photo: David Barber.

assessing the actual role of a seasonally ice-free Arctic Ocean as a future sink or source of atmospheric CO_2 required a significant improvement of our understanding of the processes and feedbacks linking freshwater and sea ice, sea ice and climate, and sea-ice, biological productivity and biogeochemical cycles on Arctic shelves (Fortier and Cochran, 2008).

Accordingly, the CASES study was developed as an international effort under Canadian leadership to study concurrently the sea ice regime, the biogeochemistry and the ecology of the Mackenzie Shelf in the Canadian Arctic Ocean. The central objective of CASES was to understand and model the complex response of the Mackenzie Shelf ecosystem to atmospheric, oceanic and continental forcing of sea ice cover variability. Given the influence of the Mackenzie River and the intense regression of sea-ice in the area in recent years, the Mackenzie Shelf represented the only North-American analogue of the immense Siberian Shelves influenced by large rivers and where sea-ice losses had been particularly severe (ACIA, 2005; Nghiem et al., 2006; Bareiss and Gorgen, 2005). The proposed scientific program of CASES was underpinned by the simple central hypothesis that the atmospheric, oceanic and hydrological forcing of sea ice variability dictates the nature and magnitude of biogeochemical carbon fluxes on and at the edge of the Mackenzie Shelf. It was subdivided in the following nine multidisciplinary subprojects:

- Atmospheric and sea ice forcing of coastal circulation
- Ice-atmosphere interactions and biological linkages
- Light, nutrients, primary and export production in ice-free waters
- Microbial communities and heterotrophy
- Pelagic food web: structure, function and contaminants
- Organic and inorganic fluxes
- Benthic processes and carbon cycling
- Millenial-decadal variability in sea ice and carbon fluxes
- Coupled bio-physical models of the carbon flows on the Canadian Arctic Shelf

Given its scope, CASES summoned an important fraction of the Canadian expertise in arctic oceanography and atmospheric sciences. This expertise was widely dispersed in Universities and Federal research institutes across Canada. The Canadian team was complemented by a host of international collaborators from the USA, Japan, Norway, Spain, Poland, the United Kingdom, Denmark and Russia. Following the national and international calls for interest, over 200 principal investigators, students and research personnel joined in the project.

Obviously, the multidisciplinary and international dimensions of this unique endeavour would require intense networking and the natural funding avenue for the Canadian component was the Research Network program of the Natural Sciences and Engineering Research Council of Canada (NSERC).

1.2 Study area and field program

The Mackenzie Shelf is typically covered with ice from October until May to early August (Barber and Hanesiak, 2004). The input of freshwater from the Mackenzie River on the shelf is present throughout the year but approximately 70% of the total annual discharge is delivered between May and September. In late summer, the nearshore zone of the ice-free shelf is dominated by the Mackenzie River plume (Macdonald et al., 1995). Usually, ice starts forming in October in shallow areas and by late fall, the freshwater plume extends immediately beneath the growing landfast ice cover. The landfast ice and the central pack ice are separated by an ice-free channel that forms a recurrent flaw lead polynya (Fig. 1.1). Throughout winter, floe rafting at the edge of the landfast ice builds the "stamukhi", a thick ice ridge parallel to the coast that forms in waters 15 to 50 m deep. In spring, the containment of the river plume by the stamukhi forms the seasonal Lake Mackenzie. Beyond the stamukhi, the flaw polynya that stretches along the entire circum-Arctic Shelf widens in summer to form the Cape Bathurst polynya in the Amundsen Gulf.

The field program proposed by CASES was designed to contrast the annual cycle of the arctic marine ecosystem in three regions: (1) the Cape Bathurst polynya; (2) the Mackenzie Shelf and (3) the edge and slope of the Shelf (Fig. 1.1). Field operations were centred on the expeditions, from 2002 to 2004, of the Canadian Coast Guard icebreakers *Radisson* and *Amundsen*, from Quebec City through the NW Passage and the icebreaker *Laurier*, from Victoria through the Bering Strait. These main expeditions were to be complemented by several missions on other vessels. The *Nahidik*, a shallow draft vessel operated by DFO, was used to access the Mackenzie Delta and the inshore waters of

the Mackenzie Shelf, either as a direct CASES operation or as part of the satellite program ARDEX (Arctic River Delta Experiment). Some CASES investigators also joined the ice-reinforced *Mirai*, a Japanese ship deployed in the offshore region of the study area from 20 September to 10 October 2002. Finally, other CASES scientist participated in the annual expedition of the *Kapitan Dranytsin*, a Polar class Russian icebreaker, to the Laptev Sea as part of an exchange program between CASES and NABOS (Nansen-Amundsen Basin Observatory Study).

In September 2002, a first 35-d mission on board the *Radisson* enabled the CASES researchers to complete

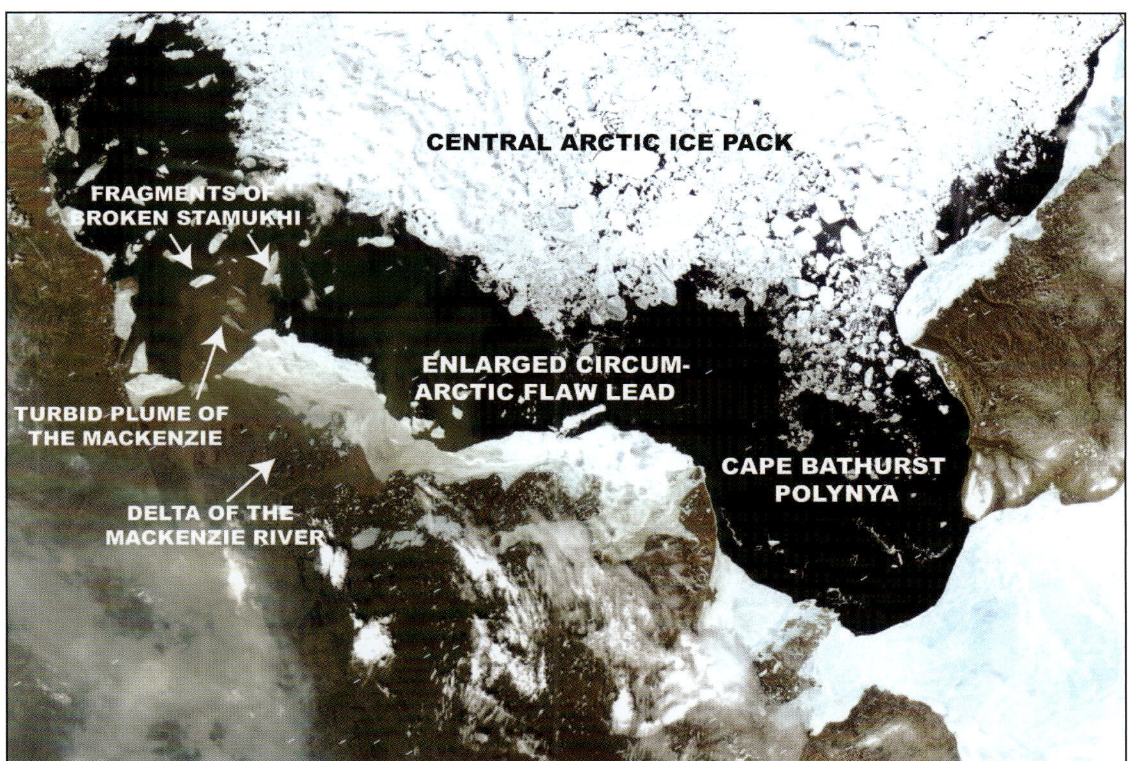

Figure 1.1

MODIS satellite image of the study area in June 2004, showing the release of the turbid waters of the Mackenzie after the break-up of the stamukhi on the Mackenzie Shelf.

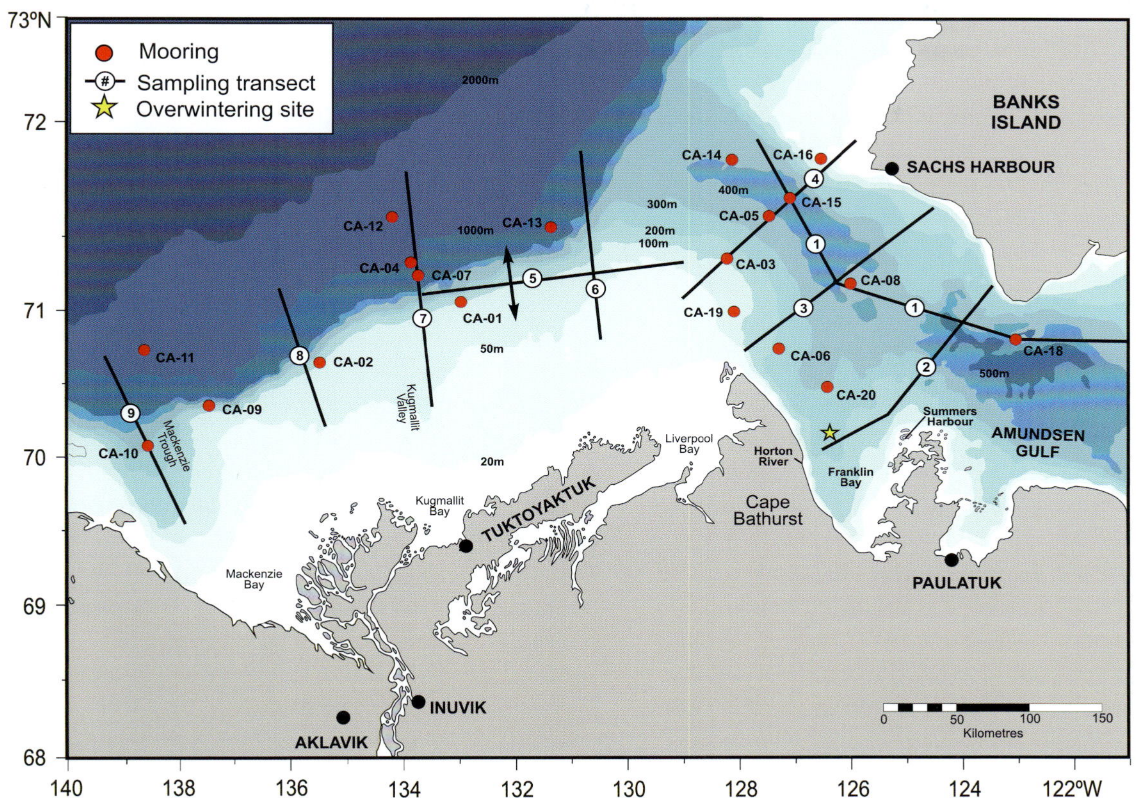

LEFT PAGE: **Figure 1.2**

The Canadian Coast Guard icebreaker Amundsen, a state-of-the-art 98m-long research vessel. The Amundsen was deployed in the CASES study area from September 2003 to August 2004. Photo: Alexandre Forest.

Figure 1.3

Bathymetric chart of the study area with position of the moorings deployed from the CCGS Radisson in September 2002 and the sampling transects covered by the CCGS Amundsen in the fall of 2003 and the spring/summer of 2004. The over-wintering position of the Amundsen in Franklin Bay is indicated by the yellow star.

a preliminary spatial survey of the oceanography, biology and biogeochemistry of the area. This was complemented by the deployment for one year of 8 oceanographic moorings during the September 2002 expedition of the Laurier to the same region. The main thrust of the planned field program was the one-year expedition of the *Amundsen* (Fig. 1.2) to the study area, starting in September 2003 (Fig. 1.3). This arctic mission of unprecedented scope comprised three major parts: (1) a fall survey covering the entire region from September to December 2003, including the recovery of the 8 moorings deployed during in 2002 and the deployment of 17 new mooring arrays; (2) the over-wintering of the ship in Franklin Bay (70° 2.73'N,

126° 18.07'W) for the monitoring of the winter evolution of the ecosystem; and (3) the spring/summer spatial survey of the region to monitor the break-up of the stamukhi, the opening of the Cape Bathurst polynya and the development of the summer ecosystem, including the recovery and redeployment in August 2004 of 7 of the 17 oceanographic moorings. Again, the annual mission of the Laurier to the study area was the opportunity to recover the remaining 10 moorings and to extend the sampling of the ecosystem into September 2004.

Overall, the CASES field program logged 543 days at sea, 377 of these days being directly chartered by

Deployment of plankton nets.
Photo: Martin Fortier/ArcticNet.

CASES and 166 being contributed by national and international partners. This corresponded to a total of 14 544 day-scientists at sea, which makes the CASES field program the largest and most comprehensive international effort thus far to decipher the functioning of the Arctic Ocean shelf ecosystem and understand the biogeochemical and ecological impacts of the present decline in arctic sea ice cover.

1.3 An overview of scientific research activities

This synthesis of the CASES research activities reviews and highlights the scientific conclusions that emerged from the extensive data sets collected and analyzed by each subproject during the program. Each chapter introduces in a general way the rationale of the subproject; provides an overview of the key findings along with scientific references for further reading; details the implication of the work; and finally makes recommendations on future research and policies. We hope this synthesis will be useful to all stakeholders, including government agencies, northern industries and Inuit communities.

Chapter 2 presents an overview of the seasonal circulation on the Beaufort Shelf. Oceanographic conditions on the Shelf are influenced by a complex range of physical processes, annual and inter-annual variations in sea ice, wind-driven circulation, upwelling events, tidal forcing and freshwater input from the Mackenzie River. During CASES, particular efforts were made to understand the shelf circulation with a focus on conditions at the outer shelf and the shelf break. This involved the annual deployment of instrumented moorings from 2002 (8 moorings) to 2004 (17 moorings), that measured ocean currents and water mass properties. On the Shelf, the general circulation observed

above 100 m was strongly seasonal and influenced by the combined effects of ice cover and discharge from the Mackenzie River. At 200m however, the wintertime circulation was generally unaffected by sea ice conditions. Schematic representations of different wind-stress influence on the shelf circulation summarize results from this chapter.

The complex exchanges at the Ocean-Sea Ice-Atmosphere interface in the CASES study area are described in Chapter 3. Results presented cover annual changes in mass, gas, energy fluxes across this interface as well as key findings of how the ocean and atmosphere affect dynamic and thermodynamic process of snow covered sea ice. Results demonstrate a declining trend in sea-ice concentration from 1979 to 2000 and that the duration of the ice break-up within the Amundsen Gulf is widely variable in comparison to the freeze-up. One of the most striking observations is the increase since the 1990's in the frequency of reversal of the normally anticyclonic Beaufort Gyre to a cyclonic mode and the spreading over the summer and fall season of this cyclonic mode. By dislocating the ice pack, cyclones accelerates the penetration of solar heat in the surface layer, which explains the accelerated regional decline in sea ice cover. The CASES study also allowed the collection of one of the most complete data sets on snow change in grain size and shape due to temperature variation within the snow as well as comprehensive measurements of the surface heat and radiation budget during winter. The observed variations in ice and snow cover will have profound impact on the environment, our policies and the development of monitoring systems.

Chapter 4 investigates the dynamics of light, nutrients and microalgae in ice-covered and ice-free waters of the Beaufort Sea throughout the year. In autumn and winter (2003-2004) the renewal of nutrients in the upper euphotic zone was small due to modest mixing by wind, convection and brine rejection. Approximately a third of the nitrate reservoir available to phytoplankton in spring 2004 appears to have been supplied by nitrification above the resilient halocline. The response of phytoplankton to the declining ice cover may depend more on the alteration of nutrient load by atmospheric forcing and freshwater input than on changes in light availability. Strikingly, deep sediments on and adjacent to the Mackenzie shelf may be a source of phosphorus to surrounding waters, rather than a sink of nitrogen. This study has shown that exopolymeric substances produced by the algae contribute directly to the carbon pool in first-year sea ice, and influence carbon cycling within the sea ice and the fate of sea-ice carbon once released to the water column. The phytoplankton species succession generally observed in the high-latitude waters, from a high diatom contribution in summer towards flagellate predominance in fall, also prevailed in the sinking material throughout the Beaufort Sea.

Microbial communities and carbon fluxes is the main topic of Chapter 5. The CASES Network and its satellite program ARDEX provided a unique opportunity to examine the microbial community structure as well as a broad suite of related processes and environmental variables in the coastal Arctic Ocean and adjacent waters. Results suggest that in a warmer climate, anticipated changes in the coloured fraction of the dissolved organic carbon and particulate organic carbon loading from eroding permafrost could lead to a reduced primary production in the freshwater and marine ecosystems. A rich diversity of bacteria, archaea, protists and viruses was observed throughout the year, even during the dark winter, though with noticeable seasonal differences in assemblage, abundance and activity. Important differences in microbial community structure existed among the riverine, the coastal and the stamukhi Lake Mackenzie habitats. One of the most exciting discoveries was that of a new group of photosynthetic picoeukaryotes in the CASES area (and elsewhere). Overall, small cells dominate the ecosystem carbon fluxes and several features make this system vulnerable to climate change.

Chapter 6 investigates the pelagic food web structure and function, as well as contaminants in relation to climate change and ice cover dynamics. The study region can be subdivided into three areas characterized by three different food webs: (1) the Mackenzie River outflow, characterized predominantly by river-based biological production; (2) the Beaufort Sea Shelf and Cape Bathurst, characterized by both river-based and marine production; and (3) the Amundsen Gulf, a principally marine-based food web. The food web proved to be very active under the ice during the dark winter and spring. The colossal abundances of Arctic cod detected by hydro-acoustics in the deep waters of Franklin Bay in 2004 seem to account for the 'missing cod' on the Beaufort Shelf required to support the yield of its numerous predators. It was also found that this fish hatch over a prolonged period in the Beaufort Sea and its survival and growth are related to ice dynamics. Mercury levels in Arctic Cod and near bottom invertebrates were higher than those of estuarine fish and this may differentially impact the health of Belugas that consume them. Theses findings must be taken into consideration in future management of the Mackenzie Shelf ecosystems.

Chapter 7 presents data on organic and inorganic fluxes in the Beaufort Sea and the Mackenzie shelf. These include CO_2 fluxes at the air-sea interface, sediment resuspension, water and sediment discharge and transport from the river, off shelf transport of particulate matter, and downward flux of organic matter. The southeastern Beaufort Sea appears to be an area of net sink for atmospheric CO_2 during the ice-free periods of the CASES study; thus this area could play a role in regulating the rising atmospheric CO_2 levels. Modelling of the Mackenzie River indicates that both water and sediment discharge may increase with increasing warming of the area. Vertical fluxes and resuspension of particulate mater during fall and winter (2003-2004) highlight the importance of resuspension and advection processes. As well, this study suggests that Franklin Bay could be a repository of particulate organic matter and that already buried particulate organic carbon can be transferred from the continental shelf to the deep sea, though its final fate remains unknown. Finally, the diagenesis of sediments in Amundsen Gulf is not driven by sedimentation of terrestrial organic matter but by the settling of marine organic matter locally produced by primary producers.

Chapter 8 focused on the calibration and the use of proxies for revealing the historical variability in paleoclimates in the CASES study area. Microfossils such as dinoflagellates, pollen grain, foraminifera, tintinnids, and thecamoebians were used has indicators of paleo-sea ice conditions, changes in physio-chemical conditions and continental climate. Comparisons were also made with previous collections done in the area in the 1970's. This work revealed that Atlantic deep water seems restricted to a narrow time interval in early postglacial times corresponding to periods of calving icebergs. Variable sea-ice cover, fresh water input and early glacial movements and more recent changes were also identified from the sediment cores. Data collected during CASES will allow establishing the glacial history of the Beaufort Shelf and to compare with other history periods, including the Holocene that this study revealed as a more climatically active time that initially thought.

Benthic communities of the Arctic shelf, presented in Chapter 9, play an important role in the cycling of carbon and the recycling of nutriment but are not always well known. CASES and the Joint Western, Arctic Climate Studies (JWACS) provided the opportunity to link benthic macrofaunal community compositions on the Beaufort Shelf to regional differences in geography, oceanography and ice processes. Differences in community compositions were observed in the Cape Bathurst polynya, the Mackenzie Canyon and the Beaufort Shelf and appeared to be indicative of the local food supply and nutrient fluxes. Spatial and seasonal variations were also observed in the benthic community respiration and production; for example, sediment-community oxygen demand was highest in late spring. The inshore fauna would be the first affected by climate warming of the Beaufort Shelf and the changes in the Mackenzie River inflow, ice dynamics, nutrients supply, sediment transport and coastal erosion and could lead to the replacement of the local community by other species better adapted to the new conditions. It is anticipated that species from Pacific origin would replace the current Arctic species.

Finally, Chapter 10 presents the data management and archiving strategy adopted by the Network to ensure that CASES data are preserved and accessible to all. Building on the success of data management within long-term environmental programs, the CASES data management strategy has an implementation plan that relies on existing infrastructures and data services. Metadata are accessible though the Polar Data Catalogue portal. The metadatabase has been structured according to National and International standard protocols. A basic search facility has been programmed that allow anyone to search the metadatabase using standardized keywords. The strategy also includes a permanent archiving component where each CASES investigator will provide datasets in a coherent format understandable to future users; these datasets should be available within three years after the end of the CASES program. Future developments of this database and hence of the CASES legacy, include a new support for geospatial tools and close collaborations with northern communities, the northern science institutes, NSERC, DFO and IPY.

1.4 Rebuilding Canada's leadership in arctic research: the legacy of CASES

In addition to their scientific impact, successful endeavours such as the Canadian Arctic Shelf Exchange Study often evolve into new national and international collaborations and projects. In addition to its superb scientific program, one of the most important contributions of CASES is no doubt to have prompted the transformation of the *CCG Amundsen* into a state-of-the-art polar research vessel thanks to a major grant from the Canada Foundation for Innovation (Fig. 1.2). The Amundsen has been the fulcrum of the success of the Network of Centres of Excellence ArcticNet and is presently supporting the most ambitious projects at the core of the Canadian International Polar Year.

Some central elements of the CASES field program are being continued and expanded as part of the scientific program of ArcticNet (http://www.arcticnet-ulaval.ca/). In particular, ArcticNet has been re-deploying annually four of the original CASES mooring arrays in the study area, thus setting up a long-term oceanographic observatory to monitor the evolution of the southeastern Beaufort Sea in response to climate change. The ArcticNet annual mission of the Amundsen to the region is the opportunity to measure key indicators of the state of the ecosystem, in the continuity of the three-year interannual comparison (2002-2004) initiated during CASES. By 2007, six long-term oceanographic observatories of this kind were operated jointly by ArcticNet and the Nansen-Amundsen Basin Observatory System (NABOS: http://nabos.iarc.uaf.edu/) in the Beaufort Sea, the North Water, Lancaster Sound, Hudson Bay, the Laptev Sea and the East-Siberian Sea.

Most importantly, the trans-disciplinary networking and the engagement of stakeholders developed during CASES have become the foundation of ArcticNet which brings together scientists and managers in the natural, human health and social sciences with their partners in Inuit organizations, northern communities, federal and provincial agencies and the private sector. The synergy among stakeholders created by ArcticNet has literally transformed the way arctic research is done in the Canadian Arctic. Inuit experts contribute at all levels of the research process, from field work to the Board of ArcticNet. The Network provides unprecedented access to arctic infrastructures (land and marine) to its researchers and their international collaborators. The ArcticNet science program is conducted in true and equal partnership with Federal agencies, and increasingly with private-sector partners. Numbers of young Canadians, including Northerners, are trained in a unique trans-sectorial context and the recruitment of arctic specialists in universities and departments has resumed.

The scientific achievements of CASES have also laid the ground for further multidisciplinary research on the Beaufort Sea ecosystem as part of the International Polar Year. The Circumpolar Flaw Lead System Study (CFL: http://www.ipy-cfl.ca/) reiterates some key elements of the CASES scientific program on sea-ice and carbon fluxes, focusing this time on the segment of the

Deploying a weather balloon.
Photo: Martin Fortier/ArcticNet.

circum-arctic flaw lead polynya that stretches over the Mackenzie Shelf and west of Banks Island (Fig.1.1). Building on the experience acquired during CASES, teams from Canada, Japan, Russia, Poland, Denmark, Norway, Germany, Italy, USA, China, Sweden, Belgium, Spain, France and the UK gathered again on the Amundsen to monitor the arctic marine ecosystem from October 2007 to August 2008. Contrary to CASES, during CFL the ship remained mobile throughout the year. This bold plan will provided scientists with a first-time opportunity to study the winter and spring ecosystem of the circum-arctic flaw lead.

Members of the CASES research network were also instrumental in development of a major scientific output in the form of a textbook on Polynya research at both poles of our planet (Massom and Barber, 2007). CASES research figured prominently in this stat-of-the-art compilation of how polynyas function both in the Arctic and in the Antarctic. The project was inspired by the International Arctic Polynya Program (IAPP) which is a committee of the Arctic Ocean Sciences Board (J. Deming Chair).

Finally, under the leadership of the Board of Directors, CASES made the archiving of metadata and data a central objective of the Network. With major input from our partners in the Departments of Fisheries and Oceans Canada and Environment Canada, CASES data management committee defined the CASES Data Management and Archive Strategy that ensures that the CASES data are: (1) integrated into comprehensive databases; (2) available to all participants of the CASES Research Network and under conditions that respect the rights of the data originators; (3) archived and ultimately accessible to the broader science community and the public in the long term; and, finally, (4) available to researchers in other national and international polar research programs such as ArcticNet and IPY. A first step in implementing the strategy was to associate with the existing Canadian Cryospheric Information Network (CCIN) which provides a portal for the archiving of Arctic data. Soon the successfully implemented CASES-CCIN metadata base was expanded to include metadata from ArcticNet. Then the Canadian IPY program decided to adopt the CASES-ArcticNet-CCIN (www.polardata.ca) database as a platform to archive the immense data sets that it presently accumulates. Hence, in a remarkable turn of events, the original CASES data management strategy could soon seed the development of one of the world's largest and most comprehensive international database for the Arctic.

In view of all these outcomes, it can be safely concluded that the Canadian Arctic Shelf Exchange Study played a defining role in enabling Canada to rebuild its leadership in the international study of the Arctic. Results from the CASES Network will provide data, information and knowledge, required to develop adaptation stratagies for the marine system in the western Canadian High Arctic. Our intention in preparing this synthesis report was to provide a digestible summary of the research results arising from the CASES network, targeted at community, local, regional and national governing bodies who hold the responsibility for management. Our scientific expertise is also available for consultation and we encourage interested parties to contact the corresponding author.

Acknowledgements

The Research Network CASES and the operation of the CCGS *Amundsen* and *Laurier* were funded by the Natural Science and Engineering Research Council of Canada (NSERC) and the Canada Foundation for Innovation, with additional support from the Department of Fisheries and Oceans Canada (Canadian Coast Guard). Special thanks to Leah Braithwaite, former NSERC Officer, for decisive support at critical times in the unfolding of CASES. International participants in CASES were funded by agencies in their respective countries, including the US National Science Foundation, the Japanese Ministry of Education, Culture, Sports, Science and Technology and the . Thanks to Dr. J. Deming, Chair of the International Arctic Polynya Program (IAPP) for coordination of the CASES project within polynya studies globally. The ambitious field program of CASES would not have been possible without the dedication and superb seamanship of the officers and crew of the Canadian Coast Guard. Thanks to Martin Fortier, Marc Ringuette and Josée Michaud for their indefatigable efforts in coordinating the Network.

References:

ACIA, 2005. Arctic Climate Impact Assessment. Cambridge University Press. 1042 p. URL: www.acia.uaf.edu.

Barber, D.G., L. Fortier, and M. Byers, 2006. The Incredible Shrinking Sea Ice. Policy Options 27(1): 66-71

Barber, D.G., and J. Hanesiak, 2004. Meteorological forcing of sea ice concentrations in the Southern Beaufort Sea over the period 1978 to 2001. J. Geophys. Res. 109, C06014, doi:10.1029/2003JC002027.

Bareiss, J., and K. Gorgen, 2005. Spatial and temporal variability of sea ice in the Laptev Sea: Analyses and review of satellite passive-microwave data and model results, 1979 to 2002. Global Planet Change 48: 28-54.

Comiso, J.C., 2002. A rapidly declining perennial sea ice cover in the Arctic. Geophys. Res. Lett. 29:1956.

Flato, G. M., G. J. Boer, and W. G. Lee , 2000. The Canadian Centre for Climate Modelling and Analysis global coupled model and its climate. Clim. Dyn. 16: 451-467.

Fortier, L., and J.K. Cochran, 2008 Introduction to special section on Annual Cycles on the Arctic Ocean Shelf, J. Geophys. Res., 113, C03S00, doi:10.1029/2007JC004457.

Johannessen, O.M., L. Bengtsson, M.W. Miles, S.I. Kuzmina, V.A. Semenov, G.V. Alekseev, A.P. Nagurnyi, V.F. Zakharov, L. Bobylev, L.H. Pettersson, K. Hasselmann, and H.P. Cattle, 2004. Arctic climate change - observed and modeled temperature and sea ice variability. Tellus, Series A: Dynamic Meteorology and Oceanography 56: 328-341.

Macdonald, R.W., D.W. Paton, E.C. Carmack, and A. Omstedt, 1995. The freshwater budget and under-ice spreading of Mackenzie River water in the Canadian Beaufort Sea based on salinity and 18O/16O measurements in water and ice. J. Geophys. Res. 100: 895–919.

Nghiem, S.V., Y. Chao, G. Neumann, P. Li, D.K. Perovich, T. Street, and P. Clemente-Colon, 2006. Depletion of perennial sea ice in the East Arctic Ocean. Geophys. Res. Lett. 33,L17501, doi:10.1029/2006GL027198.

Shindell, D.T., R.L. Miller, and G.A. Schmidt,1999. Simulation of recent northern winter climate trends by greenhouse-gas forcing. Nature 399:452-455.

Smith W.O., and D. G. Barber (eds), 2007. Polynyas: Windows to the World. Elsevier Oceanogr. Ser. 74, 458p. ISBN:978-0-444-52952-7.

Stroeve, J. C., M. C. Serreze, F. Fetterer, T. Arbetter, W. Meier, J. Maslanik, and K. Knowles, 2005. Tracking the Arctic's shrinking ice cover: Another extreme September minimum in 2004. Geophys. Res. Lett. 32 doi:10.1029/2004GL021810.

CHAPTER 2

Seasonal Circulation over the Canadian Beaufort Shelf

R. Grant Ingram[1], William J. Williams[2*],
Bon van Hardenberg[2], Jordan T. Dawe[1], Eddy C. Carmack[2]

2.1 Introduction and Rationale

The Canadian Beaufort Shelf is a broad, shallow continental shelf (~120 km wide; ~530 km long) in the southeastern Beaufort Sea which stretches from the Mackenzie Trough to Amundsen Gulf (Fig. 2.1). Its coastal boundary is defined by freshwater outflow from the Mackenzie River Delta, while its offshore boundary is the oceanic Beaufort Gyre. The region can be viewed as a large shelf estuary which draws water and associated properties (e.g., nutrients, organic carbon, plankton, sediments) from both a coastal source (i.e. the Mackenzie River; Omstedt et al., 1994; Macdonald et al., 1998) and an oceanic source (i.e. the Arctic Ocean; Carmack et al., 1989). Surface waters originate from the Mackenzie River and ice melt. Deeper waters come from both the Pacific (40-220 m) and Atlantic (> 200 m) Oceans (Carmack et al., 1989; Macdonald et al., 1989; McLaughlin et al., 1996). Given these dissimilar sources, waters on the shelf have a wide range of physical and geochemical properties. Such diversity is greatly significant to the local biota, specifically in terms of constraining primary production and providing sources for biological 'seed' populations. For example, both Atlantic and Pacific Ocean species can be found in the Beaufort Sea (Dunton, 1992); while

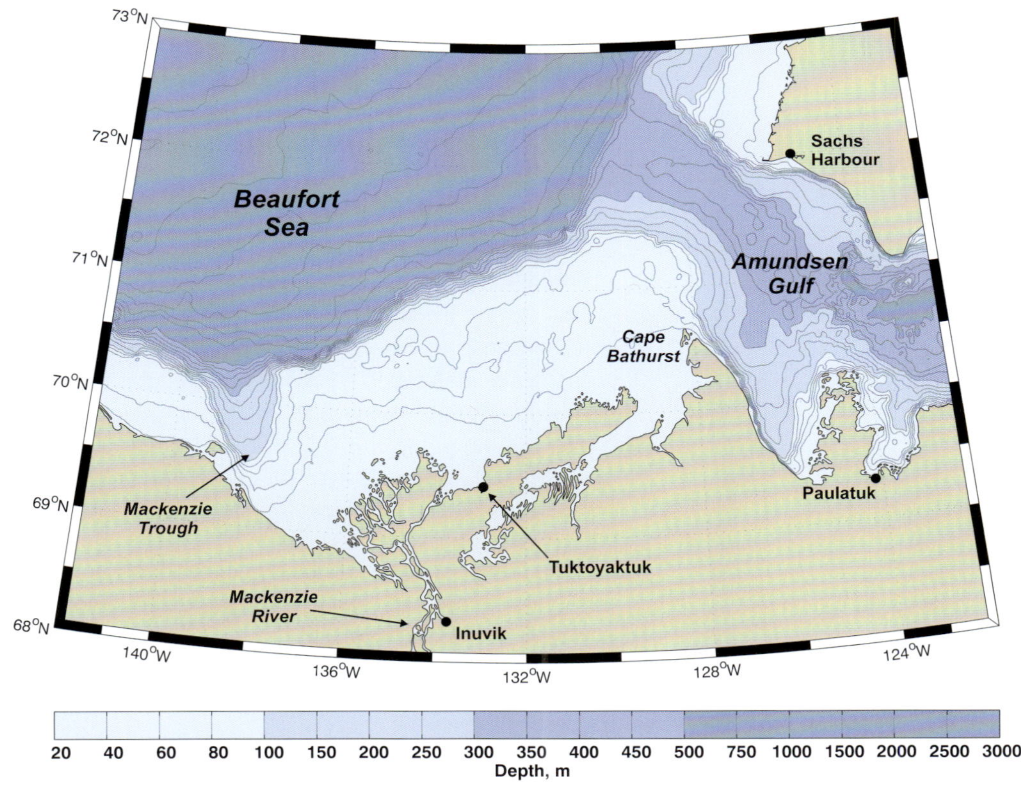

Mackenzie River outflow also ensures the presence of freshwater biota, including anadromous fish (Parsons et al., 1988 & 1989; Bodaly et al., 1989).

ABOVE: **Figure 2.1**

Map of the Canadian Beaufort Shelf.

LEFT: The CCGS Pierre Radisson breaking its way through the ice.
Photo: Martin Fortier.

[1] *Deceased: Department of Earth and Ocean Sciences, University of British Columbia, BC, V6T 1Z4, Canada*

[2] *Institute of Ocean Sciences, Sidney, British Columbia, V8L 4B2, Canada*

*Corresponding author: Bill.Williams@dfo-mpo.gc.ca

In their classic manuscript, Giovando and Herlinveaux (1981) report on "factors influencing [the] dispersion of pollutants in the Beaufort Sea" and provide a wonderful overview of historical, geographical, bathymetric and oceanographic knowledge gathered during the Beaufort Sea Project (1974-75)—including the circulation of its waters. The basic surface circulation in the offshore region is dominated by the clockwise Beaufort Gyre (Coachman and Aagaard, 1974; McLaren et al., 1987). Below the surface waters adjacent to the shelf break, however, the flow reverses to counter-clockwise, forming the so-called Beaufort Undercurrent (Aagaard, 1984). This flow moves waters of both Pacific and Atlantic origin eastward along the continental margin (Coachman and Barnes, 1961; Aagaard, 1989) and provides an offshore source of nutrients to shelf waters (Macdonald et al., 1987). The circulation on the Canadian Beaufort Shelf is highly variable, with mean velocities much smaller than the fluctuating field (Melling, 1993). Flow events related to synoptic-scale wind forcing along the Beaufort coast have been reported by Cameron (1951), Giovando and Herlinveaux (1981), Kulikov et al. (1998), and Williams et al. (2006).

The region undergoes very significant annual and inter-annual cycles in both sea ice cover and river discharge. Accordingly, oceanographic conditions on the shelf may be roughly divided into periods which reflect the interaction between these two cycles. During the ice-covered winter period, much of the direct wind forcing is dissipated by internal ice stress. Thus, shelf exchange occurs largely by density-driven flows due to brine release during ice formation (Melling and Lewis, 1982; Melling, 1993). However, wind and ice driven flaw lead events also occur during winter wherein the pack ice is forced offshore and along-shelf towards the southwest. This ice motion opens the flaw lead, causing additional ice production and brine rejection, as well as winter-time upwelling over the shelf (Williams et al., 2008). During spring, ice break-up commences in the upper Mackenzie River basin (late April and May) and river discharge peaks between mid-May and June. At this time, heat advected by the north-flowing river substantially influences the melting of river and nearshore ice. During summer, inflow from the Mackenzie River forms a shallow, turbid plume that dominates nearshore surface property distributions. At this time, the Mackenzie River is especially important for biological regimes on the shelf as it supplies nutrients, suspended sediments, and plankton. Water mass properties and density stratification (an important control over primary production) in the upper layers are very much determined by the disposition of the Mackenzie plume. The plume, in turn, is forced by synoptic wind events; moving offshore during upwelling-favorable winds (south and easterly) and onshore during downwelling-favorable winds. As river input diminishes during the fall, higher winds strongly influence shelf circulation and shelf-basin exchange. It should be noted that the timing of seasonal events (e.g. freshet, break-up, freeze-up) is more important to biota than standard quantifiable measures (e.g. annual discharge, ice thickness). Hence, climate change and other modifications brought on by human activities may impact biota by altering the timing and coupling of environmental cycles rather than by changing their magnitudes.

The Canadian Beaufort Shelf is unique among Arctic systems in that it receives large amounts of freshwater runoff throughout winter from its large headwater lakes. It also produces ventilating salty water masses by brine rejection from growing sea ice (Macdonald and Carmack, 1991). This is due to the combined effects of the offshore flaw polynya (which forms at the boundary between landfast and pack ice) and the stamukhi (which are the irregular sea ice ridges approximately parallel to the coast at the edge of the landfast ice). The Mackenzie River plume extends under landfast ice until blocked by the underside of the stamukhi. The plume containment produces a brackish area on the shelf called Lake Mackenzie (sometimes also called Lake Herlinveaux; Carmack and Macdonald, 2002). Beyond the stamukhi, there is the narrow flaw lead polynya that extends along the entire shelf region and broadens to form the Cape Bathurst polynya in Amundsen Gulf. The region is thus separated into two convective regimes by the stamukhi. Offshore of the stamukhi, the opening of the flaw polynya leads to greater ice production and brine enhancement. Onshore of the stamukhi, the regime is dominated by the impoundment of Mackenzie River water.

Significant upwelling events associated with cross-shelf topography have been recorded off the Canadian Beaufort Shelf. The shelf is cut at its southwestern end by the Mackenzie Trough, in the middle by the narrow and shallow Kugmallit Valley (immediately north of the Mackenzie Delta), and at its northeastern end by the steep slope into Amundsen Gulf at Cape Bathurst. All are recognized sites of topographically-enhanced upwelling (Macdonald et al., 1987; Carmack and Kulikov, 1998; Williams et al., 2006; Williams et al., 2008), during times of southwestward surface stress from wind or ice motion.

Tidal currents on the Canadian Beaufort Shelf are generally weak (which is typical of Arctic seas) except in the vicinity of the shelf break north of Cape Bathurst and around headlands and narrow passages in the Husky Lakes (Henry and Foreman, 1977; Kowalik and Prushutinsky, 1994; Kulikov et al., 2004). While geographically small, such areas may be of disproportionate importance to biological systems due to tidally-induced mixing and modifications to ice cover.

Knowing how Arctic Ocean shelf areas respond to variations in climate will aid our understanding of the impact of these changes on the coastal areas where marine resources are exploited by the local populations. Factors such as the presence/absence of sea ice, the timing and amplitude of freshwater input, and the direction and magnitude of wind and ice motion are thought to play major roles in carbon and nutrient transport, as well the health of marine mammals.

In order to better understand these seasonally and spatially variable processes, the CASES Research Network undertook a detailed study of the Canadian Beaufort Shelf, Amundsen Gulf, and their adjacent waters. Specifically, we present here observations for water circulation on the outer shelf and upper slope of the Canadian Beaufort Shelf obtained from long-term moorings of current meters and temperature-salinity recorders. The dataset derives from current meter moorings which were deployed at 8 sites for one year on the Canadian Beaufort Shelf and in Amundsen Gulf in the fall of 2002. It also includes data from a larger mooring program at 17 sites undertaken in 2003-2004.

2.2 Overview of Results

2.2.1 Forcing variables

The strong seasonal signal of air temperature and river discharge adjacent to the Canadian Beaufort Shelf is shown in Fig. 2.2 for 2003 and 2004. During the growing season, air temperature (Fig. 2.2a) varies over a range of almost 60 °C. They fluctuate widely (-40 to 0 °C) in April and May with the passage of synoptic weather systems, typically rise above 0 °C in early June, then drop below 0 °C in early October, signaling the onset of freeze-up. October and November are also months for strong winds and the passage of

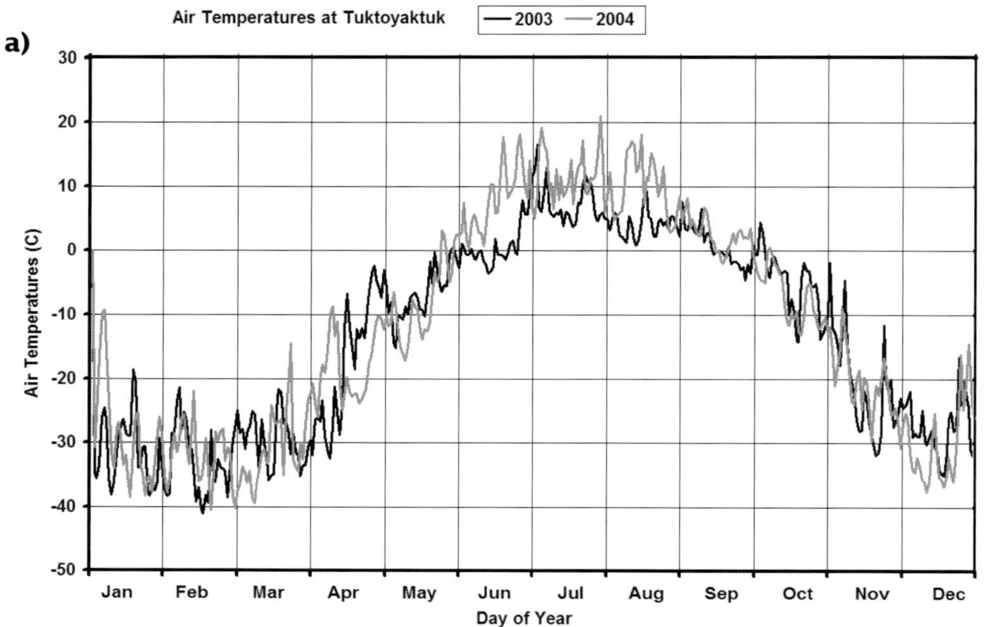

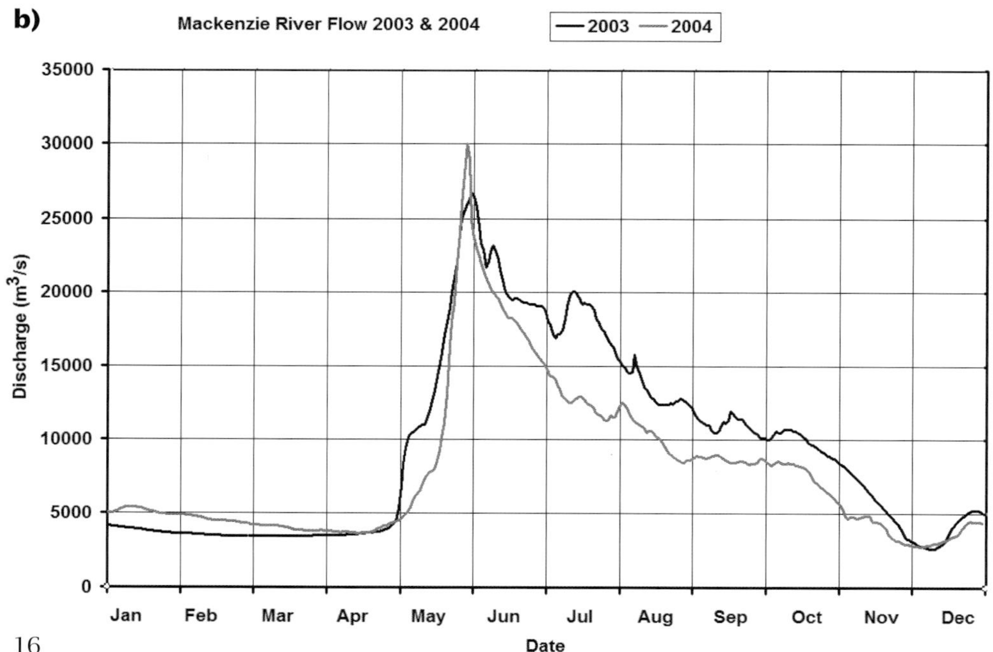

polar lows which serve to mix the water column and force shelf/basin exchange. The timing and magnitude of Mackenzie River discharge (Fig. 2.2b) influences directly phytoplankton productivity by the seasonal delivery of nutrients and seed plankton, and indirectly through the regulation of stratification and turbidity. Time series from the late 1980s of photosynthetically available radiation (PAR) (Carmack, pers. comm.) show that, even though PAR levels are relatively high by early April, most of the irradiance is reflected upwards by the high albedo of snow and sea-ice. Owing to the length of days during the Arctic growing season, the absolute values of daily-integrated PAR equal those typical of temperate latitudes.

Sea ice cover impacts underwater light and wind forcing. The historical record of sea-ice cover averaged for 1971-2000 on two dates in the Canadian Beaufort Shelf area is shown in Fig. 2.3. Typically, break-up begins in early June in the mid-shelf domain immediately seaward of the landfast ice and stamukhi, followed in early July by the rapid deterioration of the landfast ice. The outer boundary of sea ice in summer is highly variable from year to year (Melling et al., 2005) and varies on shorter time scales in response to synoptic wind events. Freeze-up typically begins in October and progresses both seaward (with the formation of landfast ice) and shoreward (with the growth of pack ice). Complete ice-cover is usually established in November (Dumas et al., 2005).

Figure 2.2

The annual cycle of (a) the surface air temperature in Tuktoyuktuk and (b) the river discharge from the Mackenzie River for 2003 and 2004.

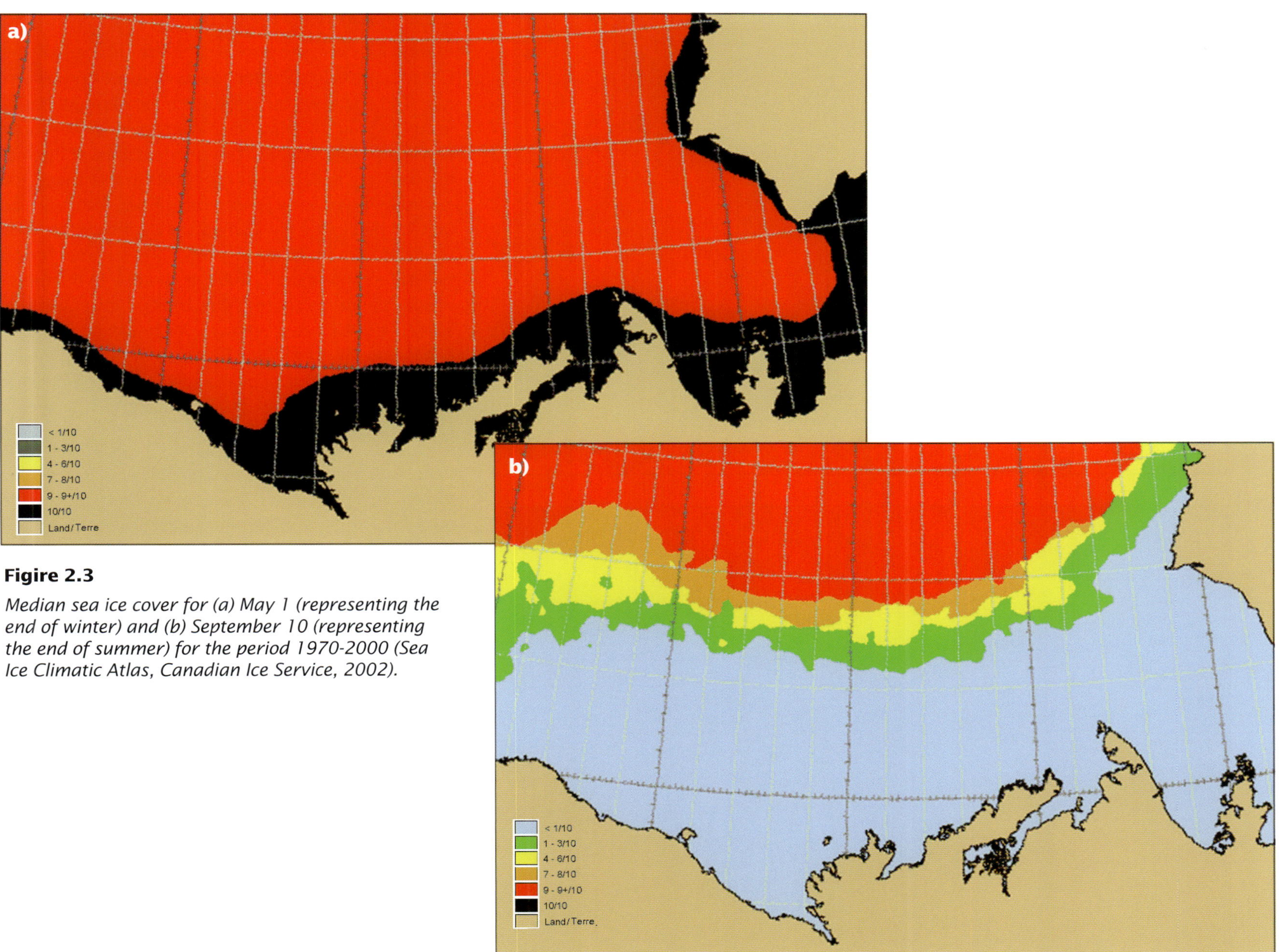

Figire 2.3

Median sea ice cover for (a) May 1 (representing the end of winter) and (b) September 10 (representing the end of summer) for the period 1970-2000 (Sea Ice Climatic Atlas, Canadian Ice Service, 2002).

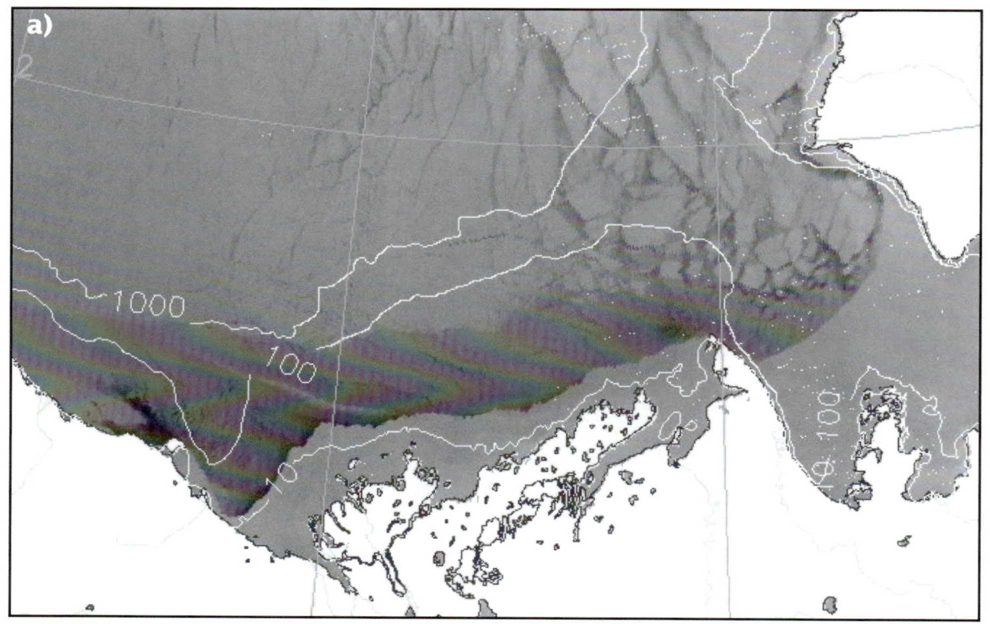

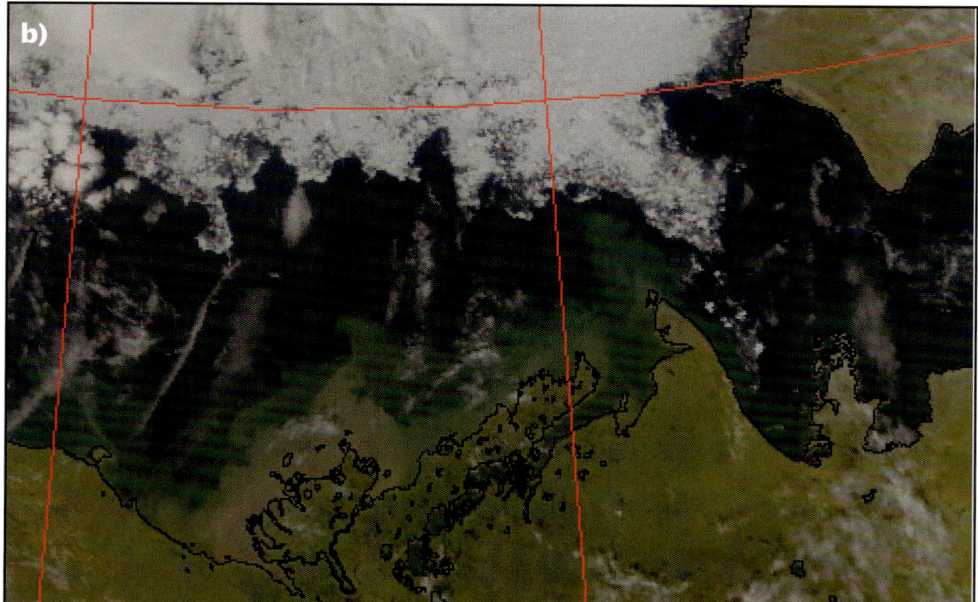

2.2.2 Seasons on the Shelf

In Fig. 2.4, we show two satellite images of the Canadian Beaufort Shelf: one in winter and one in summer. The one in winter is a surface temperature image taken on March 4, 2003, during a flaw lead event. The flaw lead, lying between the land fast ice and the mobile pack ice, appears dark grey/black in this image and is about 70 km wide. It is mostly filled with new ice that formed in the lead when the surface water was exposed to the cold atmosphere. The summer image was taken on August 7, 2004, and is one of the better images of sea ice cover in August 2004 over our study area. The shelf area is relatively ice free except for broken ice northeast of Cape Bathurst. The exceptionally turbid Mackenzie River plume can be seen dominating the region near the delta and influencing the inner shelf across most of the image.

Schematic sections of oceanographic structure across the Canadian Beaufort Shelf based on hydrographic data for winter and summer are shown in Fig. 2.5 (Carmack and Macdonald, 2002; Ingram et al., 2006). Here, it is practical to discuss the seasonality of physical and biological processes in terms of inner shelf stations (depth < 20 m), middle shelf stations (20 < depth < 80 m) and outer shelf and shelf-break stations (depth > 80 m).

Figure 2.4

Satellite images of the Canadian Beaufort Shelf: (a) A winter surface temperature image taken at 21:35UTC on 4 March 2003 during a flaw lead event. The 10, 100 and 1000-m depth contours are marked. This 1 km resolution image was prepared by Mike Schmidt and is from the Aqua satellite. (b) A visible image in summer taken on 7 August 2004.

2.2.2.1 Winter

Late winter and the beginning of oceanographic spring (~April) correspond to the time when sea-ice reaches its maximum thickness of about 2 m (Fig. 2.5a). At this time, the relatively flat land fast ice extends seaward from the coast toward approximately the 20 m isobath. The stamukhi (field of rubble ice formed by convergence of landfast and drifting ice) typically forms the outer boundary of land fast ice. Such ridges are known to gouge the bottom, and likely have a major impact on benthic habitat. The position, thickness and relative roughness of the stamukhi is highly variable from year to year, depending on the timing and intensity of winter storms. The stamukhi also acts to constrain circulation. The Mackenzie River inflow in winter, while relatively low (~ 4000 m³ s⁻¹) compared to that of summer (30 000 m³ s⁻¹), still delivers substantial quantities of freshwater to the ocean, where it is largely trapped as a floating freshwater lake (Lake Mackenzie) in the inner shelf domain behind the stamukhi dam (Macdonald and Carmack, 1991; Macdonald et al., 1995; Carmack and Macdonald, 2002). This floating freshwater lake contains about 70 km³ of winter inflow spread over an area of 12 000 km² (Macdonald et al., 1995), statistics that would rank this seasonal lake in the top 20 or 30 lakes of the world by area or volume, respectively. The interface between the near-fresh upper water and underlying seawater is a potential site of frazil ice production, owing to the differential freezing temperatures of the two water masses. The ecological functioning of this vast domain remains, surprisingly, unstudied. Furthermore, where this freshwater goes during and after break-up and the effect it has on the biological cycle of the middle and outer shelf is unknown. The land fast ice and stamukhi thus define a unique, wintertime inner shelf environment.

The mechanisms by which the river connects to the sea in late winter are not clearly understood. At that time, ice thicknesses reaching 2 m rest at the bottom of shallow areas like Kittigazuit Bay and Shallow Bay, such that river outflow requires sub-ice conduits. Tides and storm surges provide important controls on the shape and distribution of flow conduits for water passing between the ice and the bottom, which perhaps leads to pulsing events in the outflow. Where more than one river channel leads to the ocean, the ice can modulate flow between channels based on how channel geometry interacts with a progressively thickening ice cover.

The middle shelf in winter extends from the stamukhi out to the shelf break near 80 m depth. The shear zone at the boundary of land fast and pack ice is extremely dynamic and subject to the rapid opening of a wide (up to 70 km in 2003), recurrent flaw polynya running parallel to the coast. The flaw polynya is a site of increased ice production, brine release and convection. Sea ice conditions beyond the flaw polynya consist of a mixture of heavily ridged and drifting first-year and multi-year pack ice. Nutrient concentrations in the surface layers at the end of winter, which form the basis for new production in summer, are set by the degree of vertical mixing and entrainment that occurred during the preceding fall and winter. The outer shelf is recognized as a domain subject to shelf break dynamics and shelf/basin exchange.

From a biological perspective, the entire system is poised for primary production in late winter but cannot yet achieve it owing to limits set by the physical environment. Underwater light, which dictates the timing of the onset of primary production, is controlled by ice and river conditions. By April, surface layer nutrients and solar irradiance are sufficient to initiate growth,

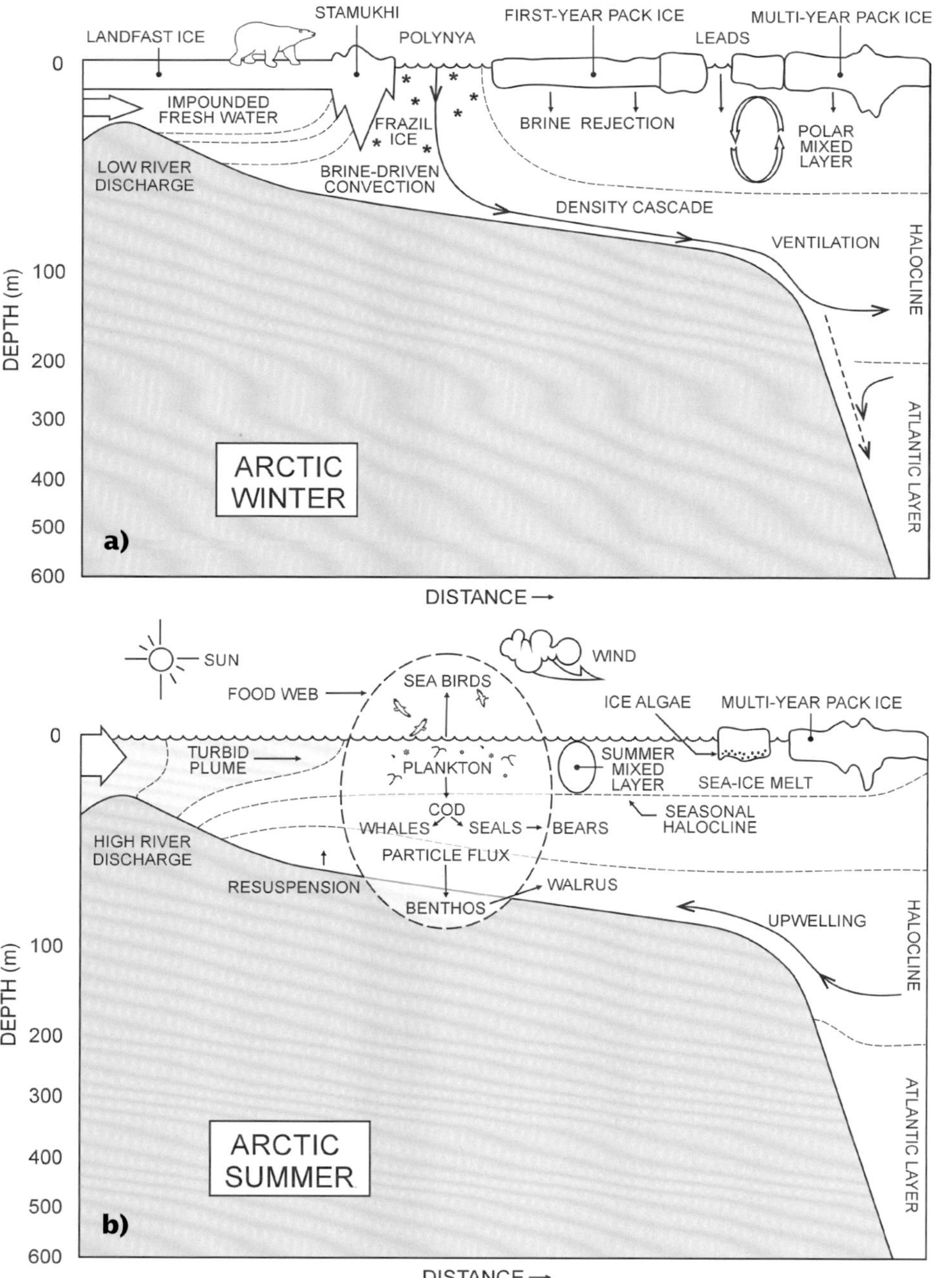

but the onset of pelagic productivity is delayed by the presence of sea ice and snow cover. At this time, algal production is likely limited to the bottom of the sea ice (Horner and Schrader, 1982). In the flaw polynya, brine-driven deep convection prohibits a stable light climate for phytoplankton growth. Farther offshore, the pack ice and its snow cover again limits light and delays the onset of productivity. With the retreat of ice cover in summer, however, light can now reach the water column.

2.2.2.2 Summer

The end of oceanographic winter (typically in late April) corresponds approximately to the time when freezing-degree-days stop accumulating (Fig. 2.5b). The transition to summer begins with the break-up of river ice in the headwaters of the Mackenzie River and the subsequent flooding of the delta and coastal area. Heat from the Mackenzie River, which overflows and underflows the land fast ice in the nearshore, accelerates ice removal in the near-shore delta region (Dean et al., 1994). Break-up of the middle and outer shelf typically spreads from existing open water in the flaw polynya, where incoming solar radiation is rapidly absorbed by the water and accelerates further melting. During break-up, the land fast ice and much of the existing pack ice melts in place. This addition of buoyant freshwater is mixed downwards by winds to form a shallow, relatively fresh mixed layer (~10 m deep) which stratifies the upper ocean. At the same time, large amounts of fresh and turbid water are delivered during the freshet of the Mackenzie River (which, in turn, becomes free of seawater intrusions landward of the

Figure 2.5

Cartoons showing typical physical processes occurring in (a) winter and (b) summer over the Canadian Beaufort Shelf (from Ingram et al., 2006).

transverse bar across Kittigazuit Bay). Plume water is distinct from sea ice melt. It is often observed to form extensive areas of highly turbid water (typical thickness: ~ 5 m) extending across the shelf and, at times, off the shelf (Macdonald et al., 1989 & 1995). At this time, the Canadian Beaufort Shelf behaves much like any estuary of a large river impinging on an open shelf; the only difference being an additional, broadly distributed freshwater source in ice melt. The dominant influence of the Mackenzie River allows the occurrence of freshwater biota, including anadromous fish (at least in the nearshore region; Parsons et al., 1988 & 1989; Bodaly et al., 1989). The plume invades the near shore and its distribution is very much affected by winds and the extent of open water. In the absence of winds or under westerly (downwelling-favourable) winds, the incoming river water will tend to flow eastward along the Tuktoyaktuk Peninsula towards Amundsen Gulf. Easterly (upwelling-favourable) winds will draw deeper (Pacific origin) waters onshore and drive plume waters offshore. Upwelling is especially evident north of Cape Bathurst, as evidenced by the frequent appearance of a tongue of cold saline water extending north along the western shelf break.

In the fall, the daylight rapidly diminishes from 12 hours at the equinox (September 21) to total darkness by mid November, and the low sun angle further limits light penetration. Intense storms can also force on-shelf upwelling. For example, Kulikov et al. (1998) observed upwelling amplitudes in Mackenzie Trough exceeding 600 m; i.e., 3 to 4 times greater than observed elsewhere along the shelf. When the wind stops, some of the dense water returns down the canyon to the ocean basin, creating a wave-like response in the offshore ocean, and some remains and mixes into shelf waters to supply production. Freeze-up commences in early to mid October after air temperatures have dropped

below the freezing point of water, the water has cooled to its freezing temperature, and freezing-degree-days start to accumulate. By then, river inflow decreases to its lowest value and, instead of entering an open estuary where it can be mixed by wind, it spreads under the near-shore (land fast) ice in a relatively quiescent environment.

2.2.3 Methods and instrumentation

A significant part of the physical oceanography component of the CASES project focused on using instrumented moorings to measure ocean currents and water mass properties. An initial array of 8 moorings was deployed in 2002 followed by a larger array of 17 moorings in 2003. Moorings were located along the edge of the shelf, within the slope region, and along the axis of Amundsen Gulf. Many of these moorings were equipped with instrumentation to measure bio-geo-chemical parameters such as fluorescence, light levels and sediment flux. Fig. 2.6 shows the 2002-03 mooring locations in black (with the deployment year added as -02) and the new deployment locations for 2003-04 in red. Tables 2.1 and 2.2 give the precise location, bottom depth, deployment period, and instrumentation associated with each mooring.

The moorings were designed with a float at the top, an anchor and acoustic releases at the bottom, and instruments with additional floatation along the mooring line. The nominal depth of the top floats used in the Beaufort Sea was ~25-30 m below the surface (in order to avoid the keel of ice ridges during deployment). The instrument directly below the top float was either a Seabird SBE-37 or Alec-ACT, measuring sea water temperature, salinity, and in some cases pressure (which indicates how much a mooring is leaning over when dragged by currents). The three shallow moorings (< 65

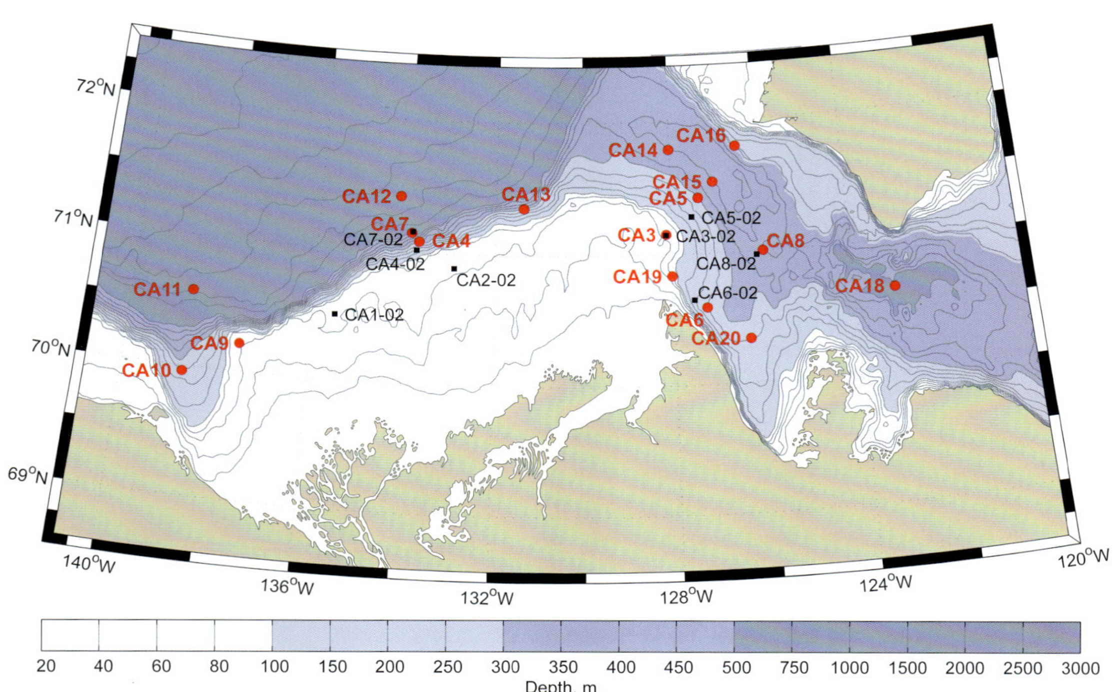

Figure 2.6

A map of the CASES study area showing the locations of the moorings deployed by the CASES project. The black moorings were deployed in the fall of 2002 and the red moorings were deployed in the fall of 2003.

m) had an Aanderaa RCM current meter located 14 m above the bottom. The other moorings were equipped with additional current meters at about 50, 100 and 200 m depth and about 15 m above the bottom. A number of moorings had ADCP Workhorse current profilers looking upward from about 100 m depth. Some moorings had two sediment traps, each at depths of about 65 and 200 m. Acoustic releases were used to recover the instruments (a single release at the three shallow moorings, double releases at all others).

The eight instrumented moorings dedicated to the first year of the CASES project were deployed from *CCGS Sir Wilfrid Laurier* in the fall of 2002. All eight moorings were successfully recovered in the fall of 2003. While most of the recovered instruments functioned properly, some of them showed damage (or failed to work) and could not be serviced for re-deployment. In the fall of 2003, the main array of instrumented moorings was deployed from *CCGS Amundsen*. The initial experiment design called for a total of twenty moorings. Much of the instrumentation was newly acquired and the remainder was recovered from moorings of the previous year. Moorings were eventually deployed at 17 locations. Moorings CA1, CA2 and CA17 were not deployed due to a lack in available instruments. Most moorings deployed in 2003 were deployed for 11 months or more. However, due to budgetary and logistical constraints, some moorings were recovered after 9-10 months. All data from recovered instruments were processed, calibrated, verified and analyzed.

TABLE 2.1

Moorings deployed in 2002: Name of mooring site, geographical location (latitude and longitude), bottom depth (m), deployment and recovery dates (UTC), and list of instrument types and their deployment depths. Data return was good from most instruments.

Mooring Name	Latitude (N)	Longitude (W)	Bottom Depth (m)	Deployment date (UTC)	Recovery date (UTC)	Instrument Type	Instr. Depth (m)
CA1-02	70° 30.000'	135° 30.000'	63.2	12-Sep-02	4-Oct-03	SBE-37 RCM-7	30 49
CA2-02	70° 53.707'	132° 54.829'	65.7	15-Sep-02	1-Oct-03	SBE-37 RCM-7	32 51
CA3-02	71° 08.996'	128° 08.021'	63.0	18-Sep-02	10-Oct-03	SBE-37 RCM-4	30 49
CA4-02	71° 01.335'	133° 46.443'	201.0	15-Sep-02	1-Oct-03	SBE-37 RCM-4 ADCP RCM-4 RCM-4	28 46 96 102 185
CA5-02	71° 16.954'	127° 32.139'	200.9	18-Sep-02	10-Oct-03	SBE-37 RCM-4 ADCP RCM-4 RCM-4	27 46 96 102 185
CA6-02	70° 38.996'	127° 32.854'	204.8	19-Sep-02	11-Oct-03	RCM-4 ADCP RCM-4 RCM-4	50 100 106 189
CA7-02	71° 09.745'	133° 52.630'	505.0	16-Sep-02	3-Oct-03	SBE-37 TRAP RCM-7 ADCP TRAP RCM7 RCM-4	31 69 75 106 199 205 484
CA8-02	70° 58.383'	126° 06.720'	389.1	19-Sep-02	12-Oct-03	SBE-37 TRAP RCM-7 ADCP TRAP RCM7 RCM-4	27 58 65 95 200 195 373

TABLE 2.2

Moorings deployed in 2003: Mooring name, location (latitude and longitude), bottom depth (m), date deployed and recovered (UTC), instrument type & sensors, and instrument depth below surface (m).

Mooring Name	Latitude (N)	Longitude (W)	Bottom Depth (m)	Deployment date (UTC)	Recovery date (UTC)	Instrument Type *	Instr. Depth (m)
CA3-03	71° 09.140'	128° 07.560'	65	11-Oct-03	29-Sep-04	ACT-HR RCM-11	32 50
CA4-03	71° 05.158'	133° 43.392'	331	5-Oct-03	7-Sep-04	ACT-HR ADCP TRAP RCM-11 TRAP	36 104 109 204 214
CA5-03	71° 25.228'	127° 22.488'	297	11-Oct-03	28-Jul-04	SBE-37 ADCP TRAP RCM-11 TRAP	33 100 105 200 210
CA6	70° 35.324'	127° 16.23'	249	11-Oct-03	5-Sep-04	SBE-37 ADCP TRAP SBE-26	30 99 105 237
CA7	71° 8.989'	133° 53.856	525	3-Oct-03	7-Sep-04	SBE-37 ADCP TRAP RCM-11 TRAP RCM-11	19 87 94 189 299 391
CA8	71° 0.049'	125° 57.852	397	13-Oct-03	28-Jul-04	ALEC-CL ADCP TRAP RCM-11 TRAP TRAP	32 101 108 202 213 309
CA9	70° 2.581'	137° 30.856	66	13-Oct-03	9-Sep-04	SBE-37 RCM-11	29 49
CA10	69° 57.307'	138° 40.434	267	6-Oct-03	8-Sep-04	ACT-HR ADCP TRAP TRAP	30 99 105 210

Mooring Name	Latitude (N)	Longitude (W)	Bottom Depth (m)	Deployment date (UTC)	Recovery date (UTC)	Instrument Type *	Instr. Depth (m)
CA11	70° 34.629'	138° 39.289	1036	5-Oct-03	9-Sep-04	SBE-37 ADCP TRAP RCM-11 TRAP RCM-11 RCM-11 RCM-11	38 107 113 208 218 411 804 980
CA12	71° 25.317'	134° 11.212	1234	2-Oct-03	27-Jul-04	ACT-HR ADCP TRAP RCM-11 TRAP RCM-11 RCM-11 RCM-11 TRAP	32 101 107 202 212 403 801 1000 1009
CA13	71° 21.356'	131° 21.814	320	9-Oct-03	4-Sep-05	SBE-37 ADCP RCM-11 TRAP	48 114 216 226
CA14	71° 47.461'	128° 0.817	399	9-Oct-03	Not Recovered		
CA15	71° 32.231'	127° 1.433	423	10-Oct-03	22-Jul-04	ACT-HR ADCP TRAP RCM-11 TRAP	36 104 109 204 214
CA16	71° 32.231'	127° 1.433	280	10-Oct-03	22-Jul-04	ACT-HR ADCP	31 101
CA18	70° 38.566'	127° 6.252	520	13-Oct-03	30-Jul-04	ACLW RCM-11 TRAP RCM-7 TRAP RCM-7 TRAP	34 103 109 205 215 406 413

TABLE 2.2 CONTINUED

Mooring Name	Latitude (N)	Longitude (W)	Bottom Depth (m)	Deployment date (UTC)	Recovery date (UTC)	Instrument Type *	Instr. Depth (m)
CA19	70° 50.180'	128° 1.210	493	11-Oct-03	6-Sep-04	ACT-HR ALEC-CL RCM-11	30 30 50
CA20	70° 20.347'	126° 21.299	280	12-Oct-03	16-Jul-04	ACLW ACT-HR ADCP PPS3- 3TRAP NIPR-TRAP	30 30 99 105 210

* Instrument Type: ACT-HR = Alec CT High Resolution, measures temperature (T) & conductivity (C); SBE-37 = Seabird Electronics CT, measures pressure (P) and T & C; ADCP = Acoustic Doppler Current Profiler, measures speed (Spd) & direction (Dir); RCM = Aanderaa Recording Current Meter, type RCM-4 measures Spd & Dir, RCM-7 measures Spd, Dir and T & C, RCM-11 measures Spd, Dir, T & C & P, turbidity and dissolved oxygen levels; ALEC-CL measures light intensity; ACLW = Alec Chlorophyll sensor; TRAP = sediment trap (one of several designs).

2.2.4 Circulation regime

When trying to capture the seasonal variability of the region's circulation regime, we used sea ice cover as our criterion for separating different oceanic conditions. Forest et al. (2007) showed the averaged Sea Ice Cover (SIC) on the Canadian Beaufort Shelf for 2003-2004, using Sea WIFS images: In mid-October, 2003, mean SIC was approximately 50%; SIC values exceeded 95% for most of the subsequent period until late April; finally, SIC values were about 50% at the end of June 2004. Using these values as a guide, we chose to group our data according to the following three-month periods: July 15 to October 15 as the *least ice cover period*; October 15 to January 15 as the *freeze-up period*; January 15- April 15 as the *maximum winter cover period*; and April 15 to July 15 as the *break-up period*. Additional *transition periods*, with SIC values between 50 and 80%, may have also demonstrated the influence of wind forcing on the waters over the shelf and the slope (unpublished model results).

To examine the 2003-04 mean circulation, ADCP and RCM current meter data were divided into three depth classes (0-30 m, 30-100 m, and 200 m) and split into these four periods. For each depth range and period, the data available at each mooring location was averaged to form a single velocity time series. The mean of the time series was then removed, and an EOF analysis was performed. The EOF analysis was used to extract the principal directions of variability for the currents, and the standard deviation of the EOF time series in each direction were used to estimate the current variance. Since many moorings were removed earlier than October 15, 2004, any mooring that had less than half of a data record for a given period was excluded from analysis. Specifically, this excluded moorings CA5-03, CA8-03, CA12-03, CA15-03, CA16-03, and CA20-30 from the 15 July – 15 October period.

The 0-30 m depth class data (Fig. 2.7) was entirely composed of ADCP measurements. Current means and

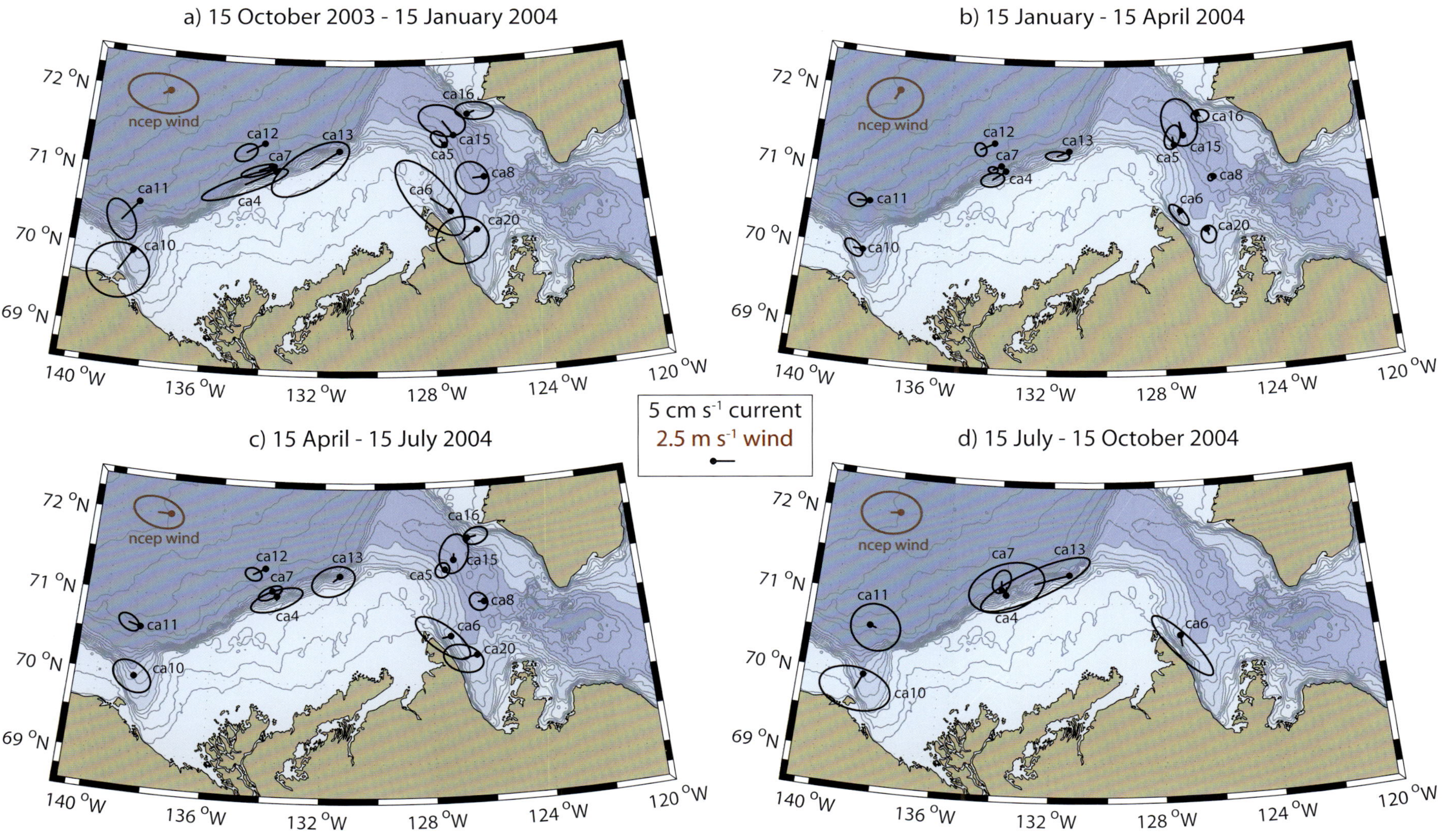

Figure 2.7

Mean and standard deviation of currents between the surface and 30 m depth measured by CASES moorings during: (a) ice formation (15 October – 15 January); (b) full ice cover (15 January – 15 April); (c) ice breakup (15 April – 15 July); and (d) summer ice (15 July – 15 October). Black dots indicate station locations; vectors indicate mean current; and ellipses centered on the vector heads indicate standard deviation of the current from the mean. NCEP winds represent an area averaged over the Mackenzie Shelf.

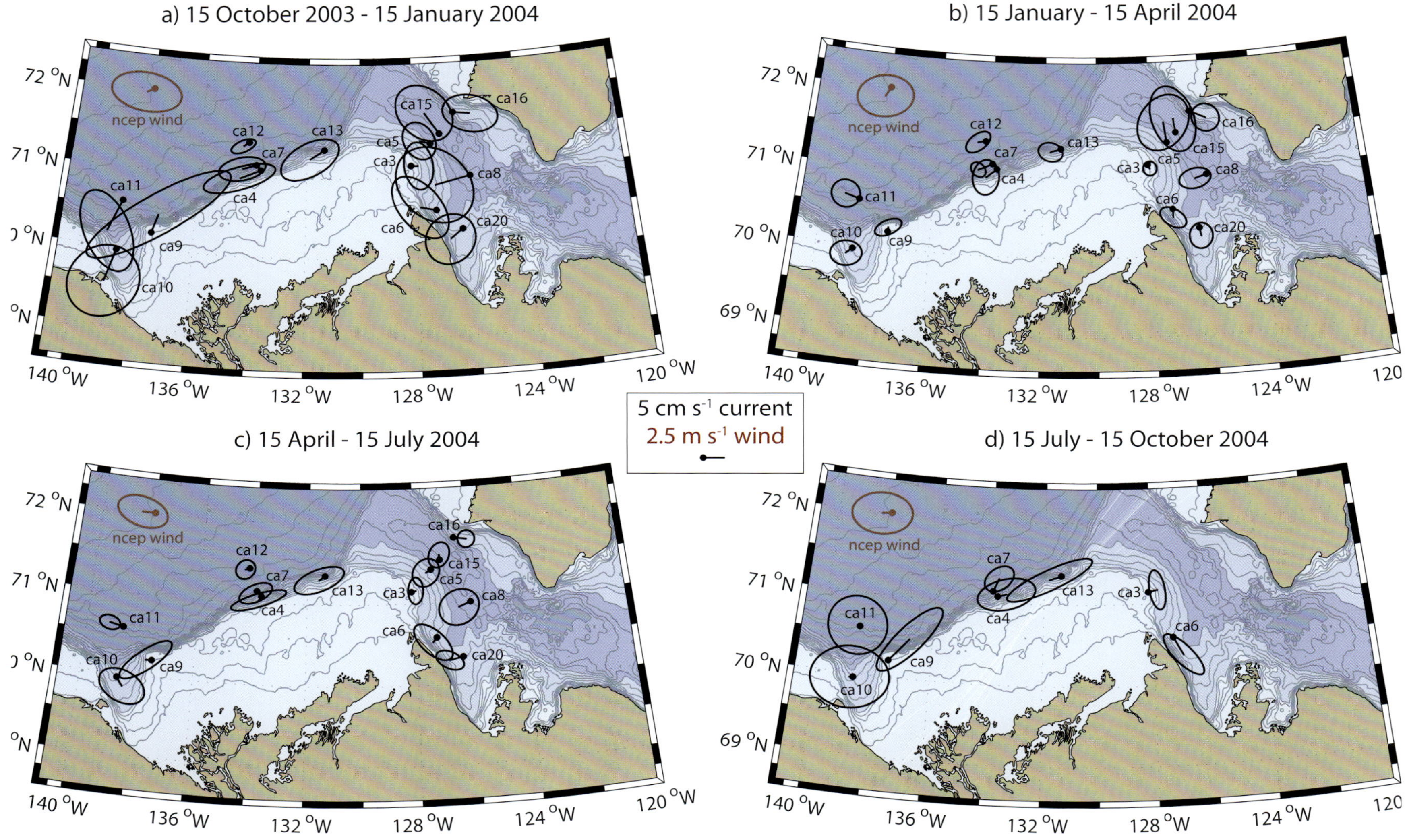

Figure 2.8

As Fig. 2.7 but for currents between 30 and 100 m depth.

variability were strong during the 2003 ice formation period (Fig. 2.7a), weakened dramatically during full ice cover (Fig. 2.7b), remained small during ice breakup (Fig. 2.7c), and became strong again as the ice opened up (Fig. 2.7d). Flow along the Beaufort Sea side of the Canadian Beaufort Shelf was generally southeastward (following the topography) with variability strongly restricted to the along-shelf direction, until a small offshore flow developed near the end of the data record. The moorings near Banks Island (CA16-03) and Cape Bathurst (CA6-03) also showed topography-following flow. Near the mouth of Amundsen Gulf, flow was generally directed from the Gulf to the open ocean. Strong on-shore flow up the Mackenzie Trough was recorded during the October 2003 – January 2004 ice formation period. However, the next two seasons showed little mean flow until another on-shore flow event occurred during the summer period.

The 30-100 m depth class data (Fig. 2.8) was composed of a mixture of ADCP and RCM current meter (on CA3-03 and CA9-03) measurements. Flow in this layer was generally similar to the flow near the surface. Significant differences between this layer and the surface included much stronger variability in the deeper currents near the Mackenzie Trough and in Amundsen Gulf. Flow in Amundsen Gulf between the Canadian Beaufort Shelf and Banks Island was still generally seaward, though the current slackened significantly during ice break up. At the eastern edge of the mouth of Mackenzie Trough (CA9-03) the flow was large and directed off the shelf during ice formation and the summer ice period, and much slacker during full ice cover and ice break-up. This area also showed extremely strong variability, again directed along the shelf topography. At the eastern edge of the Canadian Beaufort Shelf (CA3-03), a small off-shelf flow into Amundsen Gulf appeared during most of the time series.

Many of the moorings in the 2003-2004 array (with the exceptions of CA3-03, CA9-03, CA10-03, CA16-03, and CA20-03) had RCM current meters placed at ~200 m depth. These meters were used to create the 200 m circulation pattern for the region (Fig. 2.9). Flow measured at 200 m showed strong differences compared to the upper ocean. Flow weakening after the onset of the nearly continuous sea ice cover period did not occur as prominently at 200m as at the surface. Along the Beaufort Sea edge of the Canadian Beaufort Shelf the mean flow was generally reversed relative to the surface, flowing to the northeast along the shelf break. However, current variability remained strongly associated with bottom topography. Away from the shelf in the Beaufort Sea (CA11-03 and CA12-03), mean currents and variability were much weaker compared to the nearby shelf break moorings and any shallower instruments at the same locations. Flow near the mouth of Amundsen Gulf was still directed out to the sea; however, measurements deeper in the Gulf (CA18-03) showed little consistent mean flow direction.

In the introduction we discussed the importance of both river input from the Mackenzie River and seasonal sea ice cover to the study area. Our instrument depths ranged from 29 to 1000 m, i.e. different water masses of the water column than those of river-derived characteristics. A discussion of the near surface water masses over the Canadian Shelf and adjacent areas can be found in Carmack et al. (1989).

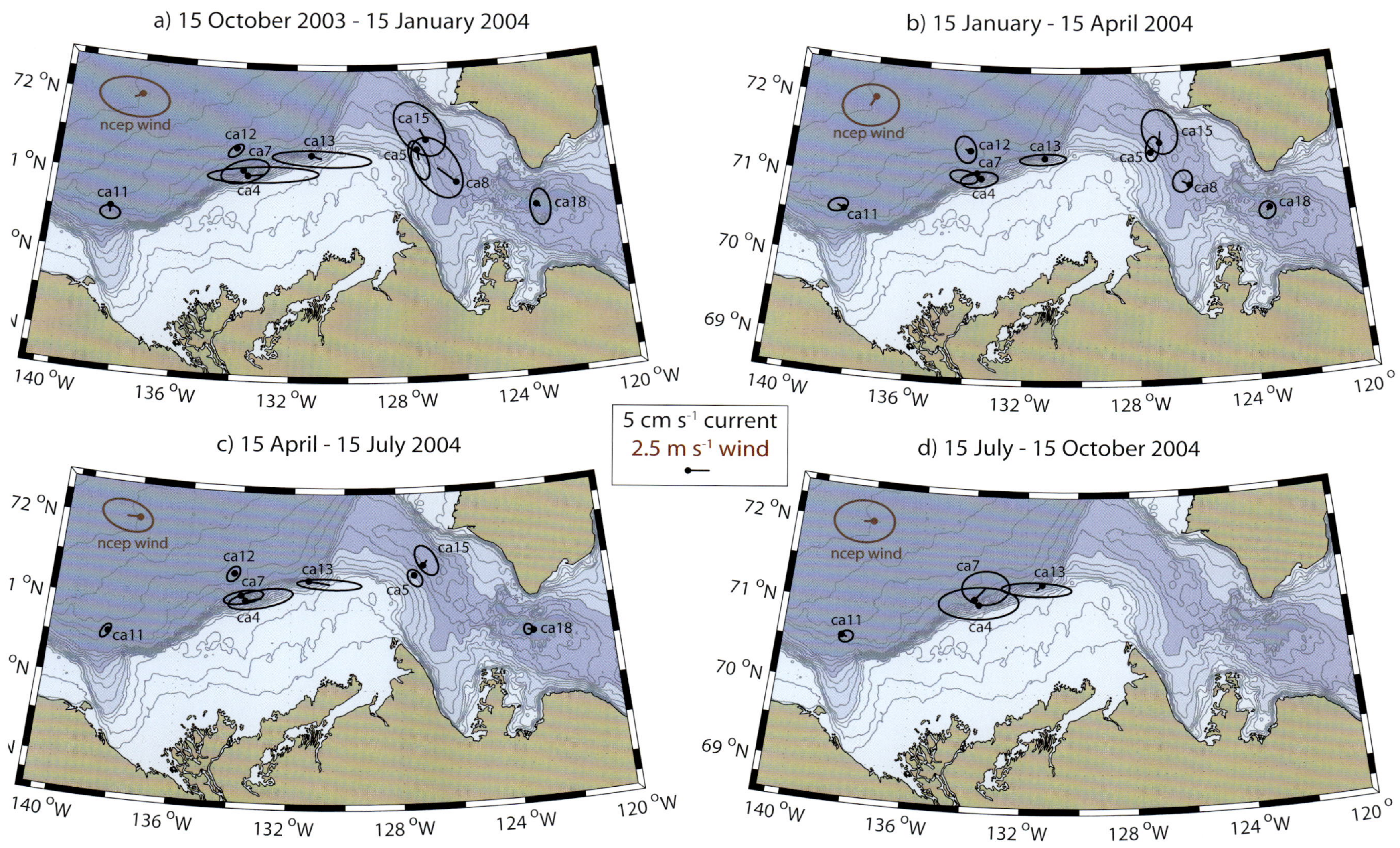

Figure 2.9

As Fig. 2.7 but for currents at 200 m depth.

2.3 Implication of this Work

The general circulation pattern observed in the upper 100 m of the water column appears to follow bottom topography and be strongly linked to sea ice cover. Mean flows are much stronger and demonstrate higher variability during reduced (<50%) ice cover and open water conditions. Weaker flows occur during complete ice cover, as well as early in the break up period. Flows are generally towards the southwest at the shelf edge and seaward in Amundsen Gulf. At 200 m depth, the influence of sea ice cover is much less apparent. Cross-shelf flows along the Mackenzie Trough are quite evident in the upper 100 m.

Synoptic wind and ice movement create stress on the surface of the ocean which produces upwelling and downwelling circulation on the shelf. A summary of our understanding of the response of the shelf to upwelling- and downwelling-favourable wind-stress is shown in Fig. 2.10. Under upwelling-favourable wind stress, the Mackenzie plume separates from the coast along the Tuktoyaktuk Peninsula and is rapidly pushed offshore such that it can extend beyond the shelf break into the Beaufort Gyre. Under downwelling-favourable wind stress, the Mackenzie plume is pushed onshore and forms a coherent coastal current that flows toward Cape Bathurst. The fate of Mackenzie River water likely depends on the predominant wind stress during the spring/summer when the river flow is highest. During upwelling-favourable wind stress, upwelling across the shelf break will occur and is topographically-enhanced at Mackenzie Trough, Kugmallit Valley and Cape Bathurst. Along-shelf flow towards the southwest during upwelling-favourable wind stress enhances the Beaufort Gyre and retards the Beaufort Undercurrent. Conversely, during downwelling-favourable wind stress, there is along-shelf flow towards the northeast, and this retards the Beaufort Gyre and enhances the Beaufort Undercurrent.

The role of the physical environment in triggering, facilitating or enhancing primary production is important. Nutrient levels and mixed layer depth at the end of winter are set by wintertime convection and circulation. The transition to summer conditions begins with increased solar radiation, river freshet, and the melting of snow on sea ice. Increasing light levels initially support ice algae production (Alexander, 1974). A distinction must be made between domains covered with land fast ice and pack ice. Break-up of both land fast and pack ice typically spreads from existing open water in a flaw polynya, where incoming solar radiation is rapidly absorbed by low-albedo open water. This buoyant meltwater is mixed downward by the wind to form a shallow, relatively fresh mixed layer. At the same time, large amounts of fresh and turbid water are delivered by river inflows. The resulting stratification of the upper ocean sets conditions for a spring bloom, the magnitude and duration of which will be limited by the initial nutrient load. The spring bloom's consumption of nutrients in the surface layer results in a subsurface chlorophyll-a maximum near the base of the mixed layer where, presumably, pelagic phytoplankton grow in an environment of relative stability (trading off the low availability of light and the high availability of nutrients; Carmack et al., 2004). Additional nutrients may be supplied in summer by upwelling and vertical turbulent mixing. The latter may be locally enhanced, for example, by tidal resonance eddies or topographically steered currents.

Any changes in the region's ice regime accompanying global climate change will likely have a determining impact on the mean and varying part of its circulation signal. There is much debate on the potential impact

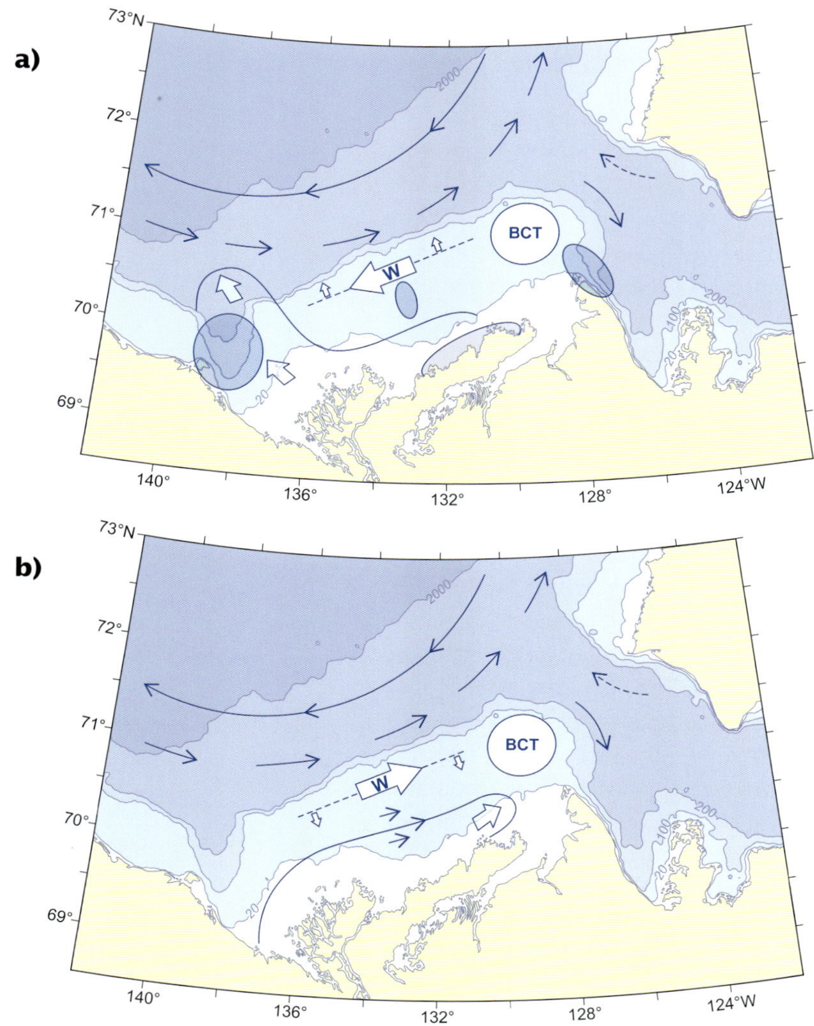

Figure 2.10

Cartoons showing circulation patterns over the Canadian Beaufort Shelf for (a) upwelling-favourable wind stress and (b) downwelling-favourable wind stress. In each diagram the along-shelf wind-stress is shown with an arrow marked 'W' and the cross shelf transport of the surface Ekman boundary layer is shown as flow perpendicular to the wind arrow. Regions of topographically-enhanced upwelling (Mackenzie Trough, Kugmallit Valley and Cape Bathurst) are shown in dark blue. Coastal upwelling, where the Mackenzie River plume has been pushed offshore, is shown as light blue. The shape and movement of the Mackenzie River plume under each kind of wind stress is shown. Near the shelf break, multiple arrows show the Beaufort Undercurrent flowing anti-clockwise; further offshore is the clockwise Beaufort Gyre. Baroclinic tides, which may perform important vertical mixing on the shelf at its northeastern end, are marked with 'BCT'.

of climate warming on sea-ice thickness (Rothrock et al., 1999; Yu et al., 2004) and certainly such changes, if realized, would greatly impact marine populations (Tynan and DeMaster, 1997; ACIA, 2004). Following Carmack and Macdonald (2002) and Michel et al. (2006), such changes could be viewed from both a bottom-up and top-down perspective. From a bottom-up perspective: increased open water would increase wind mixing and upwelling, and thus increase the availability of nutrients to phytoplankton (Carmack and Chapman, 2002); increased open water would also increase the availability of underwater light to phytoplankton and reduce ice algae production (with consequences to the pelagic and benthic food web); increased rainfall (also predicted under global warming) would increase the export of inorganic and organic terrestrial material (POC, DOC) to coastal domains; rising sea levels, combined with increased temperatures and open water areas, would accelerate coastal erosion and affect underwater light in coastal domains; Increased storms, especially in autumn, would lead to greater mixing, a greater supply of nutrients, enhanced sediment transport, and more rapid coastal erosion.

From a top-down perspective: shifting water mass fronts and currents would affect fish migration and behaviour, and the re-distribution of such grazers would impact underlying trophic structures. Any alteration in the habitat conditions of rivers and coastal seas might affect the distribution of anadromous fish, with consequences to their prey communities. Marine mammals (e.g. seals, bears) that depend on an ice platform would be seriously impacted by loss of habitat and would have to migrate elsewhere (with serious consequences to their prey and predators). Whales that depend on open water for migration would expand their habitat range in response to decreased ice concentration. Finally, a warmer climate would attract extractive, resource-based industries (e.g. oil, mining, transportation) which would have significant impacts on marine animals as well as socio-economic consequences for humans.

2.4 Recommendations

We recommend further research into upwelling and vertical mixing on the Canadian Beaufort Shelf as these processes determine the flux of nitrate to the euphotic zone and constrain primary production over the shelf. Such research would include the seasonal development of the surface mixed layer, the structure of the bottom boundary layer during upwelling events, process-oriented studies of the upwelling hot spots at Cape Bathurst and in Mackenzie Trough, and quantification of the upwelling flux across the shelf-break and to the euphotic zone. These studies need to be conducted on synoptic, seasonal, annual and interannual time scales in order to gain a deep perspective of the functioning of the ecosystem over the shelf and its variation. This will allow for better predictions of the fate and fracturing of the ecosystem under various climate change scenarios.

Ongoing research as part of the Northern Coastal Marine Studies program is using the *CCGS Nahidik* to conduct ecosystem-based research from both a monitoring and process-oriented perspective in July and August over the Canadian Beaufort Shelf. Along with providing insight on interannual variation, these cruises are allowing us a better understanding of the mixing and dispersal of the Mackenzie River plume, and of the dynamics of upwelling at Cape Bathurst and in Mackenzie Trough.

Cairn and the CCG Amundsen in the background. Photo: Alexandre Forest.

Acknowledgements

Funding for this work was provided by the NSERC Research Networks support of the CASES program, the ArcticNet program of the Canadian National Centres of Excellence program, and the Canadian Department of Fisheries and Oceans.

References

Aagaard, K., 1984. The Beaufort Undercurrent. In: P.W. Barnes, D.M. Schell, and E.Reimnitz (eds.). The Alaskan Beaufort Sea: Ecosystems and Environments. Academic Press, New York. pp. 47-71.

Aagaard, K., 1989. A synthesis of the Arctic Ocean circulation. Rapp. et Procès-verbaux des Réunions, Cons. Int. Explor. Mer. 188: 11-22.

ACIA, 2004. Impacts of a warming Arctic: Arctic Climate Impact Assessment. Cambridge University Press, Cambridge, UK.

Alexander, V., 1974. Primary productivity regimes in the nearshore Beaufort Sea, with reference to potential roles of ice biota. In: J.C. Reed and J.E. Sater (eds.). The Coast and Shelf of the Beaufort Sea. The Arctic Institute of North America, Arlington. pp. 609-632.

Bodaly, R. A., J. D. Reist, D. M. Rosenberg, P. J. McCart, and R. E. Hecky, 1989. Fish and fisheries of the Mackenzie and Churchill River basins, northern Canada. In: D.P. Lodge (ed.). Proc. Internat. Large River Symp., Canadian Special Publication of Fisheries and Aquatic Sciences, Department of Fisheries and Oceans, Ottawa, Ontario, p 128-144.

Cameron, W.M., 1951. Hydrography and oceanography of the Southeast Beaufort Sea and Amundsen Gulf; Part 1: Observations in 1951. Institute of Oceanography, University of British Columbia.

Carmack, E. C., and D. C. Chapman, 2003. Wind-driven shelf/basin exchange on an Arctic shelf: The joint roles of ice cover extent and shelf-break bathymetry. Geophys. Res. Lett., 30, 1778, doi:10.1029/2003GL017526.

Carmack E. C., and Y. A. Kulikov, 1998. Wind-forced upwelling and internal Kelvin wave generation in Mackenzie Canyon, Beaufort Sea. J. Geophys. Res., 103(C9): 18447-18458.

Carmack, E. C., and R. W. Macdonald, 2002. Oceanography of the Canadian shelf of the Beaufort Sea: A setting for marine life. Arctic, 55 (supp. 1): 29-45.

Carmack, E. C., R. W. Macdonald, and J. E. Papadakis, 1989. Water mass structure and boundaries in the Mackenzie shelf estuary. J. Geophys. Res. 94(C12): 18043-18055.

Carmack, E. C., R. W. Macdonald, and S. Jasper, 2004. Phytoplankton productivity on the Canadian Shelf of the Beaufort Sea. Mar. Ecol. Prog. Ser. 277, 37-50.

Coachman, L. K., and K. Aagaard, 1974. Physical. oceanography of arctic and subarctic seas. In: Y. Herman (ed.). Marine Geology and Oceanography of the Arctic Seas. Springer-Verlag, New York, pp.1-81.

Coachman, L.K., and C. A. Barnes, 1961. The contribution of Bering Sea water to the Arctic Ocean. Arctic 14(3): 147-161.

Dean, K. G., W. J. Stringer, K. Ahlnäs, C. Searcy, and T. Weingartner, 1994. The influence of river discharge on the thawing of sea ice, Mackenzie River Delta: albedo and temperature analyses. Polar Res. 13(1): 83-94.

Dumas, J., E. C. Carmack, and H. Melling, 2005. Climate change impacts on the Beaufort shelf landfast ice. Cold Reg. Sci. Technol., 42(1):41-51.

Dunton, K., 1992. Arctic biogeography: The paradox of the marine benthic fauna and flora. Trends Ecol. Evol. 7(6): 183-189.

Forest, A., M. Sampei, H. Hattori, R. Makabe, H. Sasaki, M. Fukuchi, P. Wassmann, and L. Fortier, 2007. Particulate organic carbon fluxes on the slope of the Mackenzie Shelf (Beaufort Sea): Physical and biological forcing of shelf-basin exchanges. J. Mar. Syst. 68, 39-54.

Giovando, L.F., and R. H. Herlinveaux, 1981. A discussion of factors influencing dispersion of pollutants in the Beaufort Sea. Report 81-4, 198 pp., Institute of Ocean Sciences, Sidney, British Columbia.

Henry, R. F., and M.G.G. Foreman, 1977. Numerical model studies of semi-diurnal tides in the southern Beaufort Sea. Pacific Marine Science Report 77-11, 71 pp., Institute of Ocean Sciences, Victoria, BC.

Horner, R. A., and G. C. Schrader, 1982. Relative contributions of ice algae, phytoplankton, and benthic microalgae to primary production in nearshore regions of the Beaufort Sea. Arctic 35: 485-503.

Ingram, R.G., E. Carmack, F. McLaughlin, and S. Nicol, 2006. Polar continental shelf boundaries. In: A. Robinson and K. Brink (eds.). The Sea, Volume 14A, Ch. 3, pp. 61-82, Harvard University Press, Cambridge, Mass.

Kowalik, Z., and A.Y. Proshutinsky, 1994. The Arctic Ocean tides. In: O.M. Johannessen, R. D. Muench, and J. E. Overland (eds.). The Polar Oceans and Their Role in Shaping the Global Environment, The Nansen Centennial Volume. Geophys. Monogr. Ser. 85:137-158, AGU, Washington, D. C.

Kulikov, E.A., E. C. Carmack, and R. W. Macdonald, 1998. Flow variability at the continental shelf break of the Mackenzie Shelf in the Beaufort Sea. J. Geophys. Res. 103(C6): 12,725-12,741.

Kulikov, E.A., A. B. Rabinovich, and E. C. Carmack, 2004. Barotropic and baroclinic tidal currents on the Mackenzie shelf break in the southeastern Beaufort Sea. J. Geophys. Res. 109(C5), C05020, doi:1029/2003JC001986.

Macdonald, R.W., and E. C. Carmack, 1991. The role of large-scale under-ice topography in separating estuary and ocean on an Arctic shelf. Atmos.-Ocean. *29*: 37-53.

Macdonald, R.W., C. S. Wong, and P. E. Erickson, 1987. The distribution of nutrients in the southeastern Beaufort Sea: Implications for water circulation and primary production. J. Geophys. Res. 92(C3): 2939-2952.

Macdonald, R.W., E. C. Carmack, F. A. McLaughlin, K. Iseki, D. M. Macdonald, and M. O. O'Brien, 1989. Composition and modification of water masses in the Mackenzie Shelf Estuary. J. Geophys. Res. 94(C12): 18,057-18,070.

Macdonald, R.W., E. C. Carmack, and D. W. Paton, 1995. Using the d18O composition in landfast ice as a record of arctic estuarine processes. Mar. Chem. 65: 3-24.

Macdonald, R. W., S. M. Solomon, R. E. Cranston, H. E. Welch, M. B. Yunker, and C. Gobeil, 1998. A sediment and organic carbon budget for the Canadian Beaufort Shelf. Mar. Geol. 144: 255-273.

McLaughlin, F. A., E. C. Carmack, R. W. Macdonald, and J. K. B. Bishop, 1996. Physical and geo-chemical properties across the Atlantic/Pacific water mass front in the southern Canada Basin. J. Geophys. Res. 101(C1): 1183-1197.

McLaren, A. S., M. C. Serreze, and R. G. Barry, 1987. Seasonal variations of sea ice motion in the Canada basin and their implications. Geophys. Res. Lett. 14, 1123-1126

Melling, H., 1993. The formation of a haline shelf front in wintertime in an ice-covered arctic sea. Cont. Shelf Res. 13: 1123-1147.

Melling, H, and E. L. Lewis, 1982. Shelf drainage flows in the Beaufort Sea and their effect on the Arctic Ocean pycnocline. Deep Sea Res. 29: 967-985.

Melling, H., D. A. Riedel, and Z. Gedalof, 2005. Trends in the draft and extent of seasonal pack ice, Canadian Beaufort Sea. Geophys. Res. Lett. 32, L24501, doi: 10.1029/2005GL024483.

Michel, C., R. G. Ingram, and L. R. Harris, 2006. Variability in oceanographic and ecological processes in the Canadian Arctic Archipelago. Progr. Oceanogr. 71: 379-401.

Omstedt, A., E. C. Carmack, and R. W. Macdonald, 1994. Modeling the seasonal cycle of salinity in the Mackenzie shelf/estuary. J. Geophys. Res. 99(C5): 10011-10021.

Parsons, T. R., D. G. Webb, H. Dovey, R. Haigh, M. Lawrence, and G. Hopkey, 1988. Production studies in the Mackenzie River-Beaufort Sea estuary. Polar Biol. 8: 235- 239.

Parsons, T. R., D. G. Webb, B. E. Rokeby, M. Lawrence, G. E. Hopkey, and D. B. Chiperzak, 1989. Autotrophic and heterotrophic production in the Mackenzie River/Beaufort Sea estuary. Polar Biol. 9: 261-266.

Rothrock, D. A., Y. Yu, and G. A. Maykut, 1999. Thinning of the Arctic sea-ice cover. Geophys. Res. Lett. 26: 3469-3472.

Tynan, C., and D. P. DeMaster, 1997. Observations and predictions of Arctic climate change: Potential effects on marine mammals. Arctic. 50: 308-322.

Williams, W. J., E. C. Carmack, K. Shimada, H. Melling, K. Aagaard, R. W. Macdonald, and R. G. Ingram, 2006. Joint effects of wind and ice motion in forcing upwelling in Mackenzie Trough, Beaufort Sea. Cont. Shelf Res. 26: 2352-2366.

Williams, W. J., H. Melling, E. C. Carmack, and R. G. Ingram, 2008. Kugmallit Valley as a conduit for cross-shelf exchange on the Mackenzie Shelf in the Beaufort Sea. J. Geophys. Res. 113, C02007, doi:10.1029/2006JC003591.

Williams, W. J., E. C. Carmack, and R. G. Ingram. Topographically induced wind-driven upwelling at Cape Bathurst, Beaufort Sea. Submitted to J. Mar. Res.

Yu, Y., G. A. Maykut, and D. A. Rothrock, 2004. Changes in the thickness distribution of Arctic sea ice between 1958-1970 and 1993-1997. J. Geophys. Res. 109(C8), C08004, doi: 10.1029/2003JC001982.

CHAPTER 3

The Ocean—Sea Ice—Atmosphere (OSA) Interface in the Southern Beaufort Sea

David.G. Barber*, Byong Jun. Hwang, Jens K. Ehn, John Iacozza, Ryan Galley, John Hanesiak, Tim Papakyriakou, Jennifer V. Lukovich

3.1 Introduction and Rationale

Sea ice conditions in the CASES region fall into three distinct 'regimes': 1) *offshore pack ice* (A in Fig. 3.1), which consists of annual and multi-year sea ice in offshore regions beyond the maximum landfast ice extent; 2) *landfast sea ice* (B in Fig. 3.1), which forms annually over the continental shelves along coastal margins; and 3) the Cape Bathurst Polynya Complex (C in Fig. 3.1), an ice regime which is composed of a series of flaw leads and a sensible/latent heat polynya. This Cape Bathurst Polynya Complex recurs annually on average at the shelf break between Amundsen Gulf and the Beaufort Sea (Fig. 3.1). It is formally described as a *recurrent* polynya, in that the flaw leads which drive its formation occur each year. This recurrence happens because the central pack ice essentially acts as an 'ice bridge'; that is, a retaining structure which prevents sea ice from advecting into the region (see Barber et al., 2001, for another example of an ice bridge in reference to the NOW polynya). The resulting anticyclonic rotation of the central pack keeps ice from advecting into the flaw lead system, promoting the formation of a latent heat polynya within Amundsen Gulf.

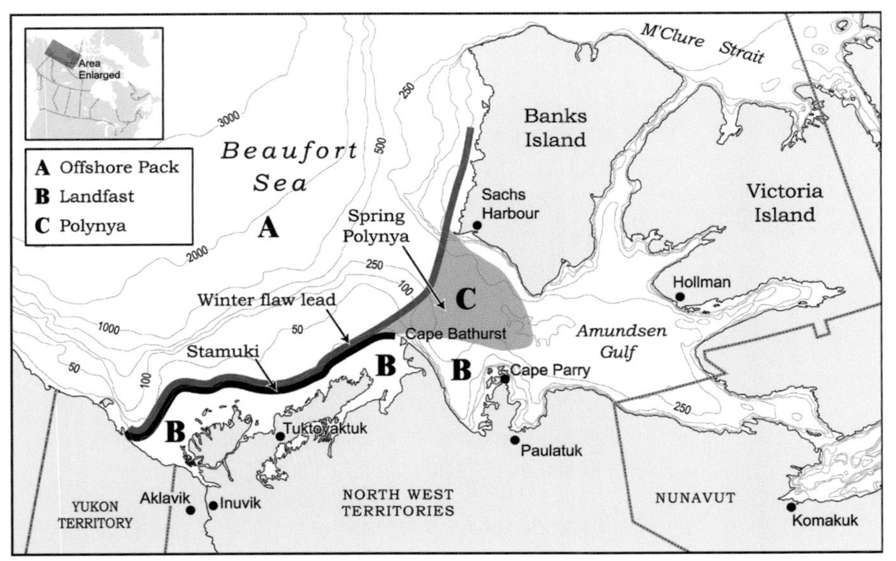

Figure 3.1

The CASES region showing the location and shape of its three ice regimes (from Barber and Hanesiak, 2004).

A significant interface occurs annually at the confluence of the offshore pack and the landfast ice regimes north of the Mackenzie estuary and the Tuktoyuktuk Peninsula (Fig. 1). Here, a highly deformed shear zone forms where the rotating offshore pack pushes against

LEFT: People working on the ice pack. Photo: Simon Belt/ ArcticNet.

Centre for Earth Observation Science, Clayton H. Riddell Faculty of Environment, Earth and Resources, University of Manitoba, Winnipeg, MB, R3T 2N2, Canada

* Corresponding author: dbarber@cc.umanitoba.ca, Clayton H. Riddell Faculty of Environment, Earth and Resources, University of Manitoba

the advancing edge of landfast sea ice, resulting in compression ridges. These ridges (termed 'stamukhi' a Russian word meaning 'grounded hummock') can be quite large, exceeding several metres in the sail and keel. They act as a barrier to freshwater from the Mackenzie River during winter, thereby forming an under-ice lake named Lake Herlinveaux (in honour of the first western scientist to document the importance of this geophysical feature; Carmack and Macdonald, 2002).

The offshore pack responds to synoptic-scale atmospheric patterns such as the Arctic Oscillation (AO) (Hurrell, 1995) and the North Atlantic Oscillation (NAO) (Thompson and Wallace, 1998) which, in turn, are features of the Northern Annular Mode (NAM). Each of these patterns (AO, NAO and NAM) represents highly correlated modes in the atmosphere over different regions. In each case, indices have been computed to show the strength of oscillatory patterns of high and low pressure systems which affect particular areas (see, for example, Proshutinsky and Johnson, 1997; Hurrell et al., 2001; Visbeck et al., 2003)

Because of its synoptic setting, the pack ice of the Beaufort Sea generally rotates in an anti-cyclonic gyre (clockwise in the northern hemisphere) centered near 78°N, 150°W. Specifically, the anti-cyclonic flow is due to a predominant surface high pressure system over the region during winter. The seasonal ice pack within the transition zone is also incorporated into the gyre during winter and rotates at about 35° per year (Thorndike and Colony, 1977). This typical circulation pattern is responsible for much of the sea ice export from the Canada Basin into the trans-polar drift (and ultimately through Fram Strait; Kwok and Rothrock, 1999) and for deformation-induced increases in ice volume observed in the Beaufort region (Melling and

Riedel, 1996). The surface waters of the Canada Basin are known to originate in the Pacific (Carmack and McDonald, 2002) whereas the deeper waters (>200 m) originate in the Atlantic. These deeper waters circulate in a counter-clockwise (cyclonic) flow opposite to the sea ice (Carmack and McDonald, 2002).

Parkinson et al. (1999) showed that between 1978 and 1996 the annual average sea ice extent over the northern hemisphere decreased by about 34,600 km^2. This reduction was spatially heterogeneous, consisting of large decreases in some locations (e.g., The Chukchi and Laptev seas) and slight increases in others (e.g., Baffin Bay). The issue of change in sea ice thickness (i.e., volume) is more controversial; observations from submarine sonar suggest a decline in thickness over the past 40 years (Rothrock et al., 1999), however this may be due to sampling issues rather than a direct reduction in the hemispheric sea ice mass balance (Holloway and Sou, 2002).

Recent evidence suggests that the areal extent of sea ice is continuing to decline. In 2005, a minimum in the extent of sea ice was observed over the Arctic Basin (since passive microwave remote sensing began in 1978). This minimum appeared to be due to anomalously strong southerly winds in spring which advected sea ice pole-ward in the Siberian sector of the Arctic. A persistent low pressure system coupled with high air temperatures also promoted divergence in the pack ice, thereby decreasing albedo and enhancing ice melt. Finally, the divergence of sea ice promoted the development of early flaw leads at continental shelf breaks along the western sector of the Arctic Basin which may have limited ice growth in late winter and promoted ablation in the spring and summer (Serreze et al., 2003). Under strong positive AO conditions, we might expect an early formation of shore flaw leads and advection of ice away from the coasts due to a predominantly cyclonic flow in the atmosphere (Rigor et al., 2002). While observations of extent are somewhat consistent with this theory, Serreze et al. (2003) note that there are differences in the interpretation of summer sea ice anomalies relative to both a weak AO index in the previous winter and in early spring.

Very little scientific literature pertains to flaw lead polynya systems or to the Cape Bathurst Polynya Complex. Work by Cavalieri and Martin (1994) illustrated the importance of the flaw lead polynya system in the formation of intermediate and deep water via ice production. They estimated based on data from 1978 to 1987 that the circumpolar flaw lead polynya system could account for about 800 km^3 of new ice production during winter. Such production would provide significant input to the Arctic Ocean's thermohaline circulation and should be included in any local estimates of shelf/basin fluxes (Serreze et al., 2003). Agnew et al. (1999) examined the formation of the flaw lead associated with Banks Island during both an episodic event in 1998 and a 10-year period between 1987 and 1997. They estimated that, over a 10-day period, ice production associated with the flaw lead was of the order of 7.36 km^3 (which is equivalent to about 22.7 x 10^{11} MJ of heat lost to the atmosphere; Agnew et al., 1999). This divergence of sea ice away from Banks Island and Prince Patrick Island likely preconditioned the sustained reduction in sea ice extent measured in the fall of 1998 (Maslanik et al., 1999).

Although imperfect, our only tools for projecting future sea ice conditions are regional, hemispheric and global circulation models. Estimates of sea ice reduction from the Canadian Centre for Climate Modelling and Analysis (CCCMA) project a reduction in sea ice for both the northern and southern hemispheres (NH, SH). Both versions of the CCCMA global model show a reduction approximately equivalent to our current passive microwave observational record. Projections suggest that NH sea ice reduction might eventually result in a seasonally ice-free Arctic (during September) as early as 2050. Maximum winter extents in 2050 might become approximately equivalent to the summer extents observed today (Flato and Boer, 2001). Recent modelling efforts by Holland et al. (2006) using high-resolution regional-coupled models confirm these predictions and projected a seasonally ice-free Arctic around 2050. If these projections are correct, we can continue to expect significant changes in the Arctic Ocean and its coupling to the rest of the planet.

The following summarizes key research results obtained by the sea ice group of the CASES program, with extensive references to peer-reviewed literature. Some of the observational data sets we collected were 'first-ever' attempts for this group. At this time, our team has translated its findings into a contribution of over 30 articles to peer-reviewed literature. The data we present are currently contained within the CASES database and copies of the original journal articles are available through www.umanitoba.ca/ceos (under 'publications').

3.2 Overview of Results

It was quite unique to spend a year documenting mass, gas and energy fluxes across the ocean—sea ice—atmosphere interface of the CASES region. What follows are highlights of some our most significant findings. We also reference the pertinent peer-reviewed literature so that the interested reader can pursue some of these studies in greater detail. We organize our synthesis into three sections: atmosphere (Section 3.2.1), sea ice (Section 3.2.2) and surface-energy balance (Section

3.2.3). Each section focuses on the processes which thermodynamically and dynamically control snow-covered sea ice, and the impact that the ocean and atmosphere have on these processes.

3.2.1 The atmosphere

The atmosphere is a major driving force for physical and biological processes operating over the ocean—sea ice—atmosphere interface. Here, we summarize six key atmospheric variables over the CASES region affecting the dynamic and thermodynamic processes associated with sea ice growth and decay.

3.2.1.1 500 hPa Circulation

The mean 500 hPa flow for the region (Fig. 3.2) is characterized by a distinct westerly (mostly zonal) pattern consisting of a polar vortex (stretching toward Hudson Bay) and a weak ridge-like pattern (extending from the Mackenzie Region into central Alaska). Seasonal variations in this flow pattern allow various types of surface air masses to intrude into the area, including Arctic-type, Pacific-type, more continental-type (from southern latitudes), and sometimes even Archipelago-type (when a brief easterly surface flow develops). The 500 hPa flow is primarily determined by two features: an upper low positioned over the central Arctic Islands during the summer (and which intensifies and shifts to northern Foxe Basin during the winter); and the Aleutian Low / Pacific High. However the polar vortex itself also appears to play a partial role in determining the local flow pattern (see, e.g., Bradley and England, 1979; Maxwell, 1981; Serreze et al., 2003).

3.2.1.2 Sea Level Pressure Pattern & Storms

The mean sea level pressure (SLP) pattern (1978-2000) for the region (Fig. 3.3) averages between 1017 and 1019 hPa and consists of a weak ridge which extends from the Arctic Ocean into the Mackenzie region. This pressure pattern suggests relatively benign weather conditions and weak annual wind averages. However, daily weather fluctuations within the region can be quite severe, particularly in the shoulder seasons (discussed later). The annual average SLP pattern is representative of the mean winter pattern which, by April, gradually becomes a quasi-stationary high pressure cell over the Arctic Ocean with a maximum central SLP of 1022 hPa. The CASES region experiences a relative annual maximum SLP in March-April (1019 hPa; Fig. 3.4a). Mean pressure gradients reach their maximum in April and May then slacken sharply by June. At that time, the central pressure of the quasi-stationary high weakens (1014 hPa) and remains over the northern part of the CASES region (whose mean June SLP is

Figure 3.2

NCEP-derived mean 500 hPa geo-potential height (m) for 1978-2000 over the CASES region.

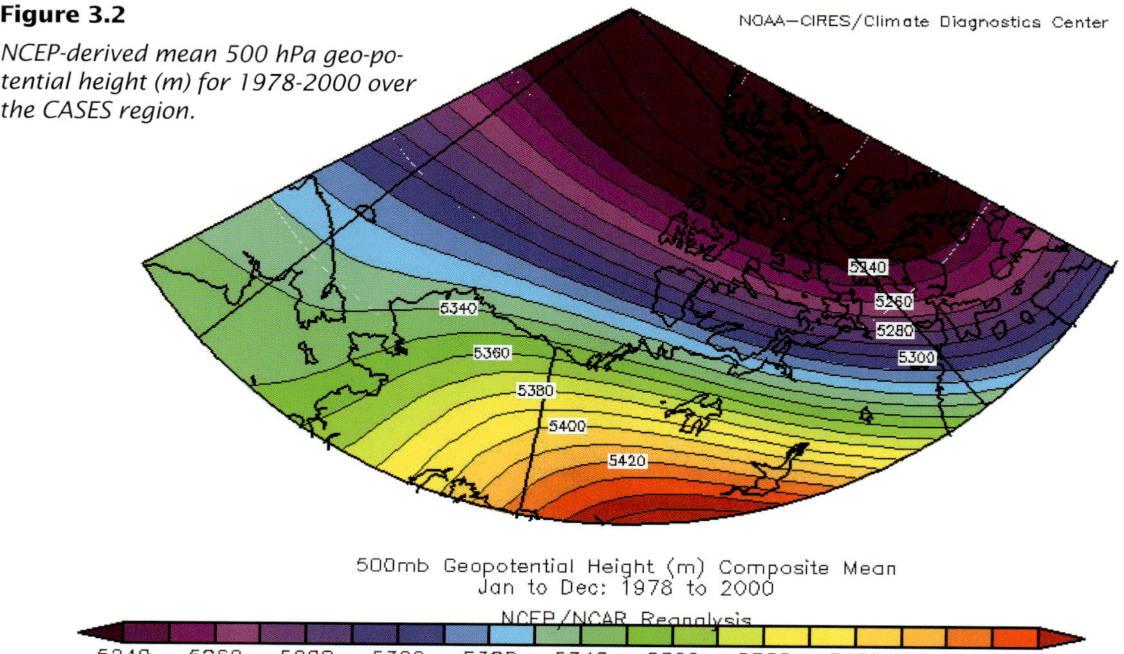

also around 1014 hPa). The climatological minimum mean SLP for the region (1012 hPa) occurs in August-September. As the ridge pattern over the Arctic Ocean and Mackenzie region begins to redevelop in the fall, the Aleutian Low intensifies and causes an increase in the baroclinicity and pressure gradient over Alaska and Yukon. However, the mean fall pressure gradients over the CASES region are weaker than in the spring.

3.2.1.3 Surface Air Temperatures

Mean surface air temperatures (1978-2000) throughout the region exhibit a north-south gradient from the south-central Arctic Ocean (-15 °C) to the Beaufort Sea coastline (about -10 °C) (see Fig. 5 in Barber and Hanesiak, 2004). These temperatures are consistent with MSC station climate normals over the 1971-2000 period (Table 3.1) and Maxwell (1981). Seasonally, mean winter air temperatures range from –24 °C (at the CASES southern limit) to –28°C (at the northern limit) and can reach +10 °C in the southern limit (+3°C in the north) during the summer (Fig. 3.4b). These figures are consistent with MSC station climate data (1971-2000; Table 3.1) and Maxwell (1981). The isotherms across the temperature gradient are mainly aligned east-to-west over the full annual cycle (due to thermal contrasts between the sea ice and terrestrial surfaces). Based on temperature measurements with respect to sea ice, winter for the CASES region is defined as the period between December and March; spring as the period between April and May; summer as the period (associated with strong melting conditions) between June and August; and fall as the period between September and November.

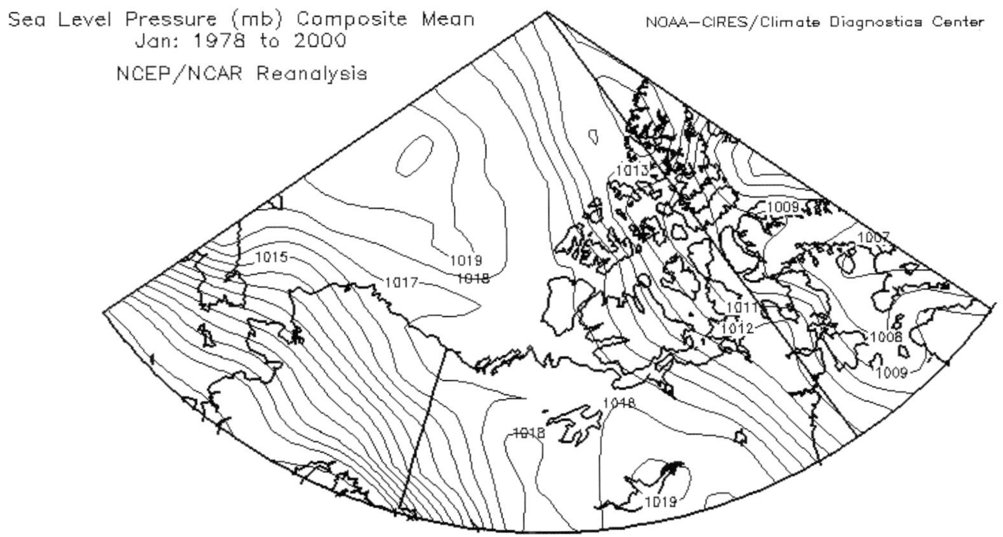

Figure 3.3

NCEP-derived mean sea level pressure (SLP) (hPa) pattern for 1978-2000 over the study region. The contour interval is 1 hPa.

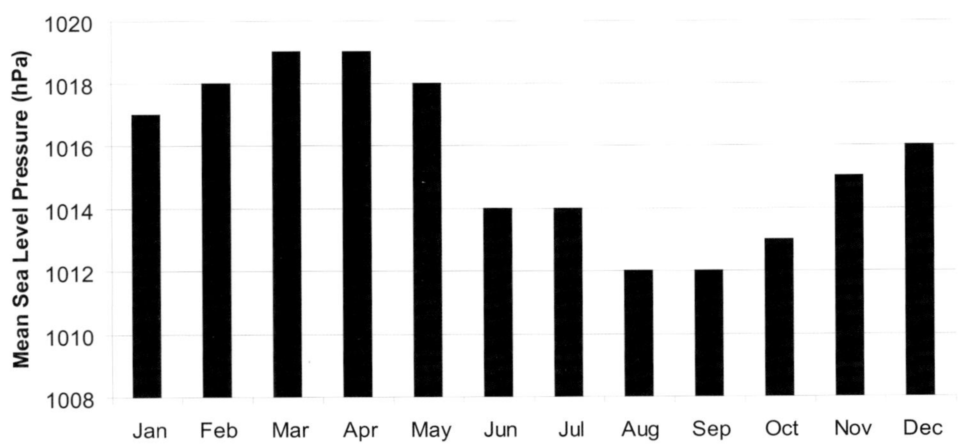

Figure 3.4a

NCEP-derived monthly (a) sea level pressure, (b) mean air temperature, and (c) winds for 1978-2000 over the CASES region. (b) and (c) have been subdivided into "north" and "south" CASES sectors.

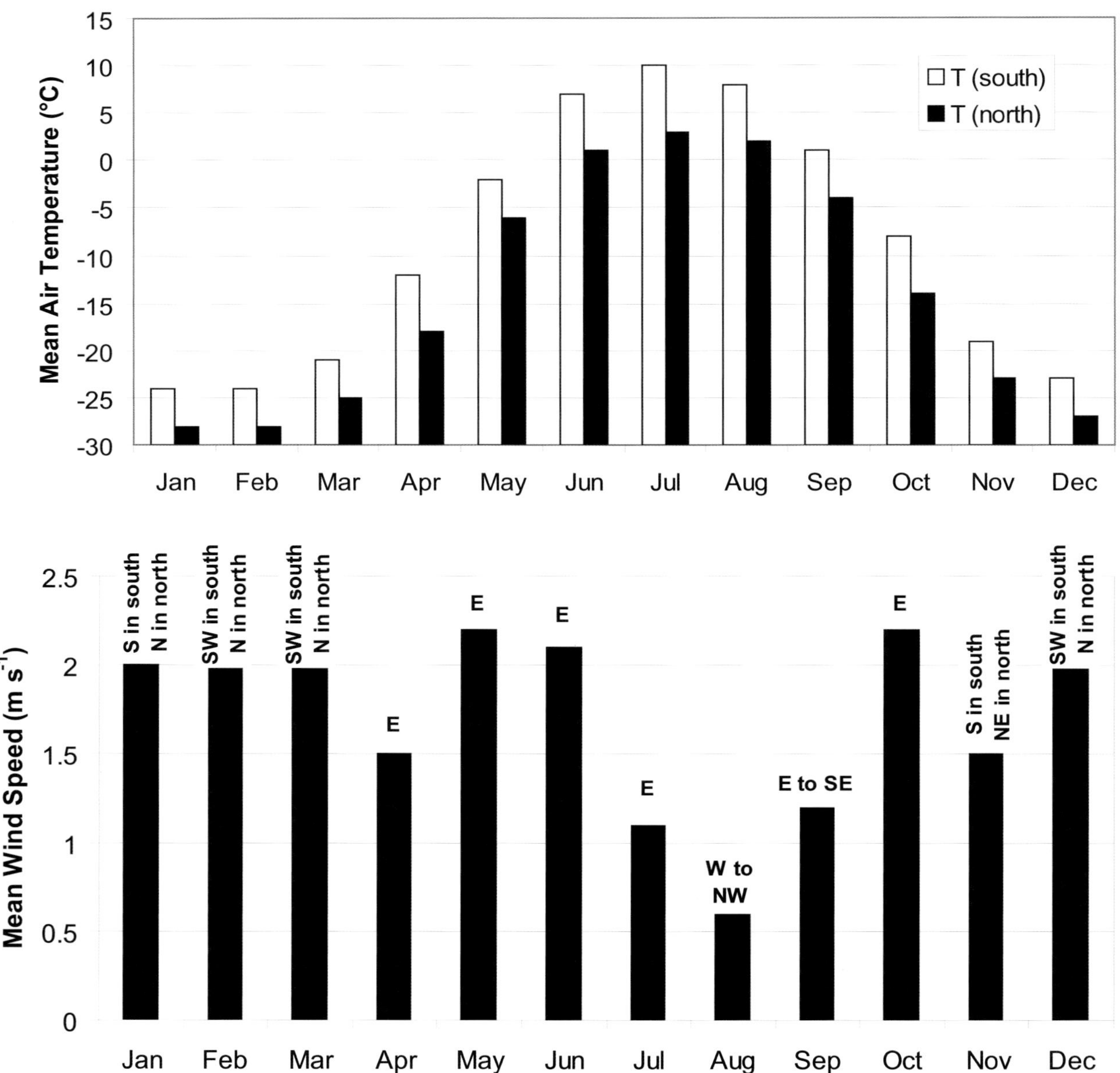

Figure 3.4
b *(TOP)* **and c** *(BOTTOM)*

TABLE 3.1

Climatology of MSC stations within and around the CASES study area. All values refer to annual means. The mean maximum monthly snow depth refers to the average maximum snow depth that occurs on any given month. FZRA is freezing rain, FZDZ is freezing drizzle, Cbs is thunderstorms, and BS is blowing snow. The superscripts refer to times of the year in which each weather element has the greatest frequency of occurrence.

Station	Avg. T (°C)	Daily max T (°C)	Daily min T (°C)	Mean winter T (°C)	Mean summer T (°C)	Rain (mm)	Snow (cm)	Total precip (mm)	Mean snow depth (cm)	Mean Max monthly snow depth (cm)	Days with FZRA/ FZDZ	Days with Cbs	Days with fog/ ice fog	Days with BS
Tuktoyaktuk	-10.2	-6.6	-13.9	-26	10	70.2	69.2	139.3	15	31	4.6[1]	0.33	33.5[3]	29.5[4]
Inuvik	-8.8	-3.9	-13.6	-26	13	117	167.9	248.4	26	57	9.4[1]	2.1	23.3[1]	10.6[4]
Sachs Harbor	-13.3	-10.0	-16.6	-29	5	52.2	105.4	149.4	9	14	8.3[1]	0.04	65.2[2]	38.5[4]
Holman	-11.7	-8.2	-15.1	-28	7.5	77.6	85.3	162.4	10	14	1.3[1]	0.48	N/A	N/A
Komakuk	-11.0	-7.2	-14.9	-25	7	83.5	77.7	161.3	14	26	1.9[1]	0.16	50.9[2]	78.1[4]
Paulatuk	-10.6	-7.2	-14.0	-25	7	90.8	77.2	168.0	11	21	2.5[5]	0.56	47.3[2]	57.3[4]
Cape Parry	-12.0	-9.0	-14.9	-28	5.5	67.1	120.9	157.2	9	18	16.5[1]	1.1	64.2[5]	69.1[4]

[1] – maximum occurs in spring / fall

[2] – maximum occurs in summer

[3] – maximum occurs in spring

[4] – maximum occurs in winter

[5] – maximum occurs in spring / summer

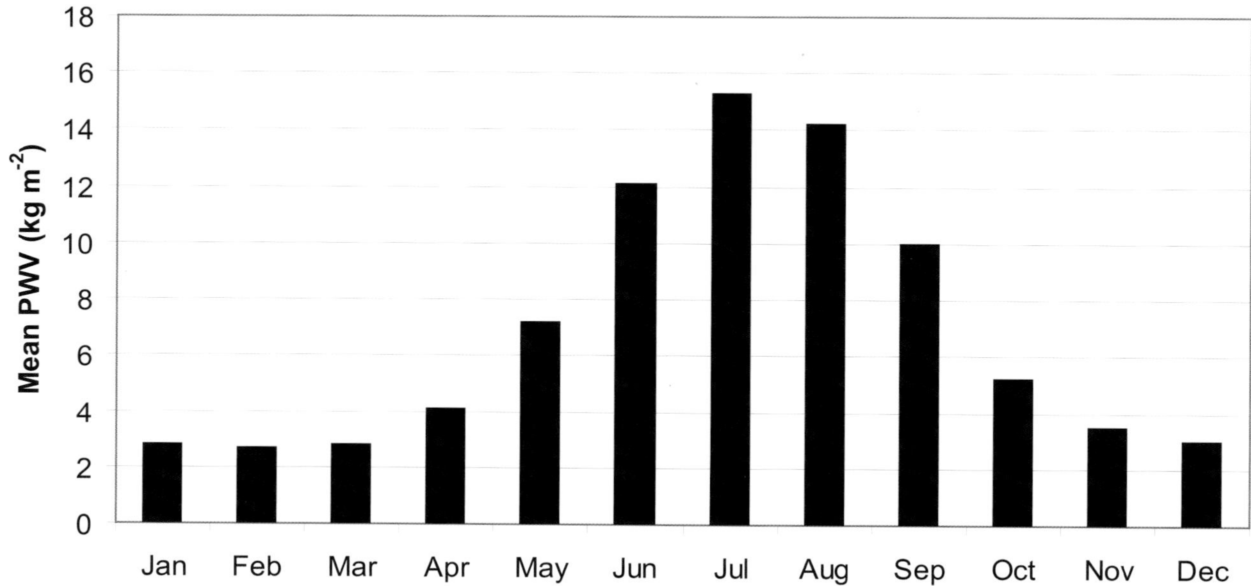

Figure 3.5

NCEP-derived monthly mean precipitable water (kg m^{-2}) for 1978-2000 over the study region.

3.2.1.4 Winds

According to NCEP re-analysis data, mean wind speeds (1978-2000) within the CASES region are around 1.0 m/s, originating from north-east and east (see Fig. 6 in Barber and Hanesiak, 2004). The strongest mean winds (> 2 m/s) generally occur in the spring and fall (primarily from the east) while the weakest ones (~ 0.5 m/s) occur in summer (Fig. 3.4c). The NCEP data also suggest a mean surface convergent flow for the region during winter, with a southerly flow in the south and a northerly flow in the north (Fig. 3.4c). This may potentially be due to low-level cloud formation combined with surface heat and moisture fluxes through local sea ice cracks and leads. Wind roses from individual MSC stations do not verify this surface flow convergence (see Fig. 7 in Barber and Hanesiak, 2004); however, this might be due to local wind effects and the sparse nature (both spatially and temporally) of the data collected in the area.

3.2.1.5 Precipitation & Adverse Weather

The mean precipitable water vapour (PWV) available to the surface ranges between 6 kg·m^{-2} in the northern portion of the CASES region and 8 kg·m^{-2} in the south (NCEP data from 1978-2000; Fig. 3.5). The spatial pattern for PWV follows that of air temperature (i.e. warmer air associated with higher PWV). PWV values are at their lowest (near 3 kg·m^{-2}) in winter and their highest (over 15 kg·m^{-2}) in July. Resulting annual average precipitation values for the coastal MSC stations of the CASES region are between 139 mm (Tuktoyaktuk) and 168 mm (Paulatuk) (see Table 3.1). Inuvik receives more precipitation (near 250 mm) due to frequent summer convective activity and snowfall during the winter. Greater snowfall in Inuvik may be due to a number of factors, including: 1) a greater probability of being affected by southern weather systems; 2) its inland location, which allows more upslope precipitation events; and 3) its proximity to a major mountain range which produces snowfalls. Annual average snow depths within the coastal region range between 9 cm (Sachs Harbour) and 15 cm (Tuktoyaktuk), while inland Inuvik experiences an average of 26 cm (Table 3.1). Coastal mean maximum snow depths for any given month in spring range between 14 cm (Holman and Sachs Harbour) and 31 cm (Tuktoyaktuk); Inuvik's mean maximum snow depth in spring is near 57 cm (Table 3.1). Such climatological information on snowfall is important to understanding the thermodynamics of sea ice as well as its associated biogeochemical cycles.

3.2.1.6 Clouds

Clouds are crucial to the Arctic climate system because they control radiative exchange with the surface (Curry et al., 1996; Walsh and Chapman, 1998; Minnett, 1999). The detection and measurement of cloud parameters, either *in situ* or from space, remains a challenge. Basic statistical properties of clouds in relation to the annual time-space cycle of sea ice are relatively poorly understood mostly due to the logistical difficulty in mounting pertinent field programs. The CASES network, however, provided an excellent vehicle for collecting information on polar clouds through an annual cycle. Here, we summarize results for time series estimates of cloud fraction range 0-1.0 (CF) over the study region.

Using CFs from manobs (manual observations of clouds) as our standard, we found an annual CF evolution which agreed well with other studies on Arctic environments; i.e., few clouds in winter and more clouds in spring (Intrieri et al., 2002; Walsh and Chapman, 1998; Schweiger et al., 1999). The major shift in CF occurred between late April and early May, with values quickly increasing from around 0.4 to > 0.7. CF values obtained using a ceilometer (a skyward looking laser profiling system) indicated that high clouds occurred

more frequently between mid-January and mid-April. This might have been due to the fact that the CF was not correctly determined from the surface when low or mid-cloud conditions were more extensive. The values obtained were similar to those of Curry and Ebert (1992), lending support to the quality of our datasets. The NCEP CF time series for the CASES region, however, showed a different pattern compared to other CF data: while all other CF datasets obtained (manobs, ceilometer, MODIS Aqua, MODIS Terra) showed a dramatic increase from late April (CF ≈ 0.4) to early May (CF ≈ 0.7), NCEP CF values remained near 0.2. We concluded that NCEP data failed to catch the seasonality of Arctic cloud cover—a finding similar to Walsh and Chapman (1998), who compared CF from three databases: Russian drifting ice stations, NCEP re-analyses data, and the European Centre for Medium-Range Weather Forecasts (ECMWF). Recently, this discrepancy in the NCEP cloud fraction product has been attributed to inaccuracy in predicting mixed-phased Arctic clouds (MPACs) during spring/fall. Several unique dynamical and thermodynamical characteristics have since been highlighted to help understand the MPAC process (Fu and Hollars, 2004; Zuidema et al., 2005; Shupe et al., 2005).

Between December and early April, CF values obtained from MODIS instruments aboard Aqua and Terra had different seasonal patterns. Specifically, in December-January MODIS-Aqua over-estimated CFs relative to MODIS-Terra (which agreed better with the manobs dataset), and from mid-February to late March, both MODIS products significantly over-estimated CF values (0.5 < CF < 0.7) relative to manobs (0.2 < CF < 0.4). Based on the small CF values observed via manobs and ceilometer, this latter period appeared to be commonly characterized by cloudless or slightly cloudy skies. Moreover, the difference between ceilometer

Using an air-iceboat to sample young consolidated pancake ice. Photo: Dave Barber.

and manobs values indicated more high clouds in this stage. It is reasonable to assume that MODIS might have misrepresented snow-covered background as clouds under clear or high cloud conditions. However, both MODIS products agreed well with manobs data after April. The monthly averaged CF deviations from the four databases compared to manobs are listed in Table 3.2.

TABLE 3.2

Month	Aqua	Terra	NCEP	Ceilometer
Dec	0.25	0.00	-0.06	-0.02
Jan	0.07	-0.13	-0.11	-0.11
Feb	0.10	0.11	-0.13	-0.16
Mar	0.31	0.29	-0.08	-0.13
Apr	0.00	-0.08	-0.16	-0.19
May	0.04	-0.05	-0.47	-0.03

Further details on CF for the CASES region, including comparison with various datasets are available in Jin et al. (2006 & 2007).

3.2.2 The Sea Ice

The annual sea ice cycle in the CASES region is a product of both thermodynamic and dynamic processes. The average cycle of sea ice formation as observed from passive microwave data (Fig. 3.6) shows that ice begins to grow southward from the summer minimum extent of the central pack around the end of September (week 39). This southward progression occurs rapidly, such that by week 43 (Fig. 3.6) the ice reaches near the coast. The Cape Bathurst Polynya Complex is the last area to form sea ice (weeks 42-44). Sea ice can remain

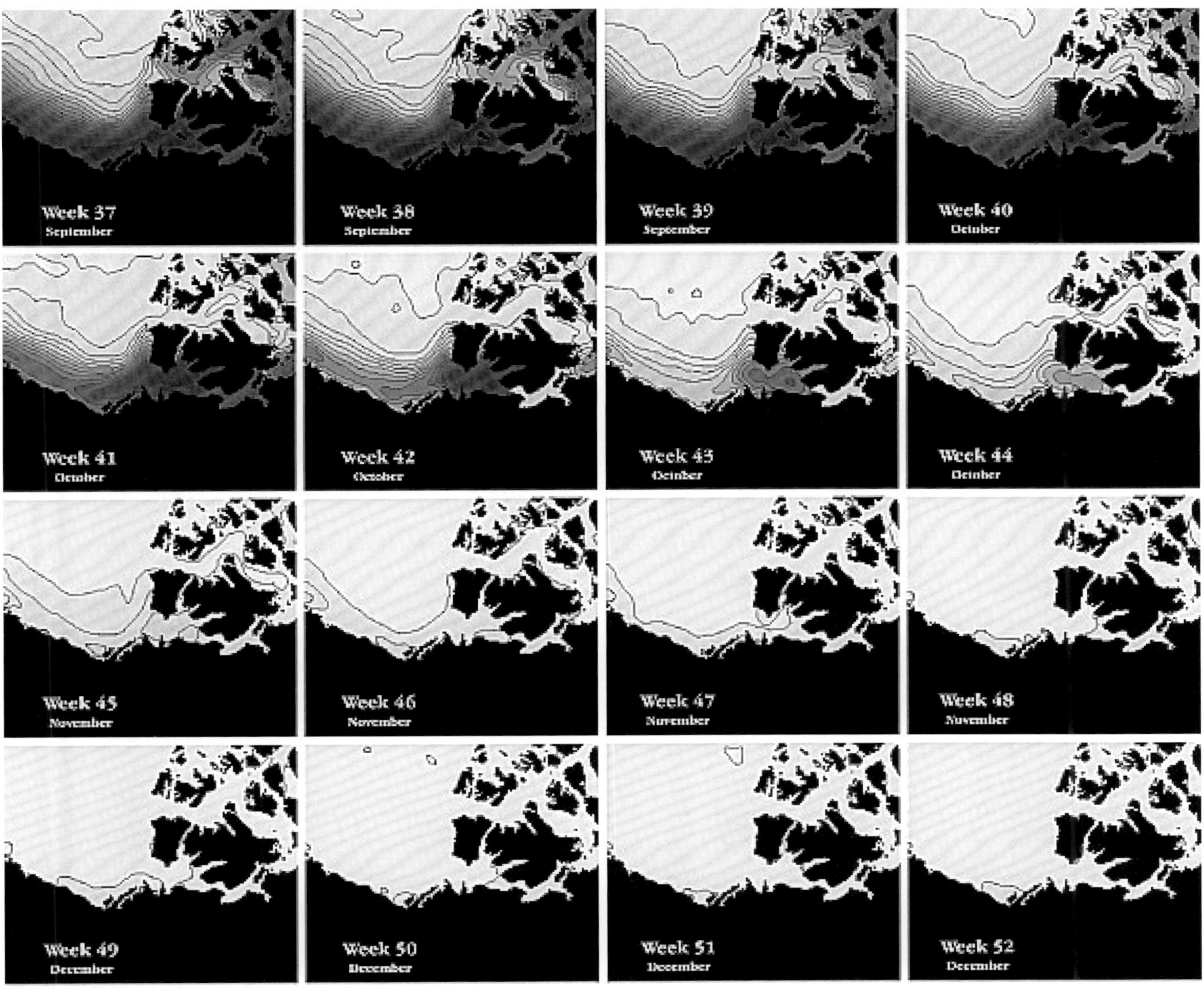

Figure 3.6

Average sea ice concentrations during freeze-up (formation) computed over the period 1979-2000 for weeks 37 to 52. Concentrations are expressed as percentages (from Barber and Hanesiak, 2004).

October to January

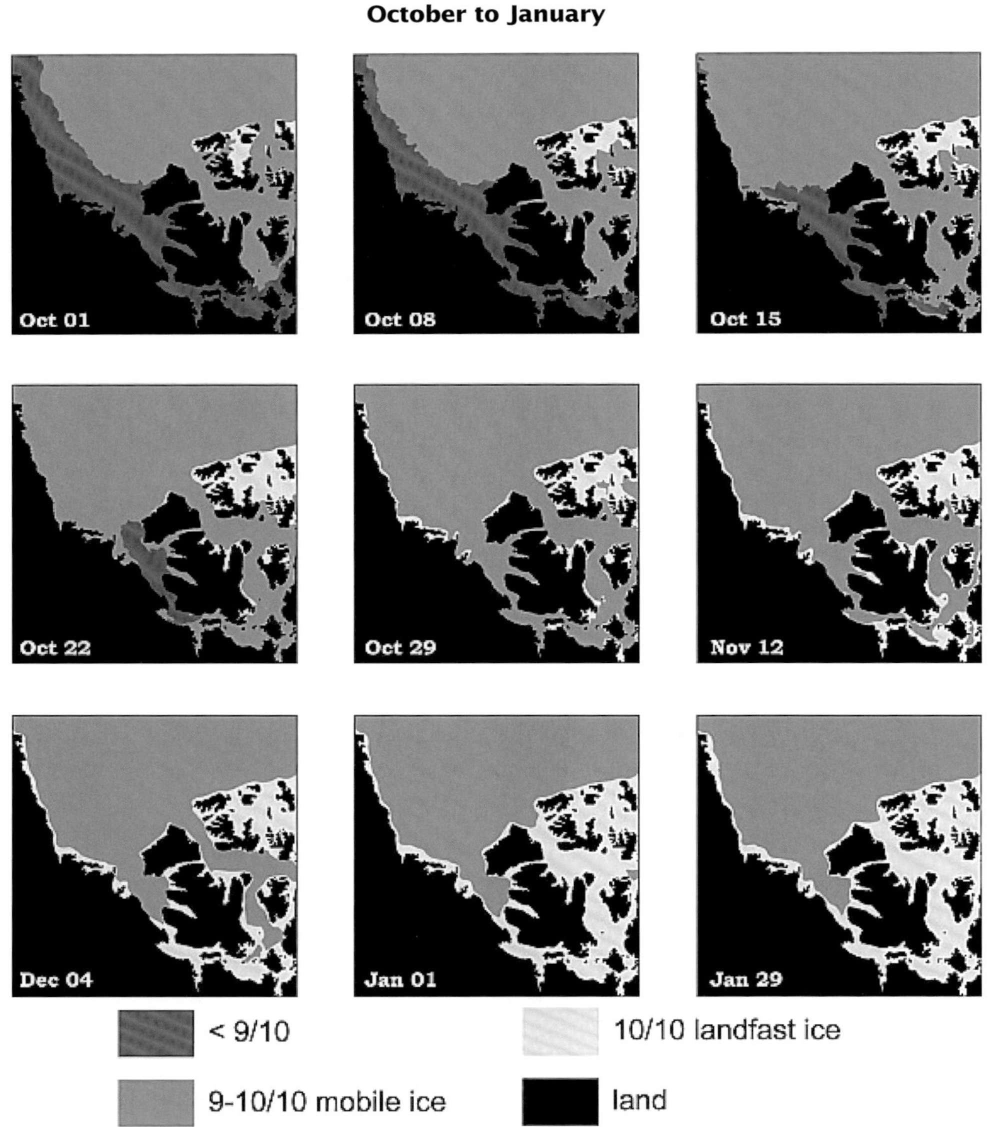

Figure 3.7

Sea ice concentrations for selected weeks from the Canadian Ice Service (CIS) digital ice chart products showing the median ice concentration (expressed in tenths) for mobile and landfast sea ice during freeze-up (formation) (from Barber and Hanesiak, 2004).

mobile in the polynya region well into the winter (approximately February); however, its concentrations are high (<9/10, 9/10, 10/10[ths]).

Data from Canadian Ice Service (CIS) digital archive provide a clear representation of the annual median ice concentrations of high-concentration mobile and fast ice (<9/10, 9/10, 10/10, Fig. 3.7). Young ice forms along the pack ice margin in early October. Both dynamic and thermodynamic processes give rise to an increase in ice volume during this period. By the end of October, high concentrations of ice are present along the Mackenzie Shelf (October 29, Fig. 3.7). By November 12, the landfast ice begins to grow in Franklin Bay and along the coast of Cape Parry. Over December and January, the fast ice regions continue to grow outwards from the coasts. These growth periods result in shear zone features (clearly visible by satellite synthetic aperture radar (SAR), not shown here) where the mobile pack ice and the landfast ice meet. CIS ice charts, such as the ones shown in Fig. 3.7, are derived from a composite of satellite and aircraft data and are thus not systematically distributed in time.

The average annual pattern of sea ice decay is quite different from that of its formation (cf. Fig. 3.6 and 3.8). Passive microwave data indicates some transient reductions in ice concentration as early as April (week 14; Fig. 3.8) in the areas where we would expect to find flaw leads. A continuous decrease in ice concentration occurs by week 16 (again near the Banks and Mackenzie Shelf flaw leads). By week 20, further ice reduction occurs in three areas: to the west of Banks Island, to the north of the Tuktoyaktuk peninsula, and in the eastern limit of Amundsen Gulf. These nodes interconnect in the subsequent weeks to produce a distinct polynya 'complex' characterized by two limbs: One along the shelf/slope break near the flaw lead polynyas, and one reaching into Amundsen Gulf (weeks 20-26). This 'tri-node' polynya complex increases in size throughout the spring melt until most of the Alaskan coast becomes ice free (week 34). Fall closure of the region (i.e. the formation of sea ice) begins with rapid southward advancement of the central ice pack (week 36 to 41; Fig. 3.9a) until the coast is reached, then continued ice formation to the south-east into Amundsen Gulf. Sea ice formation within the polynya complex occurs by week 43.

Inter-annual variability in the cycle of sea ice formation and decay presents some interesting patterns which may lead us to better understand its governing processes. Standard deviation (SD) analysis indicates that variability in the onset of ice formation (in week number) is smallest near the central ice pack and landfast ice (SD=1), twice as large in the region between the landfast ice and the offshore pack (SD=2), and largest (SD=3) in the region of the Banks Island flaw lead (Fig. 3.9b). This may be due to the fact that during ice formation, thermodynamic processes dominate over dynamic ones near the central pack and in near-shore areas (i.e. the ice grows thermodynamically more quickly than it moves). At the interface between the central pack and the landfast ice, we also expect large variability in the period of ice formation since the central pack can sometimes move across the pole in response to different hemispheric teleconnections (e.g., the two modes of the AO postulated by Proshutinsky and Johnson, 1997, and Rigor et al., 2002). Variability in the timing of landfast ice formation is highly influenced by variability in air temperature (particularly as it approaches the shelf break). In addition, variability can be caused by open water between the ice pack and the fast ice which, under favourable wind conditions, can sometimes induce upwelling (Wang and Key, 2003). High variability in the timing of ice formation in

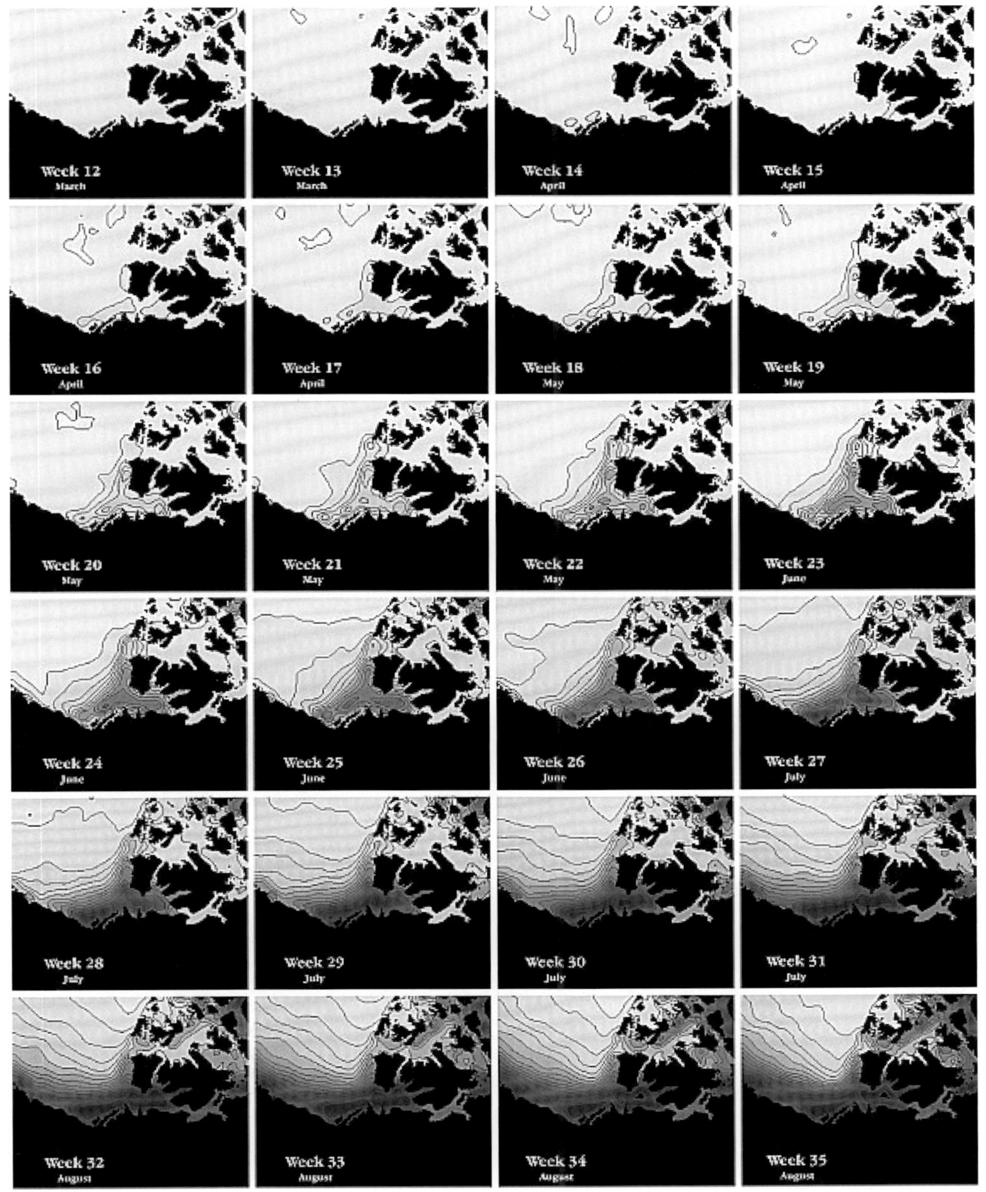

Figure 3.8 *Average sea ice concentrations during break-up (decay) averaged over the period 1979-2000 for weeks 12 to 35. Concentrations are expressed as percentages (from Barber and Hanesiak, 2004).*

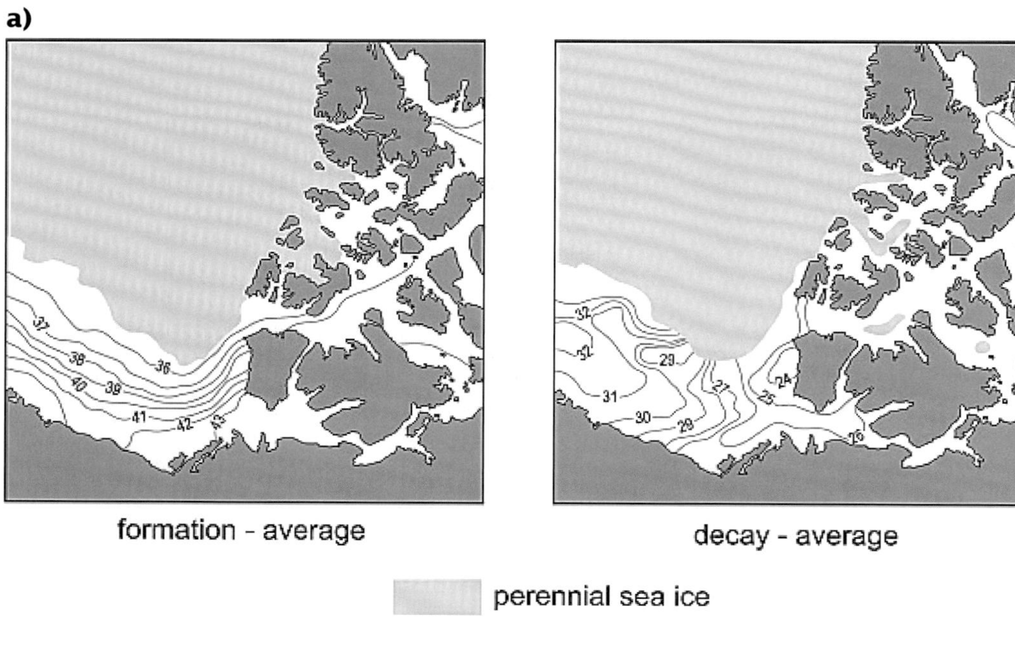

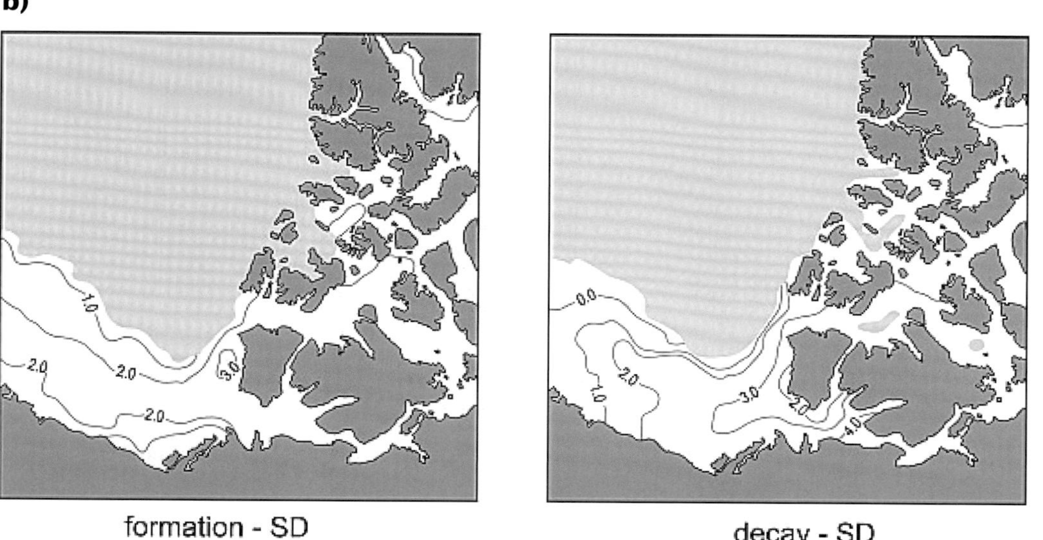

the flaw lead is likely due to periodic movement of the central ice pack towards and away from Banks Island. Standard deviation analysis also confirms that the polynya region possesses the highest variability in the timing of ice decay (Fig. 3.9b).

When one examines the full dataset of sea ice concentration anomalies (www.umanitoba.ca/ceos/iceanimations) it becomes apparent that there occurs variation in the sea ice concentration anomaly patterns in both time and space. One way to present trends in these anomalies is to compute the slopes of a least squares line through their monthly time series. Figure 3.10 provides an illustration of one such time series for a single pixel. Although this figure depicts much inter-annual variability, it is still statistically associated with a significant ($p<0.01$) slope to the least squares linear fit. The time series of the sea ice anomalies indicates a tendency toward relatively positive sea ice anomalies in the 1980's and relatively negative sea ice anomalies in the 1990's. This general observation is consistent with other studies which described various atmospheric variability time scales in the 1980's versus those of the 1990's (e.g., Proshutinsky and Johnson, 1997; Rigor et al., 2002; Drobot and Maslanik, 2003).

Figure 3.9

(a) Average date, expressed in week number, for the formation of sea ice (defined as >70 percent concentration) and decay (defined as <30 percent) averaged over the period 1979-2000, (b) The standard deviations of these average dates of formation and decay show the inter-annual variability or variance in these average weeks (from Barber and Hanesiak, 2004).

When the magnitude of the slope of each anomaly time series is plotted in the 85 by 85 pixel frame, we can visualize tendencies towards positive or negative slopes and, more importantly, the spatial coherence among (and between) points in the study area (Fig. 3.11). Between 1979 and 2000, the tendency in sea ice concentration was towards a reduction (increase in negative ice concentration anomalies). The spatial pattern for these trends showed two distinct regions of reduced concentration (Fig. 3.11). The first occurred at the interface of the central pack and the landfast sea ice north of Alaska, parallel to the coast, suggesting it was influenced by a shelf break process. The second was at the east end of the Cape Bathurst Polynya within the Amundsen Gulf. The larger of these two zones appeared to be the one at the interface of the central pack and the fast ice north of Alaska. We speculated that this was due to either a reduction in sea ice concentration within the central pack or divergence of the pack away from the coast. Local upwelling might also support the observed pattern, as well as teleconnection patterns over the Arctic and North Pacific. The observations were also broadly supported by both the AO and sea ice concentration anomaly cross-correlation results presented in Figure 9 of Barber and Hanesiak (2004).

The decrease in concentration anomalies observed in Amundsen Gulf came as a bit of a surprise as they did not directly relate to reductions in the central Arctic pack. It may be that either 1) the polynya is tending to open earlier in the year (and longer), 2) ice may be melting earlier in response to atmospheric heat advection from surrounding landmasses, or 3) local atmospheric flows are advecting ice out of the region. It is also possible that the flaw leads at the shelf break are allowing the upwelling of warmer Atlantic water to the base of the sea ice in Amundsen Gulf, thereby providing

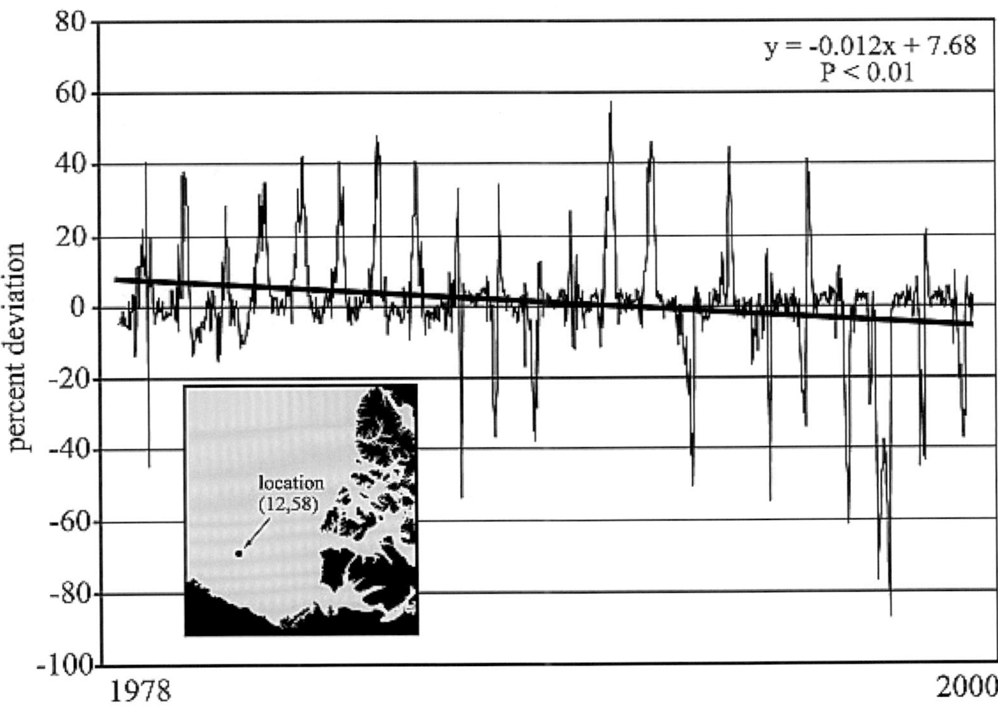

Figure 3.10

Time series profile of sea ice concentration anomalies for a single pixel located at (12r, 58c). The anomalies are expressed as percent deviation from the average computed between 1979 and 2000 and presented for each week (52) for the 22 years of the dataset (22·52) resulting in 1144 data points. The inset shows the geographic location of the pixel and the least squares best fit linear trend is shown to illustrate how the slopes in Figure 3.11 are computed (from Barber and Hanesiak, 2004).

mechanism for a sensible-heat polynya. Whatever the case, it is interesting to note that recent work (totally independent of CASES) found similar patterns in surface temperature anomalies in the same areas during 1979-2000 (Comiso, 2003).

Our most recent work on sea ice (Galley et al., 2008), based on the CIS digital archive, clearly distinguished spatial and temporal trends in sea ice types within the CASES region (e.g., multi-year, first-year, young, etc.).

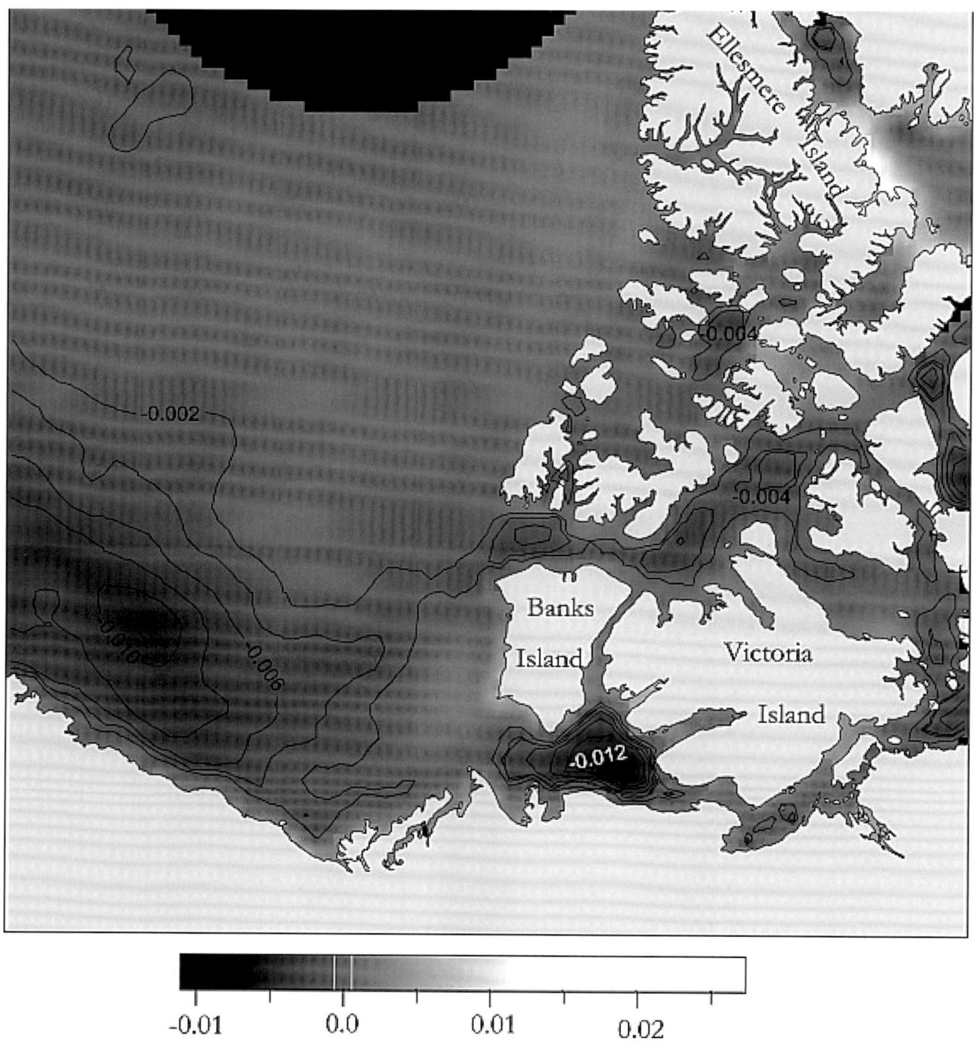

Figure 3.11

Slopes of the least squares best fit line from each pixel in the data frame based on a statistical analysis of slopes fit through each pixel in the dataset (as illustrated in Figure 3.10). The slopes show trends towards negative (dark) and positive (light) concentration anomalies as a function of spatial location over the period 1979-2000. All contours are significant at α≤0.01 (from Barber and Hanesiak, 2004).

Several of these findings supported earlier work by Barber and Hanesiak (2004). A particularly interesting finding made was that the timing and duration of ice break-up within Amundsen Gulf was widely variable compared to that of fall freeze-up. To some extent, the presence of 'old ice' in the region (i.e. ice which is re-circulated within the Beaufort Gyre and then back into Amundsen Gulf) appeared to affect the timing and duration of ice events, specifically by delaying the end of ice break-up (and therefore the beginning of the open water period) or affecting local freeze-up. We should note that ice break-up in Amundsen Gulf is dynamically-forced, while freeze-up is almost exclusively thermodynamically-forced. Break-up was observed much earlier on average during 1980-2004 compared to 1964-1974. Further details are available in Galley et al. (2008).

3.2.2.1 The Beaufort Gyre

Sea ice motion in the Beaufort Sea follows a circular rotation around the pole, which in the southern Beaufort Sea is known as the *Beaufort Sea Ice Gyre*. It generally rotates anti-cyclonically due to the persistence of a high pressure system over the pack ice in the northern hemisphere (i.e. atmospheric flow associated with the high pressure pattern moves the sea ice). The gyre has been known to reverse direction, particularly in the summer, due to the presence of low pressure systems over the sea ice.

Recent work on the dynamics of the Beaufort Sea Ice Gyre (Lukovich and Barber, 2005) has shown that the relative motion of the gyre has been changing in recent times. Ice-relative vorticity computed for the years between 1979 and 2000 demonstrated that sea ice motion within the gyre was predominantly characterized by anti-cyclonic flow with episodic reversals to cyclonic flow (Fig. 3.12). Weekly sea ice motion vector and relative vorticity fields during fall (see weeks 32 and 36 in 1983 as an example) illustrated such transitions from anti-cyclonic to cyclonic motion (see also Fig. 3 in Lukovich and Barber, 2006). It is interesting to note that the dominant reversal of cyclonic flow in the 1980's most often occurred during weeks 35-40 (August-September). This contrasted with the 1990's (Fig. 3.12), when shifts in cyclonic flow occurred over a much wider period (weeks 30-50). It is also interesting to note that during winter (weeks 40-20) there was also a tendency for the gyre to slow down and even reverse. However, the strength and frequency of these events were not as large as during summer.

3.2.2.2 Snow on sea ice

Snow-covered sea ice plays a critical role in the ocean—sea ice—atmosphere interface. Due to its low thermal conductivity and high albedo, the snow/sea ice layer acts as a barrier between the ocean and atmosphere, limiting mass and energy flow between these two areas (Kotlyakov and Grosswald, 1990; Moritz and Perovich, 1996). Snow cover also regulates the growth and decay of sea ice by impeding ice growth (via the reduction of conductive heat flux) and moderating the entry of short-wave radiation (Maykut, 1986; Eicken et al., 1995; Eicken, 2003). Finally, snow-covered sea ice is a major controlling factor in the ecology of the Arctic system. During the spring period, the distribution of snow-covered sea ice controls the spatial variability of photosynthetically active radiation (PAR) reaching the underlying ocean (Welch and Bergmann, 1989; Iacozza and Barber, 1999; Mundy et al., 2005). PAR plays an important role in algal production and provides nutrients for copepods—an important food source for Arctic cod and other Arctic fish species. The habitat selection of apex predators (i.e. bowhead whales, polar bears) in the Arctic ecosystem is also controlled by snow-covered sea ice and its associated spatial distribution

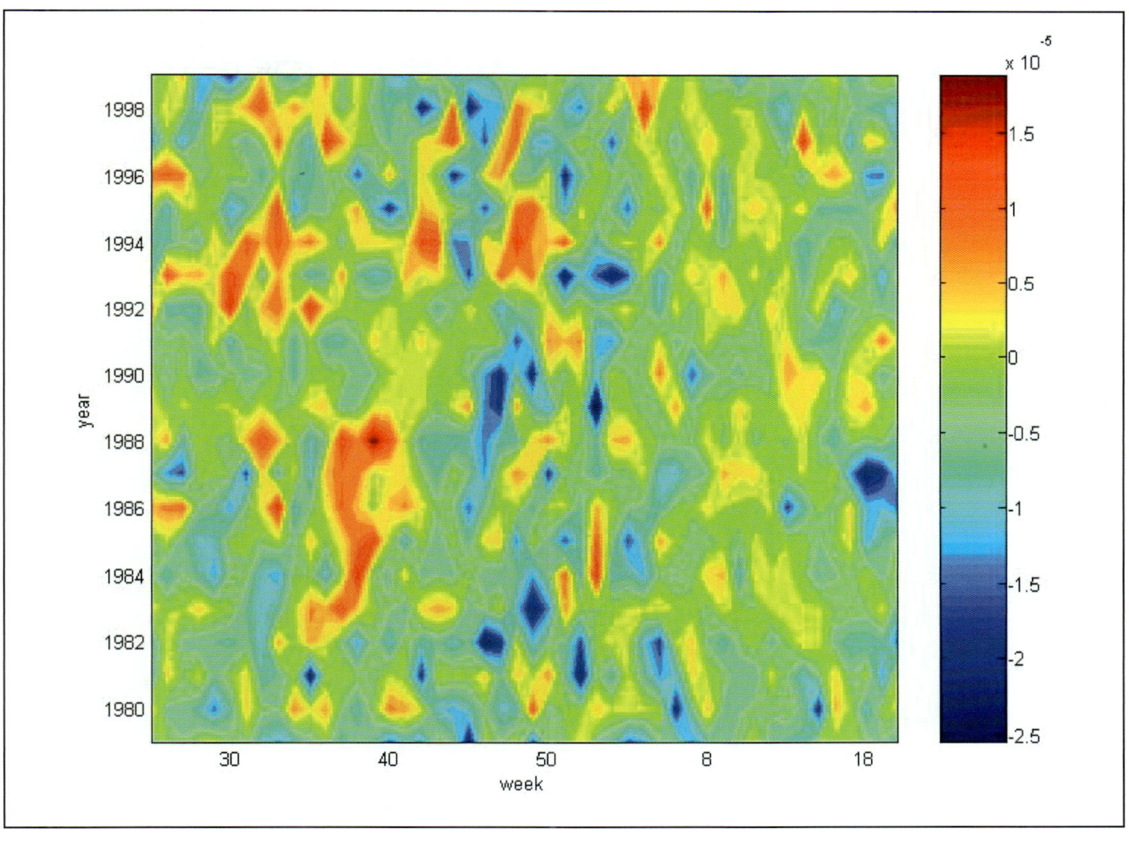

Figure 3.12

Mean relative vorticity from weeks 25 to week 20 of following year (x-axis), from 1979-2001 (y-axis). Red (Blue) shades denote cyclonic (anticyclonic) activity (from Lukovich and Barber, 2006).

(Fraker, 1979; Welch, 1991; Stirling et al., 1999; Dyke et al., 1996; Stirling and Parkinson, 2006).

The pattern of snow distribution is primarily controlled by ice surface topography (Eicken et al., 1994; Iacozza and Barber, 1999). In general, sea ice with large variations in surface topography (i.e. large ridges associated with rough sea ice and hummocks associated with multi-year sea ice) tends to exhibit large variations in snow distribution, producing patterns which mirror the surface topography and prevailing winds (Iacozza and Barber, 1999; Sturm et al., 2002). In the case of first-year sea ice with small topographic features (i.e. smooth first-year sea ice), prevailing winds will produce an undulating snow distribution pattern with dunes forming at regular intervals. In order to enhance our knowledge of physical-biological process coupling, the primary objective of this portion of our research was to investigate the relationship between ice surface roughness and snow distribution in different classes of first-year sea ice (including smooth, rough and ridged ice) at various spatial scales.

At the local scale, the best estimator for variability in snow depth proved to be the standard deviation in surface roughness (Iacozza et al., 2007). The relationship between these two indicators was non-linear and best modeled using a logarithmic function (see Fig. 6 in Iacozza et al., 2007). The model suggests that the variability in snow depth increases more rapidly for small variability in surface roughness then becomes constant or stable. This value probably depends on the amount of snow in the system that can be moved or caught by the surface roughness. The accountability of the model was significantly improved when the ice classes (i.e. ridge and rough ice) were analyzed separately. For both ice types, the resultant model was linear, as opposed to the logarithmic function used when the data were combined.

At the regional scale, a number of statistical relationships were examined (see Iacozza et al., 2007). The explained variance obtained through regression analyses ranged from 79.9% to 37.4% (see Fig. 8 in Iacozza et al., 2007). All relationships modeled were non-linear, with the dependent variable best explained by a logarithmic function. The average surface roughness measured along each transect proved to be the best estimate for snow depth (see Fig. 8 in Iacozza et al., 2007). This relationship is best fit with a logarithmic function explaining approximately 79.9% of the variability in the average snow depth for each site, while over 20% of the variability is attributed to other factors such as meteorological forcing. The candidate model is defined as:

$$x_{(sd)} = 59.6 + 19.7 \ln(x_{(SR)})$$

where $x_{(sd)}$ is the mean snow depth and $\ln(x_{(IR)})$ is the log of the mean surface roughness.

3.2.2.3 Micro-scale properties of sea ice and snow

In order to fully understand the thermodynamic processes affecting sea ice, it was important to investigate the micro-scale structure of sea ice and its snow cover. Sea ice samples obtained for this portion of our study were considered typical for landfast first-year sea ice in the Arctic. The bottom layers consisted of columnar ice, with the bottom-most 0.01-0.02 m having a porous skeletal structure. Visible coloration was evident towards the bottom of the skeletal layer due to the presence of ice algae. Analysis of ice core sections taken in May 2004, revealed that sea ice near the ice-water interface was composed of jagged crystals with a distinct platelet substructure (Fig. 3.13). The combination of crystal size, circularity, platelet width, and brine volume, gave an indication of the boundary surface area between interstitial brine and ice and could potentially be used to quantify the scattering properties of sea ice. The mean crystal size decreased with distance from the ice-water interface, with the exception of the bottom-most 0.05 m layer. Within this newly formed layer, smaller grains developed with a lesser degree of horizontal C-axis alignment. The formation of small grains during late winter/early spring, when ice growth rates are slow, was also noted by Weeks (1998) but has not been explained. The mean (and maximum) horizontal cross-sectional crystal areas observed between the ice-water interface and 0.39 m in Figure 3.13 were 18 mm² (155 mm²), 60 mm² (480 mm²), 32 mm² (350 mm²), 24 mm² (670 mm²) and, 20 mm² (490 mm²), respectively. Note that the maximum crystal areas may have been underestimated as potentially large crystals were not fully contained within the ice core area and thus excluded from our analysis.

Ice temperature, salinity, density and the partial fraction of brine ice and air are presented in Figure 5 of Ehn et al. (2008). The authors of this paper provide summary statistics on the microstructure of sea ice during the spring period. Other work by these authors provides similar information for the fall period during ice formation (Ehn et al., 2007).

The impact of snow on sea ice also required analysis at the micro-scale. During the CASES program we collected one of the most complete existing datasets on snow metamorphism (i.e. change in grain size and shape due to temperature variations within the snow). We characterized the fall-to-winter vertical profile evolution of thin and thick snow pack, and associated meteorological forcing to landfast first-year sea ice. From these data, we found that the physical and thermal properties of snow evolved according to whether the system was labelled as being in a 'cooling' or 'warming' period. During the cooling period, we only observed very small changes in the geophysical characteristics

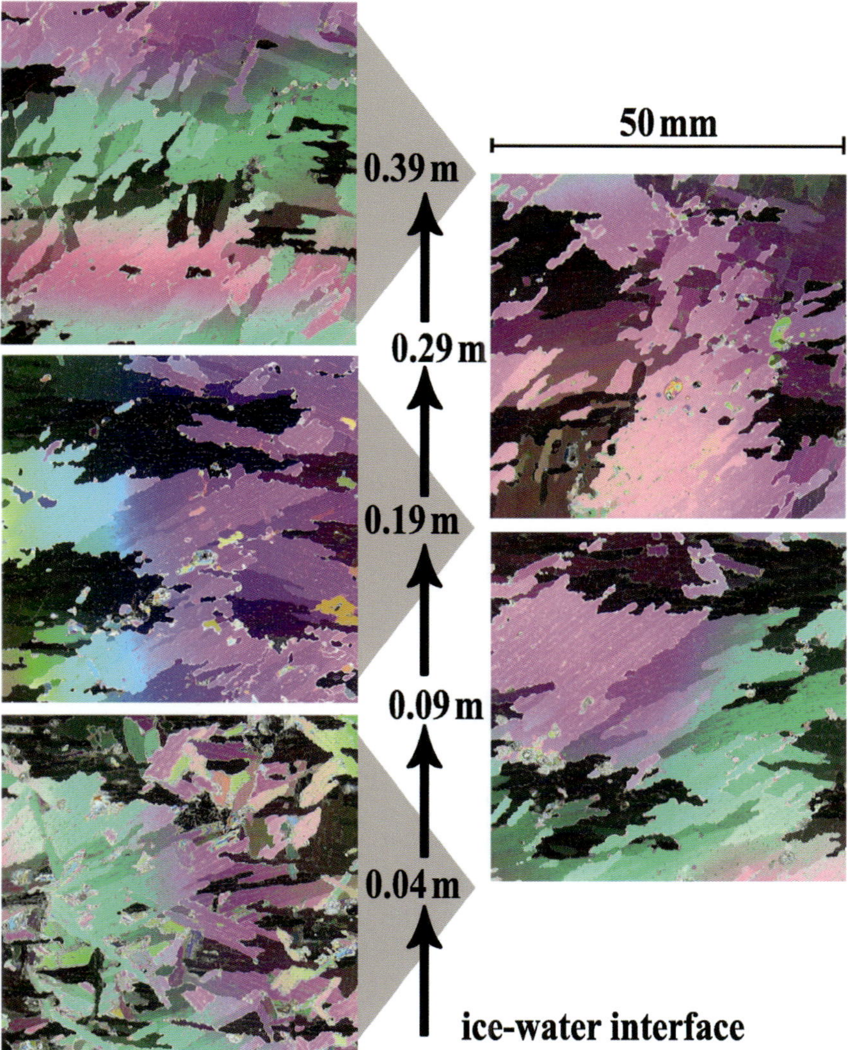

Figure 3.13

1 mm-thick thin sections showing the columnar ice structure immediately above the ice-water interface observed on 5 May 2004 when the ice thickness was 1.87 m (from Ehn et al., 2008).

of the snow-pack—with the exception of salinity, which progressively decreased. This snow desalination was stronger at the bottom of thin snow packs (–0.12 ppt day^{-1} with an R^2 of 0.52). We could not compare this rate with thick snow as that sampling started later in the season. Net shortwave and longwave radiation did not appear to have a significant influence on either thin or thick snow covers. During the warming period, significant changes were observed in the morphology of the snow grains for both thin and thick snow packs (see Fig. 5 in Langlois et al., 2006a). The rate of growth during this period was stronger in thick snow-packs (0.25 versus 0.48 mm day^{-1} for thin and thick snow, respectively) where $\delta T/\delta_L$ (vertical difference between layers L for volume temperatures) and $\delta e_{s}/\delta_L$ (vertical difference between layers L for saturation vapour pressure) were larger (i.e. stronger vapour flow) and Q* (net radiation) values near 0. Vapour flow acted as the main vehicle for brine movement within in the snow pack and this created significant changes in snow grain morphology (see Figs. 6 and 7 in Langlois et al., 2006a), snow density, and its resulting thermal properties.

3.2.3 The Surface Energy Balance

3.2.3.1 Radiative Transfer

Sea ice plays a key role in the earth's climate system partly because of its high albedo, which causes a large portion of the incident solar radiation to be reflected away from the surface (ACIA, 2005). Reflection from the sea ice surface has been extensively studied and appears to be a partial function of near surface properties (Eicken, 2003). However, much less is known about how radiation is transmitted to the bottom layers of sea ice and partitioned between reflection, absorption and transmission. The bottom-most layer of sea ice is of considerable physical and ecological interest since its porous structure creates habitat for thriving algal communities which use chloroplastic pigments to effectively absorb radiation at visible wavelengths (Horner et al., 1992).

Irradiance spectra were measured at increments within the bottom layers of landfast sea ice in Franklin Bay during early spring (April 22 to May 9, 2004). This data provided estimates of the sea ice inherent optical properties (IOPs) using the DIScrete Ordinates Radiative Transfer (DISORT) Program for a Multi-Layered Plane-Parallel Medium. To the best of our knowledge, the inverse model approach we adopted here had not been previously used for a sea ice environment. In our model, the scattering coefficient (b) was assumed to be wavelength independent within visible wavelengths; any wavelength dependence was accounted for by the absorption coefficient, $a_{tot}(\lambda)$. Resulting IOPs of the 1.8 m thick first-year sea ice cover were discussed in terms of temperature, salinity, density, and algal spectral absorption.

As expected, ice core samples and irradiance measurements revealed very low concentrations of ice algae above the bottom 0.05 m of the sea ice. For the bottom layer, chlorophyll a concentrations ranged between 16.6 and 242 mg m^{-3} (these values decreased by over an order of magnitude in overlying layers). This algal layer had a marked effect on the spectral distribution and magnitude of transmitted irradiance beneath the ice (Mundy et al., 2005).

Particulate absorption spectra, $a_{p}(\lambda)$, from melted ice samples showed that the chloroplastic pigments may have degraded during the period between sampling and measurement (i.e. the spectral shape of $K_{d}(\lambda)$ could not be explained by the measured $a_{p}(\lambda)$ for the bottom-most ice algae layer). Interior ice layers showed similar absorption spectral shapes compared to measurements made from melted samples. This provided evidence of pigment degradation within these layers. Iterations of DISORT to match the *in situ* $K_d(\lambda)$ resulted in reasonably shaped estimates of $a_p(\lambda)$. Based on the spectral shape of the modeled $a_p(\lambda)$, we concluded that the ice algae within the bottom-most layer were healthy, and that pigment degradation was responsible for modifying the spectral shape away from the ice-water interface (see Ehn et al., 2007 & 2008 for more detail).

3.2.3.2 Energy fluxes

The properties and processes associated with the atmosphere and sea ice directly impact mass and energy exchanges occurring at their interface (Maykut, 1982; Ebert and Curry 1992). This relationship underpins strong feedback processes over a wide range of spatial scales, and highlights the Arctic as an important and sensitive component of the climate system (IPCC, 2001). Surface energy budgets for the Arctic region remain poorly understood compared to other biomes, principally because of a dearth of good observational studies (Persson et al., 2002). Climatological data from the Arctic Ocean and its peripheral seas are available from numerous field experiments dating back to Nansen's expedition in 1893 (e.g., Mohn, 1905; Marshunova and Mishin, 1994; Maykut, 1982; Barber et al., 1998a; Persson et al., 2002). Most of these projects 1) were conducted over multi-year ice, 2) did not provide measurements of suitable accuracy for model development and validation, 3) included all (or most) of the important components of the energy budget at a location in conjunction with measurements of surface state variables, and 4) extended over winter conditions.

During the ice-camp phase of the CASES project, comprehensive measurements pertaining to the surface heat and radiation budget were made using sensors installed on (and around) a 6 m high antenna-type tower (located at 70° 2.516' N, 126° 15.894' W, approximately 1.5 km south east of the *CCGS Amundsen*). The experiment period extended from January 21 to May 27, 2004. Measurements were collected from a pan of uniformly consolidated seasonal sea ice that ranged in thickness from approximately 80 cm (at the start of the experiment) to 120 cm (at the experiment's end). An eddy covariance system (Baldocchi et al., 2003) provided direct measurements for the turbulent fluxes of heat, CO_2 and momentum. The four main components of net radiation were also measured, and temperature strings through the snow and sea ice provided information on the substrates' thermal state. Details of the experiment are provided in Papakyriakou et al. (2007). The following reviews some of the main features of the surface heat budget as observed during the experiment.

The energy balance for a snow layer over sea ice can be represented using:

$$Q^* - Q_H - Q_E - Q_C = \Delta Q_S + \Delta Q_M \qquad (1);$$

where Q_H and Q_E are the turbulent sensible and latent heat, respectively; Q_C is conduction to/from the underlying media; ΔQ_S accounts for sensible heat changes (i.e., temperature changes) in the volume; ΔQ_M accounts for phase changes within the volume (i.e., melting, freezing, sublimation, evaporation and condensation); and Q^* is the net radiation to the surface. This latter variable can be separated into a net shortwave (K^*) and net longwave (L^*) component:

$$= K_d - K_u + L_d - L_u \qquad (2);$$

where $_d$ and $_u$ denote down- and up-welling, and K and L denote short- and long-wave radiation, respectively. The convention here is that positive radiation terms are

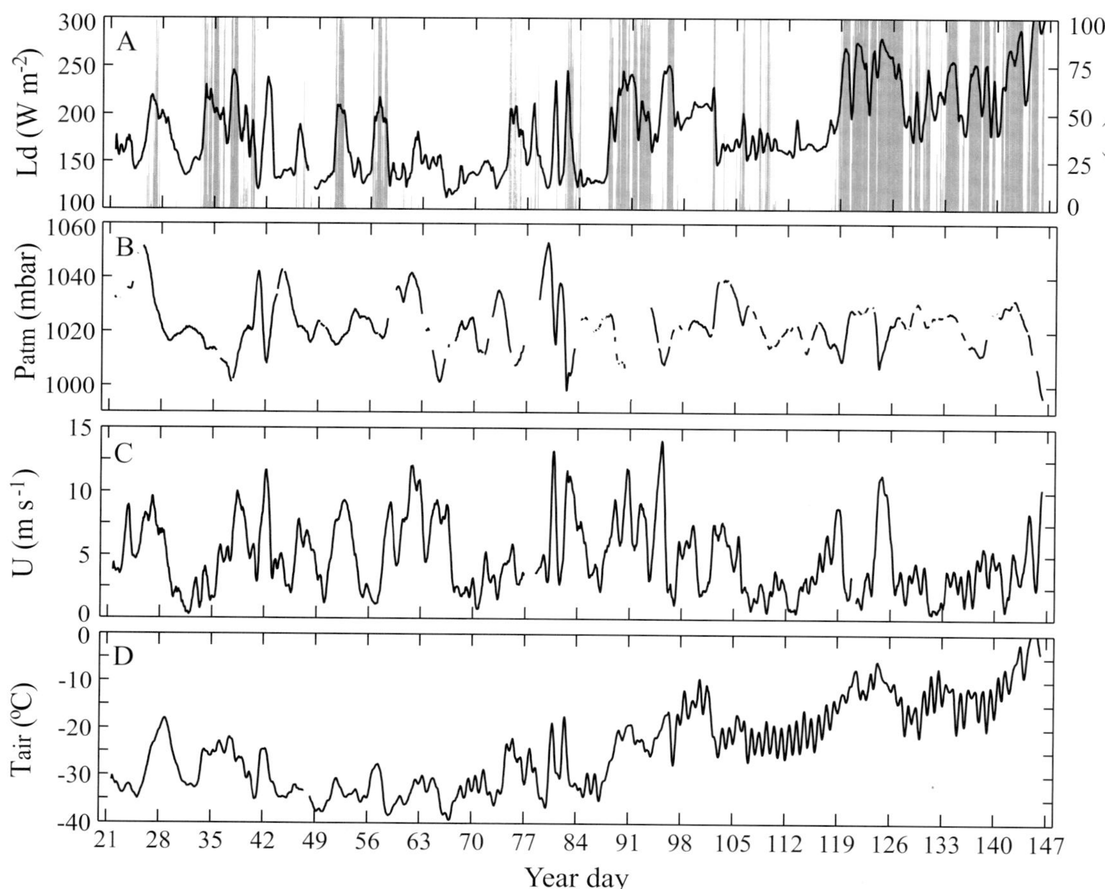

Figure 3.14

Meteorological conditions during the sampling period. CF is the cloud fraction.

directed toward the surface, while positive heat flux terms are directed away from the surface. Flux units are in watts per square meter (or W m⁻²). The ratio of outgoing to incident solar radiation is called the surface albedo (λ).

The region experienced strong variation in all basic meteorological elements. This resulted from the incursion of numerous low pressure systems combined with the seasonal rise of air temperature (Fig. 3.14). Screen height air temperature varied by 44° C over the experiment, ranging between -39° C and 5° C.

The experiment extended from polar night conditions to twenty-four hours of daylight (May 15, or year-day 135). Accordingly, the long-wave radiation exchange dominated Q* over much of the period (Fig. 3.15). The average hourly net long-wave loss was 41 W m⁻², with values exceeding 95 W m⁻². In contrast, hourly net solar radiation to the surface averaged 21 W m⁻², and peaked at 258 W m⁻². Monthly median albedo fluctuated around 0.83, although the range varied widely among months. The evolution of the heat budget showed characteristics typical of the winter-spring transition (see also Barber et al., 1998b); specifically, winter and early spring (January to March) radiative losses at the snow surface were largely offset by conductive gains to the snow layer from the underlying sea ice and ocean, and solar radiation increased its influence on net radiation during April and May (Fig. 3.16). The snow flipped from a state of net energy loss to gain on April 29 (day 119; Fig. 3.16). The latent heat flux was ineffective at removing/supplying heat throughout the experiment (hence its absence in Fig. 3.17). On average, 50% of the energy lost through net radiation was offset by conductive heat and sensible heat input to the snow prior to April 29. Thereafter, surplus net radiation approximately equalled heat gain by the snow (Fig. 3.17).

The temperature structure of the snow and sea ice (Fig. 3.18) showed a strong response to changing boundary heating, particularly during January and February (i.e., between the start of the experiment and year-day 60), and after mid-March (beyond year-day 75). Cloud cover and increased atmospheric heat content (associated with low pressure systems) tend to significantly affect surface heat loss, largely by increasing downward long-wave radiation (though on occasion significant sensible heat transfer to the surface can be observed) (Fig. 3.14). During these periods, noticeable increases in substrate temperature can occur, as less of the conductive heat input from below is lost to the atmosphere. The volume heating promotes short-term variation in the morphological, optical and electrical properties of snow (e.g., Hanesiak et al.. 1999; Langlois, et al., 2008; Hwang et al., 2007b), and a rapid and prolonged rise in net solar radiation usually follows, similar to that observed between April 6 and 9 (day 96 to 99 in Fig. 3.16). Events like these effectively promote the rapid transition to snow melt onset (see also Barber et al., 1998b). The timing and frequency of low pressure systems in the springtime therefore becomes critically important in determining the timing of *in-situ* spring melt and break-up.

3.3 Implications of this work

The observed reduction in the annual minimum sea ice extent is one of the key pieces of evidence of global warming's effect on the Arctic. The annual cycle of sea ice is being disrupted by changes in oceanic and atmospheric forcing. Because of the close connection between snow and sea ice, corresponding changes in snow distribution are also significant. These physical changes also impact most elements of the marine ecosystem (see other chapters in this CASES synthesis) and lead to profound implications in several

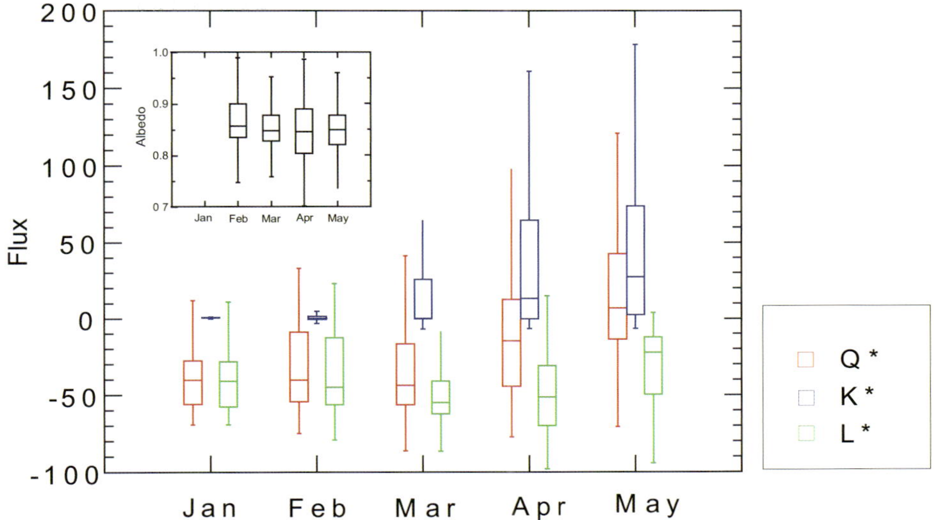

Figure 3.15

Distributions of hourly net radiative fluxes (net radiation Q, net solar K*, and net long-wave L*) by month. Albedo is shown in inset.*

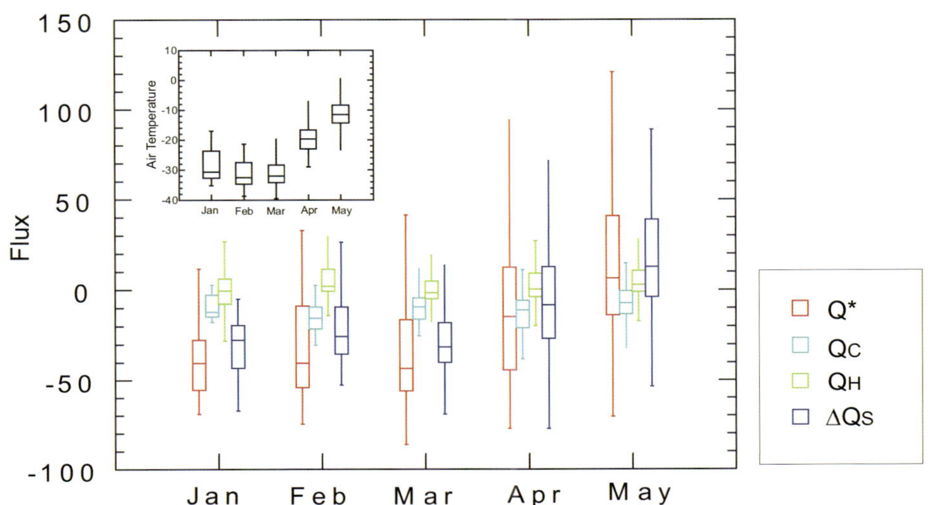

Figure 3.17

Distributions of hourly heat budget components (net radiation Q, ice surface conductive flux Q_C, sensible heat Q_H, and changing heat storage in the snow pack ΔQ_S) by month. Screen height air temperature is shown in inset.*

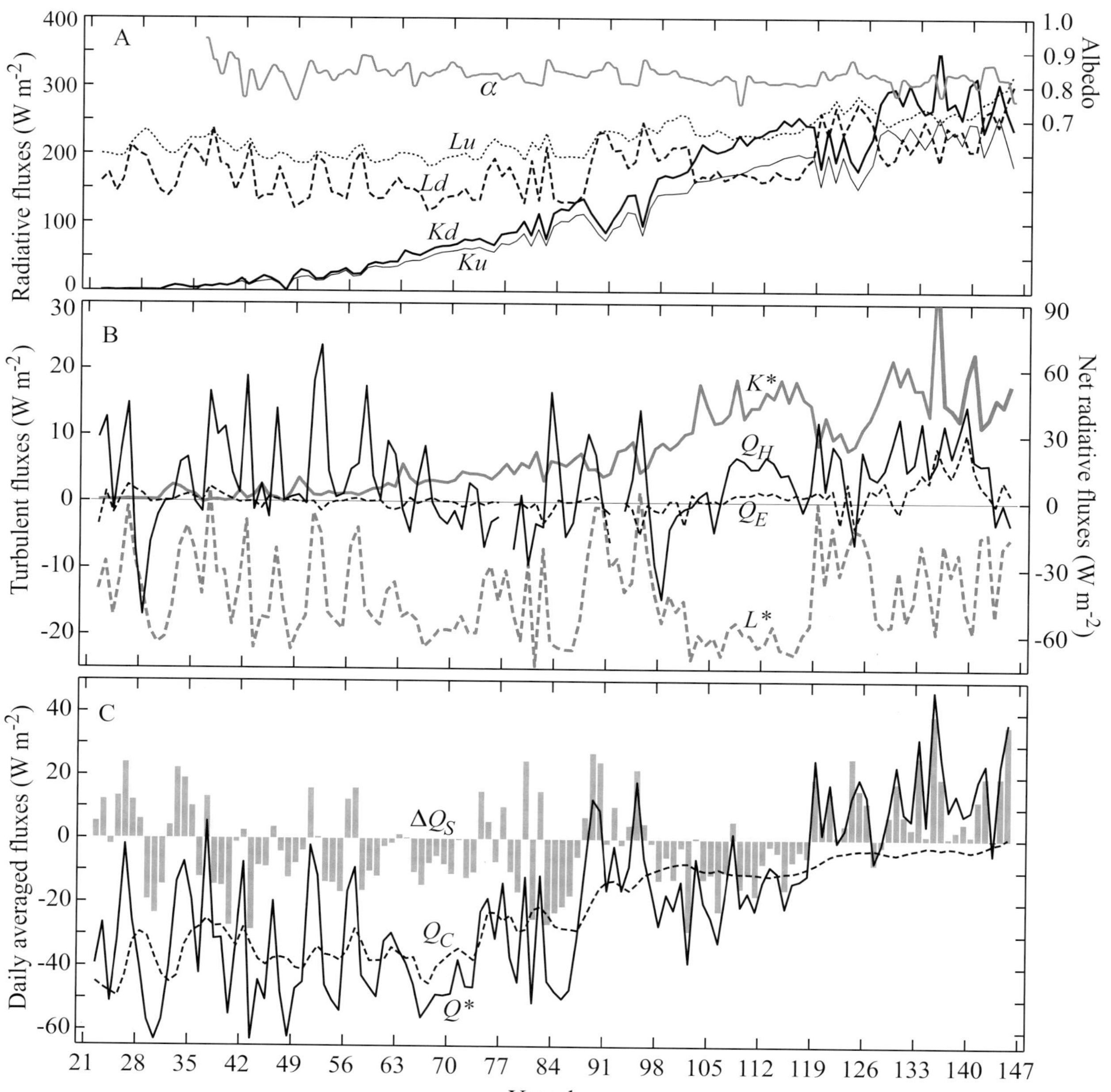

Figure 3.16

Daily averaged components of the energy budget of the snow pack during CASES. All radiation fluxes are positive towards the surface and heat fluxes (Q_H, Q_E, and Q_C) are positive away from the surface.

interrelated areas. Here we focus on three such areas: environmental security (protecting our ecosystem), political security (protecting sovereignty and security), and developing our observation systems (monitoring change to inform policy into the future).

a) Environmental Security

The current projects (including both observations and modelling) predict a seasonally ice-free Arctic around 2050 (±20yrs). The Arctic marine ecosystem has taken a long time to achieve its average annual sea ice growth cycle and has not experienced seasonally ice-free conditions for at least 1.1 million years (Barber et al., 2005). This means that organisms will have to rapidly adapt to significant changes in the timing, presence and geophysical conditions of sea ice. For the CASES region, there is also a key link between the sea ice of Amundsen Gulf and the Beaufort Sea Gyre (Galley et al., 2008). The observed reversals of the gyre may be crucially linked to observed reductions in the central Arctic pack-ice (Lukovich and Barber, 2006). Of particular relevance to the marine ecosystem are concurrent changes in the snow cover over this sea ice (Langlois et al., 2006a; Iacozza et al., 2007). This is due to the control which snow has on radiative transmission to the base of sea ice (Ehn et al., 2008) and its insulative effect on the thermodynamic growth of the sea ice. Recent work by our team and others has shown that sea ice micro-algae prefer certain habitats at the base of first-year sea ice (Mundy et al., 2007). The micro-scale control of these habitats is related to brine drainage holes at the base of the sea ice, which in turn are controlled by snow and sea ice thermodynamics. Higher up on the trophic ladder, we can expect to see changes in sub-ice grazers (e.g., copepods) and associated predator-prey relationships right through to the Arctic cod. We know that that cod use under-ice roughness as a means of avoiding predation. The dynamic processes of ice formation are in a state of flux and the consequences of this to cod are unknown. Finally, the well-known sea ice habitat of the ringed seal (and associated preference of polar bears) is also driven by snow catchment through sea ice roughness (Iacozza et al., 2007). Any changes in this habitat is therefore of considerable concern to local peoples and scientists alike (Barber and Iacozza, 2004).

The marine ecosystem is affected at all levels by climate change currently impacting sea ice and the ocean surface mixed layer. To understand the impacts of such change we must first determine how the physical environmental system operates, including what governs its underlying processes. Through this, we can eventually recognize and predict how the marine ecosystem will respond to current and oncoming changes.

b) Political Security

Current reduction in the summer minimum extent of sea ice and its thickness has received much attention from the world's media. Inuvialuit have had to deal with a fall freeze-up which occurs later in the year and an earlier spring (Barber and Hanesiak, 2004). This affects hunter practice and safety on the ice during hunts and cultural events. The observed reduction in the thickness of multi-year sea ice and the notion that we will no longer have multi-year sea ice (after about 2050) has many industry analysts looking towards the Arctic as a new ocean to explore and develop. With about 25 % of the world's known hydrocarbon reserves located in the Arctic Ocean, climate change appears to support the potential for increased development. Given the fragile nature of its ecosystem and its still challenging environmental conditions, however, the notion of 'sustainable development' in the Arctic takes on a completely new meaning. The recent workshop 'Arctic Frontiers' held on Tromso, Norway (January 2007), brought together policy makers, scientists and industry to discuss the implications of climate change in the Arctic, and to begin the process of seeking policy-related solutions to the various challenges being identified. A key element of these talks (of pertinence to the CASES region) is the opening of the Northwest Passage. Extensive natural gas finds in the Northwest Passage, passage of liquid natural gas tankers, and increased demand for natural gas are stimulating the development and use of the Northwest Passage. Major shipping companies are already looking to the north as a means for transporting goods between Europe and Asia. A likely scenario, however, might be to instead avoid the Northwest Passage and simply ship directly across the pole. This might eventually be both technically and economically possible (i.e. once climate change removes the multi-year sea ice from this region). The reduction in the extent and thickness of sea ice will also increase the amount of maritime traffic in the Canadian Arctic. This is a concern for military security, the control of border security, and immigration. The federal government has increased the presence of the Canadian Military and Canadian Rangers in the Arctic as a means of addressing this issue. Further policy linkages, pertinent to CASES are explored in Barber et al. (2005).

c) The development of observation systems

Current changes which the Arctic is experiencing have been documented using a number of different observation systems, including satellites, submarines, moorings, *in situ* studies from ice breakers and ice camps, and testimonials from local communities. It's important that we continue to improve our ability to observe the physical and biological processes operating in the

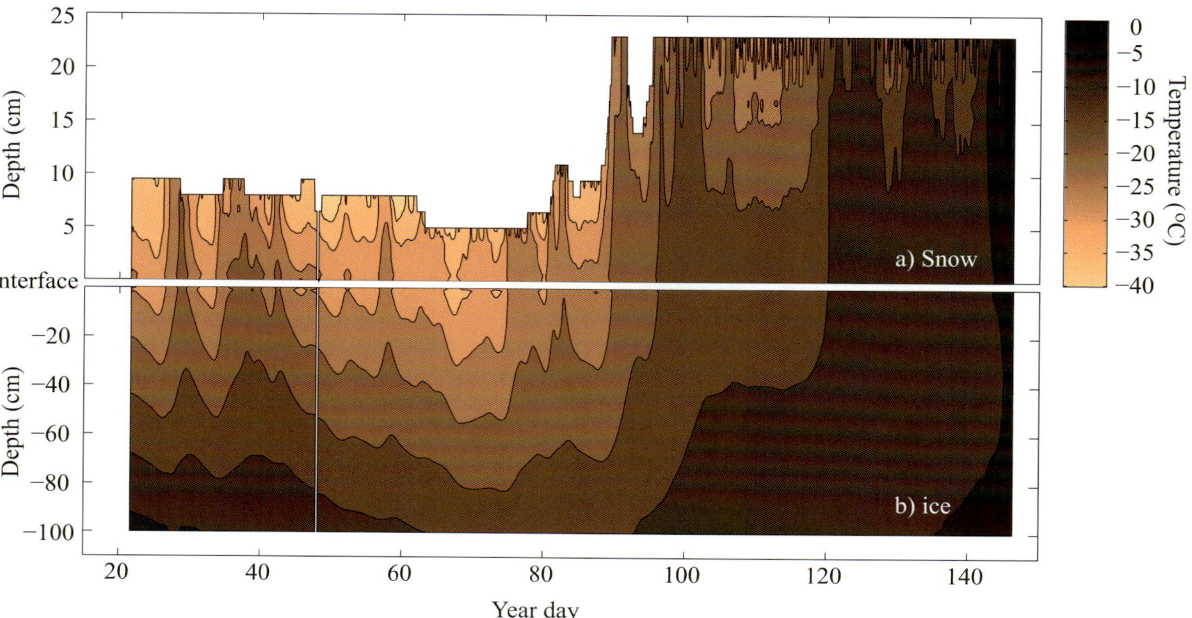

Figure 3.18

The evolution of the ice and snow temperature structure.

Arctic over a complete range of temporal and spatial scales. Several different groups within the CASES network have made contributions towards this end. We are now able to distinguish the thickness distributions of snow on sea ice based on the roughness of the ice (Iacozza et al., 2007). Perhaps we will eventually be able to use Radarsat data to estimate surface roughness (and thereby infer snow thickness) using models presented earlier in this chapter. Our use of passive microwave remote sensing data has also improved during the CASES program, and we were able to seek estimates of both the geophysical state (thickness, ice type, surface roughness, etc) and the thermodynamic state (surface temperature, temperature profile, surface albedo, etc) of sea ice. For instance, we established connections between microwave emission in young ice and the thickness and climatological albedo of the ice (Hwang et al., 2006). We were also able to improve our ability to measure young ice from space during the early part of the fall freeze-up (Hwang et al., 2007a & 2008; Ehn et al., 2007). Additional work on snow cover has also resulted in a number of significant developments. The data we collected pertaining to the annual cycle of snow grain morphology and its metamorphism, for instance, will be key in the development of better models for snow on sea ice (Langlois et al., 2006a). In addition, the development of a snow water equivalent (SWE) model may be significant to the operational monitoring of snow on sea ice (Langlois et al., 2006b). Interestingly, our team was able to show that a low pressure system during the middle of winter could create detectable changes in snow geophysics, thermodynamics and dielectrics (Langlois et al., 2008) which remain in the snow cover once the storm has passed. This finding implies a cumulative relationship between winter storminess and the overall annual cycle of sea ice. Another area of progress was the

Scientific activities around the CCGS Amundsen in Franklin Bay. Photo: Alexandre Forest.

detection and characterization of clouds using satellite data (Jin et al., 2006 & 2007) during the CASES overwintering program. This may be particularly helpful in deepening our understanding of Arctic cloud characteristics, a prerequisite for the development of better detection algorithms. Finally, the ocean mooring (biogeochemical observatories) data used during the CASES program is certainly key to the inter-annual monitoring of the changes which we see occurring within the Arctic. Such *in situ* and space-derived monitoring constitute a complete suite of Arctic-specific environmental variables which will surely aid in the development and accuracy of future biophysical models.

3.4 Recommendations

A number of important questions arise from the work our group has conducted during CASES. They relate both to changes in the physical environment and effects these changes have on the marine ecosystem:

- What role does the reversal of the Beaufort Sea Gyre play in the reduction of the summer minimum extent of sea ice?

- What role do cyclones play in local sea ice dynamic and thermodynamic processes? Are we seeing an increase in cyclogenesis and/or the advection of cyclones in the CASES region?

- What is the current climatology of cyclones in the CASES region and has this climatology changed over the past 30 years?

- How do we continue to improve our estimates of the distribution and thickness of snow over sea ice? Is there a way we can estimate sea ice surface roughness from space?

- What role does heat in the surface mixed-layer play in retarding the growth of sea ice during the fall? What is the effect of lower ice concentrations on the heat content of the ocean surface mixed layer?

- What is the future projection for fast ice in the CASES region? How do we expect ice thickness and the timing of its formation/decay to change as we move into a CO_2 enhanced atmosphere?

- How does the timing of the fast ice formation/decay relate to the accumulation of snow on the surface? If precipitation timing remains the same, more snow may fall into the ocean rather than on sea ice, creating a positive feedback to continued ice decay.

- How will ice dynamics change in the future? In particular, what roles do the atmosphere and ocean play in forming rubble, sheer zones and ridges in first-year sea ice? Does it matter if the ice forms later in the season and grows more slowly?

- What is the role of sea ice within the mobile ice zone in the seasonal evolution of the ocean surface mixed layer and its light environment?

- What is the relationship between changing sea ice dynamics and thermodynamics and the distribution of suitable ringed seal and arctic cod habitats?

- How do the thermodynamics of fast ice affect sub-ice algae development? What role does brine migration play in the selection of preferred habitat for these algae?

In order to achieve sufficient understanding of how climate change affects the ocean—sea ice—atmosphere system, we need a combination of scientific and local knowledge. The local knowledge component is required both to structure scientific studies and inform/interpret results from these studies. Although difficult to achieve, we suggest that communities of the North establish community-based monitoring (CBM) projects. CBM should be initiated, conducted and evaluated exclusively by northern communities and their local government organizations (i.e., hamlet offices, hunters and trappers committees etc). Science can inform the design of these CBM projects, but the management and responsibility for running them must lie within the community.

References

ACIA, 2005. Future Climate Change: Modeling and Scenarios for the Arctic. Chapter 4 in *Arctic Climate Impact Assesment*, URL: www.acia.uaf.edu.

Agnew, T.A., H. Le, and M. Shokr, 1999. Characteristics of Large Winter Leads over the Arctic Basin from 85.5Ghz DMSP SSM/I and NOAA AVHRR imagery. Can. J. Rem. Sens. 25(1): 12-20.

Baldocchi, D.D., 2003. Assessing the eddy covariance technique for evaluating carbon dioxide exchange rates of ecosystem: past, present and future. Clim. Change Biol. 9: 479-492

Barber, D.G., A.K. Fung, T.C. Grenfell, S.V. Nghiem, R.G. Onstott, V.Lytle, D.K. Perovich, and A.J. Gow, 1998a. The Role of Snow on Microwave Emission and Scattering over First-Year Sea Ice. IEEE Transactions on Geoscience and Remote Sensing. ONR ARI special issue. 36(5):1750-1763.

Barber, D.G., J. Yackel, R. Wolfe, and W. Lumsden, 1998b. Estimating the thermodynamic state of snow covered sea ice using time series synthetic aperture radar (SAR) data. Proceedings of the Eighth International Offshore and Polar Engineering Conference, Montreal Canada, May, 1998. International Society of Offshore and Polar Engineers. Vol III: 50-54.

Barber, D.G., R. Marsden, P. Minnett, G. Ingram, and L. Fortier, 2001. Physical Processes within the North Water (NOW) Polynya. Atm.-Ocean 39(3), 163-166.

Barber, D.G., and J. Hanesiak, 2004. Meteorological forcing of sea ice concentrations in the Southern Beaufort Sea over the period 1978 to 2001. J. Geophys. Res. 109(C06014) doi:10.1029/2003JC002027.

Barber, D.G., and J. Iacozza, 2004. Historical analysis of sea ice conditions in M'Clintock Channel and Gulf of Boothia, Nunavut; Implications for Ringed Seal and Polar Bear Habitat. Arctic 57(1): 1-14.

Barber, D.G., L. Fortier, and M. Byers, 2005. The Incredible Shrinking Sea Ice. Policy Options 27(1): 66-71.

Bradley, R.S., and J. England, 1979. Synoptic climatology of the Canadian high Arctic, Geogr. Ann. 61A (3-4):187-201.

Carmack, E.C., and R.W. Macdonald, 2002. Oceanography of the Canadian Shelf of the Beaufort Sea: A setting for Marine Life. Arctic 55: 29-45.

Cavalieri, D.J., and S. Martin, 1994. The contribution of Alaskan, Siberian, and Canadian coastal polynyas to the cold halocline layer of the Arctic Ocean, J. Geophys. Res. 99(C9): 18343-18362.

Comiso, J.C., 2003. Warming trends in the Arctic from clear sky satellite observations. J. Clim. 16: 3498-3510.

Curry, J.A., and E.E. Ebert, 1992. Annual cycle of radiation fluxes over the Arctic Ocean: Sensitivity to cloud optical properties. J. Clim. 5(11): 1267-1280.

Curry, J.A., D. Randall, W.B. Rossow, and J.L. Schramm, 1996. Overview of arctic cloud and radiation characteristics. J. Clim. 9: 1731-1764.

Drobot, S.D., and J.A. Maslanik, 2003. Interannual variability in summer Beaufort Sea ice conditions: Relationship to winter and summer surface and atmospheric variability. J. Geophys. Res. 108(C7) doi:10.1029/2002JC001537.

Dyke, A.S., J. Hooper, and J.M. Savelle, 1996. A history of sea ice in the Canadian Arctic Archipelago based on postglacial remains of the bowhead whale (*Balaena mysticetus*), Arctic 49(3): 235-255.

Ebert, E.E., and J.A. Curry, 1992. An intermediate one-dimensional thermodynamic sea ice model for investigating ice-atmosphere interactions. J. Geophys. Res. 98: 10,085-10,109.

Ehn, J., B.J. Hwang, R. Galley, and D.G. Barber. 2007. Investigations of newly formed sea ice in the Cape Bathurst polynya: Part 1. Structural, physical and optical properties. J. Geophys. Res. 112(C05002) doi:10.1029/2006JC003702.

Ehn, J. K., C. J. Mundy, and D. G. Barber, 2008. Bio-optical and structural properties inferred from irradiance measurements within the bottom-most layers in an Arctic landfast sea ice cover. J. Geophys. Res. 113, C03S03, doi:10.1029/2007JC004194.

Eicken, H., M.A. Lange, H.-W. Hubberten, and P. Wadhams, 1994. Characteristics and distribution patterns of snow and meteoric ice in the Weddell Sea and their contribution to the mass balance of sea ice. Annales Geophysicae, 12(1): 80-93. Fiscus, C.H. and W.M. Marquette, 1975. National Marine Fisheries Service field studies relating to the bowhead whale harvest in Alaska, 1974. Natl. Mar. Fish. Serv. Seattle. Northwest Fisheries Centre. Processed Rep. 23 pp.

Eicken, H., H. Fischer, and P. Lemke, 1995. Effects of the snow cover on Antarctic sea ice and potential modulation of its response to climate change, Ann. Glaciol. 21: 369-376.

Eicken, H., 2003. From the microscopic to the macroscopic to the regional scale: Growth, microstructure and properties of sea ice. Pages 22-81 in Sea Ice: An Introduction to its Physics, Biology, Chemistry and Geology D.N. Thomas & G.S. Dieckmann (eds). Blackwell Scientific Ltd, London.

Flato, G., and G.J. Boer, 2001. Warming Asymmetry in Climate Change Simulations. Geophys. Res. Lett. 28(1): 195-198.

Fraker, M.A., 1979. Spring migration of bowhead (*Balaena mysticetus*) and white whales (*Delphinapterus leucas*) in the Beaufort Sea. Canadian Fisheries and Marine Services. Tech. Rep. 859. 36 pp.

Fu, Q., and S. Hollars, 2004. Testing mixed-phase cloud water vapour parameterizations using SHEBA/FIRE-ACE observations. J. Atmos. Sci. 61: 2083-2091.

Galley, R.J., E. Key, D.G. Barber, B.J. Hwang, and J.K. Ehn. 2008. Spatial and Temporal Variability of Sea Ice in the Southern Beaufort Sea and Amundsen Gulf: 1980 – 2004. J. Geophys. Res. 113, doi:10.1029/2007JC004553.

Hanesiak, J.M., D.G. Barber, and G.M. Flato, 1999. The role of diurnal heating in the seasonal evolution of sea ice and its snow cover. J. Geophys. Res. 104(6): 13,593-13,603.

Holland, M.M., C.M. Bitz, and B. Tremblay, 2006. Future abrupt reductions in the summer Arctic sea ice, Geophys. Res. Lett., 33(L23503) doi:10.1029/2006GL028024.

Holloway, G., and T. Sou. 2002. Has Arctic sea ice rapidly thinned? J. Clim. 15: 1691-1701.

Horner, R., S.F. Ackley, G.S. Dieckmann, B. Gulliksen, T. Hoshiai, L. Legendre, I.A. Melnikov, W.S. Reeburgh, M. Spindler, and C.W. Sullivan, 1992. Ecology of sea ice biota. Polar Biol. 12(3,4): 417-427.

Hurrell, J.W. (1995). Decadal trends in the North Atlantic Oscillation: Regional temperatures and precipitation. Science 269(5224):676-679.

Hurrell, J.W., Y. Kushnir, and M. Visbeck, 2001. The North Atlantic Oscillation. Science 291:603-605.

Hwang, B.J., J.K. Ehn, and D.G. Barber. 2006. Relationships between sea ice albedo and microwave emissions during fall freeze-up: An *in-situ* study, Geophys. Res. Lett. 33 (L17503) doi:10.1029/2006GL027300.

Hwang. B.J., J.K. Ehn, D.G. Barber, R. Galley, and T.C. Grenfell, 2007a. Investigation of newly formed sea ice in the Cape Bathurst polynya: 2 Microwave emission. J. Geophys. Res. 112 (C05003) doi:10.1029/2006JC003703.

Hwang, B.J., A. Langlois, D.G. Barber and T.N. Papakyriakou. 2007b. On detection of the thermophysical state of landfast first-year sea ice using *in-situ* microwave emission during spring melt, Rem. Sens. Environ. doi:10.1016/j.rse.2007.02.033.

Hwang, B.J., J.K. Ehn, and D.G. Barber, 2008. Impact of ice temperature on microwave emissivity of thin newly formed sea ice during fall freeze-up. J. Geophys. Res. 113 C02021 doi:10.1029/2006JC003930.

Iacozza, J. and D.G. Barber, 1999. An examination of the distribution of snow on sea-ice, Atmos.-Ocean 37(1): 21-51.

Iacozza, J., D.G. Barber, and S.J. Prinsenberg, 2007. On the relationship between snow distribution and sea ice surface roughness in the Canadian Arctic. J. Geophysi. Res. submitted.

Intrieri, J., C. Fairall, M. Shupe, P. Persson, E. Andreas, P. Guest and R. Moritz, 2002. An annual cycle of Arctic surface cloud forcing at SHEBA. J. Geophys. Res. 107(C10) 803910.1029/2000 JC000439.

IPCC, 2001. Climate Change 2001: The Scientific Basis. Contribution of Working Group I to the Third Assessment Report of the Intergovernmental Panel on Climate Change, Houghton, J.T., Y. Ding, D.J. Griggs, N. Noguer, P.J van der Linden, X. Dai, K. Maskell, and C.A. Johnson (eds). Cambridge University Press, Cambridge, United Kingdom and New York, NY, USA, 881 pp.

Jin, X., J. Hanesiak, and D.G. Barber, 2006. Detecting cloud vertical structures from radiosondes and MODIS over Arctic first-year sea ice. Atmos. Res. 83(2007): 64-76.

Jin, X., J. Hanesiak, and D.G. Barber, 2007. Time series of daily-averaged cloud fractions over landfast first year sea ice from multiple data sources. J. Appl. Meteor. Climatol. 46: 1818-1827.

Kotlyakov, V.M., and M.G. Grosswald, 1990. Interaction of sea ice, snow and glaciers with the atmosphere and ocean, Polar Geography and Geology 14(1): 3-248.

Kwok, R., and D.A. Rothrock, 1999. Variability of Fram Strait ice flux and North Atlantic Oscillation. J. Geophys. Res. 104: 5177-5189.

Langlois, A., T. Fisico, D.G. Barber, and T.N. Papakyriakou, 2008. The response of snow thermophysical processes to the passage of a polar low-pressure system and its impact on in situ passive microwave radiometry: A case study. J. Geophys. Res. 113, C03S04, doi:10.1029/2007JC004197.

Langlois, A., C.J. Mundy, and D.G. Barber, 2006a. On the winter evolution of snow thermophysical properties over landfast first-year sea ice. Hydrol. Process. 21(6): 705-716, doi: 10.1002/hyp.6407.

Langlois, A., D.G. Barber and B.J. Hwang, 2006b. Development of a winter snow water equivalent algorithm using in situ passive microwave radiometry over snow covered first-year sea ice. Rem. Sens. Environ. 106(1): 75-88, doi:10.1016/j.rse.2006.07.018.

Lukovich, J.V., and D.G. Barber, 2005. On sea ice concentration anomaly coherence in the southern Beaufort Sea. Geophys. Res. Lett. 32, L10705, doi:10.1029/2005GL022737.

Lukovich, J., and D.G. Barber, 2006. Atmospheric controls on sea ice motion in the southern Beaufort Sea. J. Geophys. Res. 111, D18103, doi:10.1029/2005JD006408.

Marshunova, M.S., and A.A. Mishin, 1994. Handbook of the Radiation Regime of the Arctic Basin (Results From the Drifting Stations), 69 pp., Applied Physics Laboratory, University of Washington, Seattle, Wash.

Maslanik, J.A., M.C. Serreze and T. Agnew, 1999. On the record reduction in 1998 western Arctic sea-ice cover. Geophys. Res. Lett. 26: 1905-1908.

Maxwell, J.B., 1981. Climatic regions of the Canadian Arctic islands. Arctic 34: 225-240.

Maykut, G.A., 1986. The surface heat and mass balance. Pages 395-463 in The Geophysics of Sea Ice, N. Untersteiner (ed). Martinus Nijhoff Publishers. Dordrecht.

Maykut, G.A., 1982. Large-scale heat exchange and ice production in the Central Arctic. J. Geophys. Res. 87: 7971-7984.

Melling, H., and D.A. Riedel, 1996. Development of seasonal pack ice in the Beaufort Sea during the winter of 1991-1992: A view from below. J. Geophys. Res. 101(C5): 11975-11991.

Minnett, P.J., 1999. The influence of solar zenith angle and cloud type on cloud radiative forcing at the surface in the Arctic. J. Clim. 12: 147-158.

Mohn, H., 1905. 1995 The Norwegian North Pole Expedition 1893-1896, Scientific Results, vol. VI, Meteorology, edited by F. Nansen, F. Nansen Fund for Advancemetn of Science, Christiania, Oslo.

Moritz, R.E. and D.K. Perovich, 1996. Surface Heat Budget of the Arctic Ocean Science Plan, ARCHSS/OAII report number 5. University of Washington, Seattle. 64 pp.

Mundy, C.J., D. Barber and C. Michel, 2005. Variability of snow and ice thermal, physical and optical properties pertinent to sea ice algae biomass during spring. J. Mar. Syst. 58: 107-120.

Mundy, C.J., J.K. Ehn, D. G. Barber and C. Michel, 2007. Influence of snow cover and algae on the spectral dependence of transmitted irradiance through Arctic landfast first-year sea ice. J. Geophys. Res. 112(C03007) doi:10.1029/2006JC003683.

Papakyriakou, T.N., J.K. Ehn, and D.G. Barber, 2007. Winter and springtime fluxes of heat, radiation and CO_2 over Arctic first-year sea ice during CASES03-04. J. Geophys. Res. In preparation.

Parkinson, C. D., P. Cavalieri, P. Gloersen, H. Zwally, and J. Comiso, 1999. Arctic sea ice extents, areas, and trends, 1978-1996. J. Geophys. Res. 104(C9): 20837-20856.

Persson, P.O.G., C.W. Fairall, E.L. Andreas, P.S. Guest, and D.K. Perovich, 2002. Measurements near the atmospheric surface flux group tower at SHEBA: Near-surface conditions and surface energy budget. J. Geophys. Res. 107(C10): 8045 doi:10.1029/2000JC000705.

Proshutinsky, A. Y., and M.A. Johnson, 1997. Two circulation regimes of the wind-driven Arctic Ocean. J. Geophys. Res. 102(C6): 12493-12514.

Rigor. I. G., J.M. Wallace, and R.L. Colony, 2002. Response of sea ice to the Arctic Oscillation. J. Clim.15: 2648-2663.

Rothrock, D.A, Y. Yu, and G.A. Maykut, 1999. Thinning of the Arctic Sea ice Cover. Geophys. Res. Lett. 26(23): 3469-3472.

Schweiger A.J., R.W. Lindsay, J.R. Key, and J.A. Francis, 1999. Arctic clouds in multi-year datasets. Geophys. Res. Lett. 26(13): 1845 – 1848.

Serreze, M.C., J.A. Maslanik, T.A. Scambos, F. Fetterer, J. Stroeve, K. Knowles, C. Fowler, S. Drobot, R.G. Barry, and T.M. Haran, 2003. A record minimum arctic sea ice extent and area in 2002. Geophys. Res. Lett. 30, doi:10.1029/2002GL016406.

Shupe, M.D., T. Uttal, and S.Y. Matrosov, 2005. Arctic Cloud Microphysics Retrievals from Surface-Based Remote Sensors at SHEBA. *J. App. Met.* 44(10): 1544-1562.

Stirling, I., N.J. Lunn, and J. Iacozza, 1999. Long-term trends in the population ecology of polar bears in western Hudson Bay in relation to climate change. Arctic. 52(3): 294-306.

Stirling, I., and C.L. Parkinson, 2006. Possible effects of climate warming on selected populations of polar bears (*Ursus maritimus*) in the Canadian Arctic. Arctic. 59(3): 261-275.

Sturm, M., J. Holmgren and D.K. Perovich 2002. The winter snow cover on the sea ice of the Arctic Ocean at the Surface Heat Budget of the Arctic Ocean (SHEBA): temporal evolution and spatial variability. J. Geophys. Res. 107(C10): 8047 doi:10.1029/2000JC000400.

Thompson, D.W.J. and J.M. Wallace (1998). The Arctic Oscillation signature in the wintertime geopotential height and temperature fields. Geophys. Res. Lett., 25(9):1297-1300.

Thorndike, A.S., and R.Colony, 1977. Large scale Ice Motion in the Beaufort Sea during Aidjex, April 1975-76. Sea Ice Processes and Models, Univ. Wash.

Visbeck, M. E., Chasignet, R. Curry, T. Delwroth, R. Dickson, and G. Krahmann, 2003. The ocean's response to North Atlantic Variability. Pages 113-145 in The North Atlantic Oscillation: Climate Significance and Environmental Impact. Geophys.Monogr. 134. American Geophysical Union.

Walsh, J. E., and W. L. Chapman, 1998. Arctic Cloud–Radiation–Temperature Associations in Observational Data and Atmospheric Reanalyses. J. Clim. 11: 3030-3045.

Wang, X., and J.R. Key, 2003. Recent trends in Arctic surface, cloud and radiation properties from space. Science 299(5613):1725-1728.

Weeks, W.F., 1998. Growth conditions and the structure and properties of sea ice. In: Lepparanta, M. Ed. , Physics of Ice-Covered Seas, vol. 1, Univ. of Helsinki, Helsinki, pp. 25–104.

Welch, H.E., 1991. Comparison between lakes and seas during the Arctic winter. Arctic Alpine Res. 23(1): 11-23.

Welch, H.E., and M.A. Bergmann, 1989. Seasonal development of ice algae and its prediction from environmental factors near Resolute, N.W.T., Canada, Can. J. Fish. Aquat. Sci. 46: 1793-1804.

Zuidema, P., B. Baker, Y. Han, J. Intrieri, J. Key, P. Lawson, S. Matrosov, M. Shupe, R. Stone, and T. Uttal, 2005. An Arctic Springtime Mixed-Phase Cloudy Boundary Layer Observed during SHEBA. J. Atmos. Sci. 62(1): 160–176

CHAPTER 4

Light, Nutrients and Primary Production

Michel Gosselin[1*], Sonia Brugel[1], Serge Demers[1], Thomas Juul-Pedersen[1], Pierre Larouche[2], Bernard LeBlanc[3], Christine Michel[3], Christian Nozais[4], Michel Poulin[5], Neil Price[6], Andrea Riedel[1,3], Magdalena Różańska[1,5], Kyle Simpson[6], Jean-Éric Tremblay[6,7]

4.1 Introduction and Rationale

The production and export of biogenic carbon on Arctic shelves (which represent 20 % of the world's continental shelves) may account for a large fraction of the overall sequestration of carbon in the Arctic Ocean (Legendre et al., 1992; Stein & Macdonald, 2004). Diatoms play an important role in the vertical and trophic export of organic carbon and biogenic silica from the euphotic zone (e.g. Dugdale & Wilkerson, 1998). Their large size makes them prone to both rapid sinking and grazing by large zooplankton that produce fast-sinking fecal pellets. Rapid sinking of diatoms at the end of blooms may reflect the life cycle of some species, as observed in the North Water polynya (Michel et al., 2002). In Arctic waters, low winter irradiance and thick ice cover throughout most of the year limit phytoplankton production to a short period during the summer. The diatom bloom usually follows ice cover melt in late July or August (e.g., Hsiao, 1992), although it may occur earlier under the ice in response to early snow melt (Fortier et al., 2002). In ice-free polynyas, diatom blooms can occur almost immediately after the end of the polar night in April-May (Klein et al., 2002), 2-3 months earlier than in ice-covered areas at the same latitude.

Phytoplankton, ice algae, and benthic microalgae are the three sources of primary production in nearshore Arctic waters. Horner and Schrader (1982) studied these three autotrophic components during winter and spring in the western Beaufort Sea. In winter, they observed low phytoplankton abundance and a chlorophyll$a$ (chl a) biomass near the limit of detection. Microflagellates were the most abundant autotrophs in the water column along with some diatoms. Low cell numbers, low chl a, and weak primary production persisted until the ice break up in July-August. Algal cells were found in sea ice from the time it formed in the fall. They were generally scattered throughout the ice layer, with microflagellates dominating in number, but with pennate diatoms representing the bulk of the biomass.

[1] *Institut des sciences de la mer (ISMER), Université du Québec à Rimouski, 310 Allée des Ursulines, Rimouski, QC, G5L 3A1, Canada*

[2] *Institut Maurice-Lamontagne, Pêches et Océans Canada, C.P. 1000, Mont-Joli, QC, G5H 3Z4, Canada*

[3] *Freshwater Institute, Fisheries and Oceans Canada, 501 University Crescent, Winnipeg, MB, R3T 2N6, Canada*

[4] *Département de biologie, Université du Québec à Rimouski, 300 Allée des Ursulines, Rimouski, QC, G5L 3A1, Canada*

[5] *Research Division, Canadian Museum of Nature, P.O. Box 3443, Station D, Ottawa, ON, K1P 6P4, Canada*

[6] *Department of Biology, McGill University, 1205 Dr. Penfield, Montréal, QC, H3A 1B1, Canada*

[7] *Present address: Québec-Océan, Département de biologie, Université Laval, Québec City, QC, G1V 0A6, Canada*

* Corresponding author: michel_gosselin@uqar.qc.ca

LEFT: Arctic Sunset. Photo: Ramon Terrado.

These microalgae concentrated in the bottom few centimeters of the ice column in response to increasing light levels in March-April. The growth continued until late May/early-June when maximum production and the standing stock occurred. Benthic microalgal production was barely detectable in spring, although chl *a* concentrations were high, representing perhaps the biomass from the previous production season.

Light is apparently the major factor controlling primary production during spring, with ice algae being able to take advantage of increasing light levels at the end of the polar night. The ice algal assemblage shades both phytoplankton and benthic microalgae, such that production in these habitats does not increase until after the ice algae has been released from the ice, typically in early June or July. During the spring period studied by Horner and Schrader (1982), ice algae provided about two-thirds of the primary production; phytoplankton provided one-third, and the contribution of the benthic assemblage was found to be negligible (see also Macdonald et al., 1998). Hsiao et al. (1977) studied the phytoplankton species composition, abundance, biomass and production from both open water and ice stations in the southern Beaufort Sea (69-71.2°N; 130-138.5°W) during three consecutive summers (July-August). The phytoplankton biomass and production generally decreased with increasing distance from shore and the Mackenzie River mouth. This pattern was attributed primarily to higher nutrient concentrations and warmer temperatures in coastal waters.

On the Mackenzie Shelf, the total primary production (PT) and new production (Pnew, i.e. the part of PT fuelled by allochthonous nutrients—mainly nitrate (NO_3)), were estimated at 12 and 8 g C m^{-2} y^{-1}, respectively (Carmack et al., 2004). On the inner shelf, the dirty landfast ice and turbid surface waters limited primary production most of the year. On the outer shelf, reduced water turbidity and enhanced stratification by ice melt and solar heating favored microalgal growth from early spring to summer, depending on the timing of the opening of the flaw lead system (Arrigo & van Dijken, 2004; Carmack et al., 2004). In years of early opening of the flaw lead, the primary production could be limited by nutrient supply sooner in the season. According to satellite imagery, the timing and intensity of the phytoplankton summer bloom on the Canadian Beaufort Sea Shelf are strongly affected by the sea ice dynamics and stratification in the water column (Arrigo & van Dijken, 2004).

The Canadian Arctic Shelf Exchange Study (CASES) has given us a unique opportunity to investigate in detail the annual dynamics of nutrients and microalgae in the ice-covered and ice-free waters of the Canadian Beaufort Sea. All three main oceanographic provinces of the study area were visited, i.e., the Mackenzie Shelf, the Amundsen Gulf / Cape Bathurst polynya region, and Franklin Bay.

4.2 Overview of Results

4.2.1 Optical properties and remote sensing of phytoplankton biomass

Remote sensing is often seen as an important tool for understanding the large-scale distribution and evolution of various marine properties. The estimation of phytoplankton biomass and its relation to physical characteristics (e.g. salinity, temperature, and circulation) was the main objective of the remote sensing group. In Arctic regions, it is well known that the operational remote sensing algorithms used to estimate phytoplankton biomasses are not performing optimally. Such operational algorithms were originally developed for low latitude phytoplankton species which have different light absorption properties than high latitude species. In the south-eastern Beaufort Sea and Amundsen Gulf, the situation is even more complex given the Mackenzie River's massive input of freshwater which is enriched in dissolved organic material. Due to its strong absorption in the blue wavelength, this material contaminates the optical signal coming out of the ocean and results in an overestimation of phytoplankton biomass.

In order to build more reliable remote sensing algorithms, we measured the light absorption properties of the three main optical marine components in the CASES study region: phytoplankton, non-algal suspended matter (NAP), and colored dissolved organic matter (CDOM). This will hopefully permit both a better understanding of local physical/biological interactions and better estimates of local primary production. In the fall of 2003, CDOM clearly dominated the total non-water light absorption—reflecting the strong influence of Mackenzie River outflow over the study area—followed by the NAP and the phytoplankton absorption (Matsuoka et al., 2008). West of the Mackenzie River, NAP and phytoplankton were the two main contributors to the total non-water absorption (Matsuoka et al., 2007). In general, an enhanced light absorption capacity was observed in the phytoplankton as light levels decreased with time. The phytoplankton tended to be smaller in size and more efficient in their light absorption. The change in the phytoplankton community was attributed to a change in the light level caused by decreasing sun elevation and sea-ice formation.

The inherent optical properties (IOP) of phytoplankton, CDOM and NAP affect the spectral signature of water measured by remote sensing systems. The particular nature of these properties measured during CASES indicated that the operational algorithms were not performing well in the southern coastal Beaufort Sea and Amundsen Gulf. Ben Mustapha et al. (2006) showed that both the SeaWiFS and MODIS operational algorithms (OC4v4 and OC3M) overestimated the *in situ* chl *a* by a factor of 3.64. The Arctic versions of the SeaWiFS algorithm (OC4L and OC4P), designed to take into account CDOM absorption by Mackenzie River freshwater outflow, provided slightly better results. However, it still greatly overestimated chl *a* levels, which is unacceptable for the Cape Bathurst polynya region. Based on these results, the phytoplankton biomass and production estimated by Arrigo & van Dijken (2004) were seriously biased towards higher values. According to our SeaWiFS corrected data, the daily phytoplankton production in the Cape Bathurst polynya was lower than in the Northeast Water (NEW) and the North Water (NOW) polynyas.

In addition, new algorithms for the SeaWiFS and MODIS sensors were built using *in situ* light attenuation measurements performed between June and August 2004. These algorithms are better adapted to regional optical conditions and will eventually allow more precise estimations of phytoplankton production for ecosystem studies.

4.2.2. Nutrient dynamics and cycling

An annual study of nutrient distribution and cycling was carried out throughout the Mackenzie Shelf, Amundsen Gulf and Franklin Bay from September 2003 to August 2004 (Simpson et al., 2008a). In vertical profiles of NO_3, phosphate (PO_4) and silicic acid ($Si(OH)_4$), concentrations were highest at intermediate depths in water of 32–33.1 salinity, as recorded elsewhere in the Canadian Basin. The concentrations of dissolved organic phosphorus, measured for the first time in this region, also showed a subsurface enrichment that was coincident with the major nutrients. Temperature-salinity characteristics of this water mass suggested that it was derived from the Pacific Ocean, modified as it flowed over Chukchi and adjacent shelves. None of the organic or regenerated nitrogen forms showed any significant vertical structure and were generally most abundant in the surface mixed layer in summer. In averaged profiles, however, we found some evidence of subsurface peaks in dissolved organic nitrogen (DON) and urea concentration at 50 m and 250-300 m. These high concentrations may be caused by aggregations of bacteria, zooplankton and/or fish; organisms that are known to produce these compounds during remineralization/metabolism.

A quasi-time series of measurements in the surface layer identified a seasonal cycle between the inorganic and organic forms of nitrogen. Consumption of NO_3 and $Si(OH)_4$ occurred in surface waters throughout the region immediately after ice breakup and during the growth of pelagic algae. Much of the NO_3 taken up was apparently converted to DON as DON concentration increased during the bloom and post bloom periods. Microbial decomposition of the surface water DON during winter is hypothesized to reduce its concentrations to springtime levels. Nitrite (NO_2) and ammonium (NH_4) concentrations also increased coincidentally with NO_3 drawdown, but to a much lesser extent than DON. Higher average concentrations of DOP were also present in surface waters following the phytoplankton bloom, but reciprocal decreases in PO_4 could not be detected at that time.

Based on deep water measurements of nutrient and oxygen concentrations, regeneration ratios were computed as a function of depth and region. In the shallow Pacific-derived waters (halocline layer), the values were approximately 9.0 mol O_2/mol NO_3 and 122.5 mol O_2/mol PO_4 while in the deep Atlantic waters (Atlantic layer) they were 17.4 and 193.5, respectively. Below the nutrient maxima in the Amundsen Gulf, higher concentrations of PO_4, NO_3 and $Si(OH)_4$ were detected relative to source waters in the Beaufort Sea. The excess nutrient calculated by difference was 0.12 µM PO_4, 1.7 µM NO_3, and 6.2 µM $Si(OH)_4$, and varied in proportions consistent with the remineralization ratios derived from oxygen-nutrient regressions. Oxygen concentration was also relatively lower in the deep Amundsen Gulf waters. The enhanced export of particulate matter from the overlying polynya and subsequent remineralization at depth is believed to account for this deep water nutrient enrichment.

The dynamics of nutrients and chl maximum was studied in Franklin Bay and adjacent waters from October 2003 to August 2004 (Tremblay et al., 2008). The time series showed that vertical mixing in fall 2003 was modest and resulted in a weak upward supply of nutrients to the surface. In this context a large fraction of the NO_3 renewal during winter was presumably driven by nitrification within the upper mixed layer. In 2004, the biological consumption of nutrients began under the ice during May so that waters were already impoverished at the time of ice melt. This situation led to the rapid establishment of a subsurface chl maximum, which drove a significant portion of the seasonal NO_3 drawdown and generated the primary NO_2 maximum. In the upper mixed layer the consumption of dissolved inorganic carbon and PO_4 was in excess of the cumulative depletion of inorganic nitrogen. This pattern shows that the net productivity of the pelagic community was higher than new production based on the allochthonous supply of NO_3. The nitrogen source required to drive the supplemental net production has not been identified, but could be of atmospheric or riverine origin. Overall, this study suggests that the ongoing increase in river discharge to the coastal Arctic may lower NO_3-based new production in nearshore environments due to increased vertical stratification. Exceptions to this general pattern occur at the shelf break were wind-driven upwelling may inject substantial amounts of nutrients in the upper euphotic zone.

Estimates of phytoplankton new production in the Amundsen Gulf and the Cape Bathurst polynya were derived from time series analysis of changes in nutrient concentration (Simpson et al., 2008b) and from uptake of stable isotope tracers (Simpson et al., 2008c). In the Cape Bathurst polynya, ice retreat was quickly followed by the depletion in the surface mixed layer of dissolved NO_3, PO_4 and $Si(OH)_4$ concentration as phytoplankton bloomed. Within 18 days, all of the NO_3 was depleted from the surface to a depth of about 30-50 m, while PO_4 and $Si(OH)_4$ concentrations were still detectable. Consumption of the PO_4 and $Si(OH)_4$ continued for an additional 10 days after NO_3 exhaustion.

Using a logistic model we calculated the maximum consumption rates and total integrated drawdown of each of the nutrients. The ratio of Si:N uptake (1.8:1 mol/mol) indicated that the phytoplankton bloom was predominately composed of diatoms. Published values of the phytoplankton C:N ratio were used to determine the amount of new production associated with the disappearance of NO_3. This method yielded an estimate of spring new production of 16.6 ± 1.5 g C m^{-2}. Horizontal advection of NO_3-enriched water could potentially increase this estimate by roughly 20 %. Similar amounts of new production were computed from the net PO_4 and $Si(OH)_4$ consumption (15.8 ± 2.1 and 22.5 ± 3.1 g C m^{-2}, respectively). ^{15}N-tracer measurements of NO_3 uptake during the spring bloom agreed remarkably well with the estimates of NO_3

consumption from the logistic model. Spring new production determined by the tracer was 13.2 g C m^{-2} and it represented roughly 43 % of annual new production (25–36.7 g C m^{-2}). Uptake of NH_4 and urea fueled most of the phytoplankton growth annually and urea in particular was extremely important in supplying 2/3 of the regenerated N demand. The average f-ratio for the region was correspondingly low.

4.2.3 Phytoplankton production and photosynthetic properties

Phytoplankton production in the euphotic zone was measured using the ^{14}C-assimilation method. Differences in the production regime were observed between the Mackenzie Shelf and Amundsen Gulf during early October 2002 and 2003. Phytoplankton production, chl *a* biomass and cell size were highly variable on the Mackenzie Shelf, presumably because of the influence of the Mackenzie River. Phytoplankton production and biomass on the Shelf, which were dominated by small algal cells (<5 µm), were lower than in the Gulf. During early October 2002 and 2003, primary production and biomass values on the Shelf were similar (i.e., ca. 66 mg C m^{-2} d^{-1} and 12 mg chl *a* m^{-2}, respectively). In contrast, the phytoplankton biomass in the Gulf was higher in 2003 than in 2002. In 2003, the phytoplankton biomass in the Gulf averaged 26 mg m^{-2}, and large algal cells (>5 µm) contributed for 50 % of the total algal biomass. These results suggested that we sampled the end of a fall bloom. The variable nature of the fall bloom in Amundsen Gulf was previously detected by Arrigo & van Dijken (2004) using satellite observations. By mid-October, the primary production was reduced to ca. 25 mg C m^{-2} d^{-1}, both in the Shelf and Gulf areas, due to decreasing solar irradiance and ice freeze-up. The phytoplankton biomass then decreased throughout the end of the fall (9 mg m^{-2} in November)

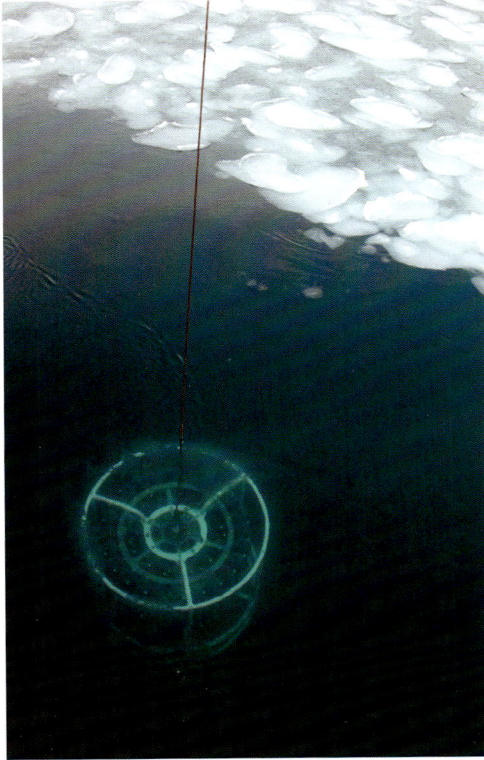

Rosette sampling of sea water.
Photo: Martin Fortier/ArcticNet.

and the beginning of the polar night, and exhibited a constant dominance of small algal cells (about 75 %). Minimal biomass values were reached in January at the end of the polar night (< 2 mg m^{-2} in the surface layer). Phytoplankton production and biomass started to increase again as early as February under the ice cover. In April-May, phytoplankton production reached ca. 7 mg C m^{-2} d^{-1} and its biomass was constrained to the upper meters of the water column. In June, phytoplankton production and biomass were highly variable in the Amundsen Gulf, probably due to the opening of the Cape Bathurst polynya. The centre of the Amundsen Gulf was already experiencing post-bloom conditions and exhibited a low phytoplankton biomass dominated by small algal cells. The western part of the Amundsen Gulf, however, was characterized by a high algal biomass (286 mg m^{-2}) dominated by large cells (mainly diatoms). This spring bloom was probably induced by upwelling events. In July, the phytoplankton biomass in the Shelf was highly variable due to the Mackenzie River plume. The primary production was ca. 200 mg C m^{-2} d^{-1} and large algal cells contributed for 65 % of the total algal biomass (110 ± 137 mg m^{-2}). In July-August, the phytoplankton production was low (105 mg C m^{-2} d^{-1}) and the biomass dominated by small cells (>70 %) in Amundsen Gulf.

The photosynthetic properties of psychrophilic phytoplankton and ice algae collected in Franklin Bay in late May 2004 were determined at 4°C using a pulse amplitude modulation fluorometer (Phyto-PAM) (Ban et al., 2006). Rapid light curve measurements allowed the assessment of the photosynthetic efficiency (α), maximal electron transport rate (rETR$_{max}$), and minimum saturating irradiance (E$_k$) in the samples. The values of α in phytoplankton (0.63–0.68) were much larger than those in ice algae (0.10–0.51), and the values of rETR$_{max}$ in phytoplankton (4.6–6.7) were relatively larger

than those in ice algae (1.8–4.3). However, E$_k$ showed similar values in both samples and were around 10 µmol photons m^{-2} s^{-1}. These values were systematically compared to those obtained from mesophilic marine diatoms grown under various irradiances in the laboratory. The highly shade-adapted features of ice algae and phytoplankton were disclosed through this comparative analysis. It was also found that the non-photochemical quenching was much higher in psychrophilic samples than in mesophilic diatoms grown under moderate irradiance. Furthermore, in ice algae and phytoplankton, the decrease in rETR at high irradiances was prominent, showing that they were highly susceptible to photoinhibition. The comparative analysis using psychrophilic phytoplankton, ice algae, and two strains of mesophilic diatoms also revealed that the dependency on the xanthophyll cycle for the protection mechanisms of photosystems were remarkably different between the groups, indicating that the acclimation strategies to growth irradiances were variable between species. Such variable acclimation strategies could be one of the driving forces for the diverse algal flora that enables the region around Franklin Bay to be a productive, even though the psychrophilic phytoplankton and ice algae are highly shade-adapted.

4.2.4 Sea ice microbial dynamics

Community dynamics and trophic interactions (including nutrient cycling) within newly formed sea ice are still poorly understood, especially those surrounding the very thin sea ice formed on Arctic shelves. Therefore, the fall and winter contribution of sea ice assemblages to Arctic carbon cycling is largely unknown. Newly formed sea ice was sampled on the Mackenzie Shelf, in Amundsen Gulf and in Franklin Bay between 30 September and 19 November 2003 (Riedel, 2006; Riedel et al., 2007a; Rózanska et al., 2008). Protist abundance and taxonomic composition were determined for four development stages of newly formed sea ice (new ice, nilas, young ice, and thin first-year ice) and in the underlying surface waters (Różańska et al., 2008). Picoalgae and nanoalgae were counted by flow cytometry whereas photosynthetic and heterotrophic protists ≥ 4 µm were identified and counted by inverted microscopy. Protists were always present in sea ice and surface water samples throughout the study period. The most abundant protists in sea ice and surface waters were cells <4 µm. They were less abundant in sea ice ($418–3051 \times 10^3$ cells L^{-1}) than in surface waters ($1393–5373 \times 10^3$ cells L^{-1}). In contrast, larger protists (≥ 4 µm) were more abundant in sea ice ($59–821 \times 10^3$ cells L^{-1}) than in surface waters ($22–256 \times 10^3$ cells L^{-1}). These results suggested a selective incorporation of larger cells into sea ice. The ≥ 4 µm protist assemblage was composed of a total number of 73 taxa, including 12 centric diatom species, 7 pennate diatoms, 11 dinoflagellates and 16 flagellates. The taxonomic composition in the early stage of ice formation (i.e., new ice) was very similar to that observed in the surface waters and was composed of a mixed population of nanoflagellates (Prasinophyceae and Prymnesiophyceae), diatoms (mainly *Chaetoceros* species) and dinoflagellates. In older stages of sea ice (i.e., young ice and thin first-year ice), the taxonomic composition became markedly different from that of the surface waters. These older ice samples contained relatively fewer Prasinophyceae and more unidentified nanoflagellates than the younger ice. Diatom resting spores and dinoflagellate cysts were generally more abundant in sea ice than in surface waters. However, further studies are needed to determine the importance of this winter survival strategy in Arctic sea ice. Our investigation clearly showed the selective incorporation of large cells (≥ 4 µm) in newly formed sea

ice, and the change in the taxonomic composition of protists between sea ice and surface waters as the fall season progresses.

Newly formed sea ice and its underlying surface waters were also analyzed to assess the concentration and enrichment of nutrients, exopolymeric substances (EPS, measured with Alcian blue), chl *a*, autotrophic and heterotrophic protists, and heterotrophic bacteria (Riedel et al., 2007a). Dark incubations were also conducted to estimate the net heterotrophic NH_4 regeneration rates in sea ice <5 cm thick. Large (≥ 5 µm) autotrophs were selectively enriched during sea-ice formation, having the highest average enrichment index (IS = 62), although heterotrophic protists (IS = 19), EPS (IS = 17), bacteria (IS = 6) and dissolved inorganic nitrogen (IS = 3 to 5) were also significantly enriched in the sea ice. Significant relationships were observed between sea ice EPS and total chl *a* concentrations (Pearson's r = 0.59, p < 0.001) and between sea ice EPS and ≥ 5 µm autotroph enrichment indices (r = 0.48, p < 0.01), suggesting that EPS were actively produced by algae entrapped in the sea ice. These relationships also suggest that the presence of EPS may enhance the selective enrichment of large autotrophs. Heterotrophic regeneration contributed to the observed enrichment of NH_4 in the sea ice, with an average regeneration rate of 0.48 µM d^{-1}, contributing 67 % to the sea-ice NH_4 concentrations. In the newly formed ice, the NH_4 regeneration was coupled to the NO_3 and $Si(OH)_4$ consumption and was significantly related to EPS concentrations (r = 0.87, p < 0.05). Our data suggest that EPS enhance the NH_4 regeneration by acting as a carbon source for sea-ice heterotrophs or a substrate for sea-ice bacteria.

The seasonal dynamics of the ice community were investigated in the first-year landfast sea ice of Franklin

Sunset on ice covered sea.
Photo: Thomas Juul-Pedersen.

Bay from 24 February and 20 June 2004 (Riedel et al., 2006). Bottom sea ice (from under high and low snow cover) and surface water samples were collected on 21 occasions and analyzed for EPS, particulate organic carbon (POC) and chl *a*. The concentrations of EPS were measured using Alcian blue staining of melted ice samples. Chl *a* and bacterial sinking velocities were also assessed with settling columns, to determine the potential role of EPS in the transport of sea-ice biomass. EPS concentrations in the bottom ice were consistently low in March (avg. 185 µg xanthan equivalents L^{-1}), after which they increased to maximum values of 4930 and 10500 µg xanthan equivalents L^{-1} under high and low snow cover, respectively. EPS concentrations in the surface water were consistently 2 orders of magnitude lower than in the sea ice. Sea ice EPS concentrations were significantly correlated with sea-ice chl *a* biomass (Kendall's τ = 0.70, p < 0.01). Sea ice algae were primarily responsible for EPS production within the sea ice, whereas bacteria produced insignificant amounts of sea ice EPS. EPS-carbon contributed, on average, 23 % of POC concentrations within the sea ice, with maximum values reaching 72 % during the melt period. Median chl *a* sinking velocities were 0.11 and 0.44 m d^{-1} under high and low snow cover, respectively. EPS had little effect on chl *a* sinking velocities. However, bacterial sinking velocities did appear to be influenced by diatom-associated and free EPS within the sea ice. Diatom-associated EPS could facilitate the attachment of bacteria to algae thereby increasing bacterial sinking velocities, whereas the sinking velocities of bacteria associated with positively buoyant, free EPS, could be reduced. EPS contributed significantly to the sea-ice carbon pool and influenced the sedimentation of sea-ice biomass, which emphasizes the important role of EPS in carbon cycling on Arctic shelves.

Heterotrophic bacterial dynamics in the sea ice and surface waters were determined at the same station from 5 March to 3 May 2004 (Riedel et al., 2007b). On 11 occasions, heterotrophic protist bacterivory was assessed from the disappearance of fluorescently labeled bacteria (FLB) in sea ice samples collected from areas of high and low snow cover. Concurrently, sea ice and surface water samples were analyzed for dissolved organic carbon (DOC), EPS and chl *a* concentrations, and protist and bacterial abundances. Total bacterial abundances were significantly higher in the sea ice than in surface waters. However, DOC concentrations and abundances of large (≥0.7 µm) bacteria were not significantly higher in the sea ice compared to surface waters. This suggested that DOC was being released from the sea ice, potentially supporting the growth of large-sized bacteria at the ice-water interface. Heterotrophic protist (HP) bacterivory averaged 57 % d^{-1} of large-sized bacterial standing stocks in the sea ice with ingestion rates averaging 768 and 441 bacteria HP^{-1} d^{-1}, under high and low snow cover, respectively. High concentrations of EPS during the sea ice algal bloom may have interfered with the grazing activities of heterotrophic protists as indicated by the significant negative correlations between ingestion rates and EPS-carbon concentrations under high (τ = -0.57, p < 0.05) and low (τ = -0.56, p < 0.05) snow cover. Bacterivory satisfied heterotrophic protist carbon requirements prior to (but not during) the sea-ice algal bloom, under high and low snow cover. EPS may have been an additional carbon source for the heterotrophs, especially during the sea ice algal bloom period. This study provided evidence of an active heterotrophic microbial food web in first-year sea ice, prior to and during the sea ice algal bloom. It also highlighted the significance of DOC and EPS as integral components of the microbial food web within the sea ice and surface waters of Arctic shelves.

An understanding of microbial interactions in first-year sea ice on Arctic shelves is essential for identifying potential responses of the Arctic Ocean carbon cycle to changing sea ice conditions. The study of Riedel et al. (2008) assessed the concentrations of DOC and POC, EPS and chl *a*, and the biomass of bacteria and protists in the sea ice and underlying surface waters of Franklin Bay from February to June 2004. The dynamics of and relationships between different sea ice carbon pools were investigated for the periods prior to, during, and following the sea-ice algal bloom, under high and low snow cover. A predominantly heterotrophic sea ice community was observed prior to the ice algal bloom under high snow cover only. However, the heterotrophic community persisted throughout the study with bacteria accounting for, on average, 44 % of the non-diatom particulate carbon biomass overall the study period. There was an extensive accumulation of sea ice organic carbon following the onset of the ice algal bloom, with diatoms driving seasonal and spatial trends in particulate sea-ice biomass. DOC and EPS were also significant sea-ice carbon contributors such that sea-ice DOC concentrations were higher than, or equivalent to, sea-ice algal carbon concentrations prior to and following the algal bloom, respectively. Sea-ice algal carbon, DOC and EPS-carbon concentrations were significantly interrelated under high and low snow cover during the algal bloom (r values ≥ 0.74, p < 0.01). These relationships suggest that algae were primarily responsible for the large pools of DOC and EPS-carbon and that similar stressors and/or processes could be involved in regulating their release. This study demonstrated that DOC can play a major role in organic carbon cycling on Arctic shelves.

4.2.5. Photodegradation of sea ice carbon

Photoprocesses accelerate dissolved organic matter (DOM) cycling in the ocean by transforming biologically refractory DOM into labile DOM and by degrading DOC to small molecules, mainly CO_2 and carbon monoxide (CO). The photoproduction of CO in sea ice was studied in Franklin Bay in May 2004 (Xie & Gosselin, 2005). The concentration of CO in the sea ice was 40 times higher than in the underlying seawater and 15 times higher than in the adjacent open water. The CO concentration in the sea ice decreased with increasing depth and increased rapidly at the bottom where there was an abundance of ice microalgae. The depth distribution of CO concentration was consistent with a photochemical source of this compound in sea ice, which was further inferred from the vertical profiles of the DOM absorption coefficients and directly verified by incubation of ice samples refrozen from melted ice. Results from this study suggested that a substantial photooxidation of the organic matter occurs in sea ice. This process may affect the organic carbon cycle in the Arctic Ocean.

4.2.6 Microalgal settling

The sinking export of biogenic carbon was studied in ice-covered and open waters of the Canadian Beaufort Sea using short-term particle interceptor traps (Juul-Pedersen, 2007). The sinking export of particulate material under first-year landfast sea ice was studied from the winter period to spring melt in Franklin Bay (Juul-Pedersen et al., 2008b). Traps were deployed at 1, 15 and 25 m under the ice on 16 consecutive occasions from February 23 to June 20, 2004. The sinking material was analyzed for chl *a*, phaeopigments, total particulate carbon (TPC), POC, particulate organic nitrogen (PON) and for biogenic silica (BioSi). The sinking fluxes of chl *a* and BioSi increased steadily after March 19 and until the onset of the spring melt (May 26), after which these sinking fluxes increased considerably (maxima of 2.0 and 90.4 mg m^{-2} d^{-1}, respectively). The contribution of large algal cells (>5 μm) to the total chl *a* sinking flux also increased after March 19 (from ca. 60 % to 90 %), reflecting an increasing contribution of diatoms to the sinking export of algal material. Accordingly, the chl *a* sinking fluxes at 1 m showed a significant linear relationship with the bottom ice chl *a* biomass. On average, almost half (46 %) of the chl *a* exported at 1 m was lost in the upper 25 m. POC was the main component of the TPC sinking fluxes (91 %) throughout the study. POC sinking fluxes remained fairly stable until the onset of the spring melt (median values of 21.0 and 24.6 mg m^{-2} d^{-1} at 1 m), after which a considerable increase was observed (maximum of 522 mg m^{-2} d^{-1}). High POC:chl *a* ratios (ranging from 75.8 to 3474 g:g at 1 m) indicated a significant contribution of non-algal material to the sinking POC. The daily sinking loss rates of chl *a*, POC and PON from the sea ice and interfacial layer (top 1 m of the water column) varied seasonally and were highest during the winter period. Over the 4-month duration of this study, under ice sinking fluxes of chl *a*, POC and PON at 1 m were 31.3 mg m^{-2}, 7.2 g C m^{-2} and 1.2 g N m^{-2}, respectively. These results illustrated the continuous downward sinking export of organic material under landfast ice, from winter throughout late spring.

The study of Juul-Pedersen et al. (2008a) examined the influence of the Mackenzie River plume on the sinking fluxes of particulate organic and inorganic material on the Mackenzie Shelf. Short-term particle interceptor traps were deployed under the halocline at 3 stations across the shelf during fall 2002 and at 3 stations along the shelf edge during summer 2004. During the two sampling periods, the horizontal pattern

in sinking fluxes of particulate organic carbon (POC) and chl *a* paralleled those in chl *a* biomass within the plume. Highest sinking fluxes of particulate organic material occurred at stations strongly influenced by the river plume (maximum POC sinking fluxes at 25 m of 98 mg C m^{-2} d^{-1} and 197 mg C m^{-2} d^{-1} in 2002 and 2004, respectively). The biogeochemical composition of the sinking material varied seasonally with phytoplankton and fecal pellets, contributing considerably to the sinking flux in summer, while amorphous detritus dominated in the fall. Also, the sinking phytoplankton assemblage showed a seasonal succession from a dominance of diatoms in summer to flagellates and dinoflagellates in the fall. The presence of the freshwater diatom *Eunotia* sp. in the sinking assemblage directly underneath the river plume indicated the contribution of a phytoplankton community carried by the plume to the sinking export of organic material. Yet, increasing chl *a* and BioSi sinking fluxes with depth indicated an export of phytoplankton from the water column below the river plume during summer and fall. Grazing activity, mostly by copepods, and to a lesser extent by appendicularians, appeared to occur in a well-defined stratum underneath the river plume, particularly during summer. These results showed that the Mackenzie River influences the magnitude and composition of the sinking material on the shelf in summer and fall, but does not constitute the only source of material sinking to depth at stations influenced by the river plume.

Juul-Pedersen (2007) presented an extensive spatial and seasonal coverage of the sinking export of particulate organic material below the euphotic zone in the Canadian Beaufort Sea. Free-drifting short-term particle interceptor traps were deployed, generally at 50 m, during fall 2002 and 2003, and summer 2004. The different regions of the sampling area (i.e. the Cape Bathurst polynya and the Mackenzie shelf and slope) showed comparable ranges for the sinking export of chl *a* and POC in fall, while regional differences were observed in summer. The two regions showed a general decreasing trend in sinking fluxes towards fall. The highest chl *a* and POC sinking fluxes during this study were therefore recorded during summer (3.6 and 258.4 mg m^{-2} d^{-1}, respectively). A high retention of suspended biomass was observed throughout this study, i.e. low daily loss rates of suspended chl *a* and POC (both averaging ca. 1 % d^{-1}) were observed. Still, the POC sinking export accounted for, on average, half of the particulate primary production throughout this study. Zooplankton, primarily copepods, played an important role in the sinking export of particulate organic material, particularly in the Cape Bathurst polynya. A cluster-based analysis of the sinking protist cell assemblage revealed a seasonal succession that prevailed over spatial and interannual differences between the stations sampled in the eastern Beaufort Sea. Flagellates dominated throughout the study area, while diatoms, dominated by *Fragilariopsis cylindrus*, showed a decreasing contribution to the sinking protist cell assemblage towards fall. The presence of the sea ice related pennate diatoms *Nitzschia frigida* and *Navicula vanhoeffenii* in the material collected during summer reflected an input of organic material from the sea ice. Results from particle interceptor traps deployed at a landfast sea ice station during ice-covered and ice-free conditions showed the importance of taking into account underice sinking fluxes (up to 115 mg C m^{-2} d^{-1} for POC) for sinking export estimates on Arctic shelves.

4.2.7 Sea ice-pelagic-benthic coupling

Finally, we investigated the relationship between ice algal standing stock and benthic respiration between January and July 2004 at the time series station of Franklin Bay (Renaud et al., 2007). Both ice algal chl *a* and benthic sediment oxygen demand showed >10-fold increases between March and April. While some of the increase in oxygen demand could be attributed to bacteria and meiofauna, most was due to the activities of the macroinfauna. We also observed a trend toward lower sediment pigment content during the pulse in benthic carbon remineralization. While the chl *a* sedimentation also increased by a factor of 7 during this period, fluxes were not sufficient to account for the increased carbon demand. We suggest that the sedimenting ice algae provided a cue for increased benthic activity, and that the direct consumption of ice algae and increased oxygen availability in the sediment due to bioturbation by epifaunal organisms led to the enhancement in respiration rates. Seasonal patterns in primary productivity and the activity of resident epifaunal and infaunal communities are, thus, important factors in determining carbon cycling patterns on Arctic shelves.

4.3 Implications of this work

During this investigation, light spectral intensity, nutrient availability, nutrient uptake, size-fractionated microalgal production and biomass, microbial distribution, taxonomic composition of protists, and vertical fluxes of biogenic carbon were assessed in ice-covered and ice-free waters of the three oceanographic provinces of the eastern Beaufort Sea.

Continuous measurements from September 2003 to August 2004 allowed us to monitor nutrient/microalgal dynamics and sinking exports throughout the annual physical and chemical forcing cycles characteristic to this region (and other Arctic shelves).

Until now, the concentration of nutrients available to microalgae at the end of winter in the Beaufort Sea was a matter of speculation. We have shown that during fall 2003 and winter 2004 the renewal of nutrients in the upper euphotic zone was small due to modest mixing by wind, convection, and brine rejection. We hypothesize that other mechanisms of N and P supply may also play determining roles in this context. Approximately, one third of the NO_3 reservoir available to phytoplankton in spring 2004 was likely supplied by nitrification above the resilient halocline. However, this remains to be confirmed experimentally.

Irradiance clearly affects the timing and the rate of primary production in seasonally ice-covered Arctic waters, however our analysis supports the notion that cumulative new and net production are driven primarily by nitrogen loading. The response of phytoplankton to declining ice cover may depend more on the alteration of nutrient loads by atmospheric forcing and freshwater input than on changes in light availability.

DOP concentrations were greatly enriched in Pacific derived waters and represent an important phosphorus input to the Arctic that has previously been overlooked. The deep water of the Amundsen Gulf / Cape Bathurst polynya exhibited a marked divergence in nutrient and oxygen content compared to waters of similar depth in the Beaufort Sea. Our analysis suggests that the residence time of these waters in the Amundsen Gulf is sufficiently long to allow the accumulation of nutrients from the remineralization of exported diatom material. The enrichment confirms the importance of the polynya as a pump for transferring nutrients (and ultimately carbon) to depth.

The Cape Bathurst polynya appears to be remarkably similar to other polynyas in the Arctic and Antarctic oceans in that the spring new production represents 40 % of the annual new production and occurs over a relatively short period of time following ice retreat.

The variable acclimation strategies among microalgal species could be one of the driving forces for creating a diverse flora and enabling the Arctic Ocean to be a productive area.

Newly formed sea ice is significantly enriched with large photosynthetic cells (>5 μm) in the fall. Microorganisms in newly formed sea ice are active, taking up NO_3 and $Si(OH)_4$ and producing NH_4.

This study has shown that EPS not only contributes directly to the carbon pool in first-year sea ice, but also influences carbon cycling within the sea ice and the fate of sea ice carbon once released to the water column.

Photoprocesses are primarily responsible for elevated carbon monoxide levels in sea ice, and photoremineralization in sea ice may influence the organic carbon cycle of the Arctic Ocean.

The Mackenzie River plume has a strong influence on the sinking export of particulate material on the Beaufort Sea shelf and slope.

Regardless of the spatial and interannual variability between sampling stations, a strong seasonal signature in the species composition of the sinking protist assemblages emerged for the study area. These results showed that the phytoplankton species succession generally observed in the high latitude waters—from a high diatom contribution in summer towards a higher proportion of flagellates in the fall—prevailed in the sinking material throughout the Canadian Beaufort Sea.

The impacts of climate change on carbon cycling in Arctic marine communities are difficult to predict. However, this study suggests that factors changing the patterns of ice algal production on Arctic shelves may have significant consequences for carbon processing and storage in benthic sediments. It is increasingly clear that both the structure and seasonal activity of benthic faunal communities interact with local primary productivity regimes to determine carbon preservation and regeneration on Arctic shelves.

Many species of consumers that have fixed life history strategies may be unable to adjust to changes in the timing of the phytoplankton bloom such that earlier ice retreat in a warmer Arctic Ocean may alter the coupling between benthic and pelagic processes.

Phytoplankton communities adapt to the changing light conditions so that a changing ice cover may alter the productivity of the system.

4.4 Recommendations

Future information is required on the importance of non-nitrate forms of allochthonous nitrogen for primary producers, the dynamics of nitrification near the surface, and the ecology of SCMs and their significance to annual primary production, food webs, vertical flux and elemental cycling.

Future studies are needed to determine the presence of potentially harmful algae in a changing Arctic environment.

Further studies are needed to determine the coupling between physical forcing and phytoplankton production at the scale of the ecosystem, with the objective of identifying biological hot spots.

Our results challenge the view that there is little export of ice material prior to ice melt. The validation of an early coupling between ice algal biomass and the sinking export of algal material therefore urges that future studies on shallow sinking export to include the ice algal growth period.

Acknowledgements

This project was supported by grants from the Natural Sciences and Engineering Research Council (NSERC) of Canada (Research Network grant; Individual and Northern Research Supplement Discovery grants), the Fisheries and Oceans Canada (DFO) Academic Science Subvention Program, the DFO Science Strategic Fund and the Fonds québécois de la recherche sur la nature et les technologies (FQRNT) through Québec-Océan and by financial support from the Canadian Museum of Nature. Graduate students received scholarships from NSERC (A.R.N), the Institut des sciences de la mer de Rimouski (S.B., A.R.N., T.J.-P., M.R.), Université du Québec à Rimouski (UQAR; A.R.N., T.J.-P., M.R.) and the Fondation de l'UQAR (bourse Estelle-Laberge to A.R.N.) and financial support from Indian and Northern Affairs Canada for fieldwork (A.R.N). We sincerely thank the officers and crew of the CCGS *Pierre Radisson* and CCGS *Amundsen* for their invaluable support during the expeditions; M. Simard, C. Jose, J. Ferland, A. Tatarek, J. Wiktor, M. Lizotte and others for assistance in the field and/or laboratory; D. Bérubé for CHN analyses; C. Belzile for flow cytometry analyses; C. Lovejoy and S. Lessard for assistance in some taxonomic identification.

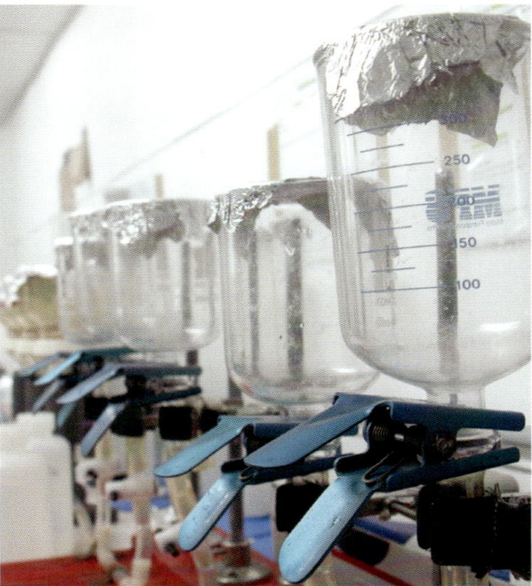

Filtration bench. Photo: Martin Fortier/ ArcticNet.

Collecting new sea-ice. Photo: Martin Fortier/ArcticNet.

References

Arrigo, K.R., and G.L.V. Dijken, 2004. Annual cycles of sea ice and phytoplankton in Cape Bathurst polynya, southeastern Beaufort Sea, Canadian Arctic. Geophys. Res. Lett. 31, L08304, doi:10.1029/2003GL018978.

Ban, A, S. Aikawa, H. Hattori, H. Sasaki, M. Sampei, S. Kudoh, M. Fukuchi, K. Satoh, and Y. Kashino, 2006. Comparative analysis of photosynthetic properties in ice algae and phytoplankton inhabiting Franklin Bay, the Canadian Arctic, with those in mesophilic diatoms during CASES. Polar Biosci. 19:11–28.

Ben Mustapha, S., P. Larouche, S. Bélanger, and M. Babin, 2006. Evaluation of ocean color algorithms in the Cape Bathurst polynya using MODIS and SeaWiFS spectral bands. Proceedings of Ocean Optics XVIII, Montréal, Canada, 9-13 October 2006.

Carmack, E.C., R.W. Macdonald, and S. Jasper, 2004. Phytoplankton productivity on the Canadian Shelf of the Beaufort Sea. Mar. Ecol. Prog. Ser. 277:37–50.

Dugdale, R.C., and F.P. Wilkerson, 1998. Silicate regulation of new production in the equatorial Pacific upwelling. Nature 391:270–273.

Fortier, M., L. Fortier, C. Michel, and L. Legendre, 2002. Climatic and biological forcing of the vertical flux of biogenic particles under seasonal Arctic sea ice. Mar. Ecol. Prog. Ser. 225:1–16.

Horner, R., and G.C. Schrader, 1982. Relative contributions of ice algae, phytoplankton and benthic microalgae to primary production in nearshore regions of the Beaufort Sea. Arctic 35:485-503.

Hsiao, S.I.C., M.G. Foy, and D.W. Kittle, 1977. Standing stock, community structure, species composition, distribution, and primary production of natural populations of phytoplankton in southern Beaufort Sea. Can. J. Bot. 55:685–694.

Hsiao, S.I.C., 1992. Dynamics of ice algae and phytoplankton in Frobisher Bay. Polar Biol. 12:645-651.

Juul-Pedersen, T., 2007. Variations spatiales et temporelles de la sédimentation sous la zone euphotique dans le secteur canadien de la mer de Beaufort. Thèse (Ph.D.), Université du Québec à Rimouski, 142 p.

Juul-Pedersen, T., C. Michel, and M. Gosselin, 2008a. Influence of the Mackenzie River plume on the sinking export of particulate material on the shelf. J. Mar. Syst. 74:810-824.

Juul-Pedersen, T, C. Michel, M. Gosselin, and L. Seuthe, 2008b. Seasonal changes in the sinking export of particulate material under first-year sea ice on the Mackenzie Shelf (western Canadian Arctic). Mar. Ecol. Prog. Ser. 353:13–25.

Klein, B., B. LeBlanc, Z.P. Mei, R. Beret, J. Michaud, C.-J. Mundy, C.H. von Quillfeldt, M.-È. Garneau, S. Roy, Y. Gratton, J.K. Cochran, S. Bélanger, P. Larouche, J.D. Pakulski, R.B. Rivkin, and L. Legendre, 2002. Phytoplankton biomass, production and potential export in the North Water. Deep-Sea Res. II 49:4983–5002.

Legendre, L., S.F. Ackley, G.S. Dieckmann, B. Gulliksen, R. Horner, T. Hoshiai, I.A. Melnikov, W.S. Reeburgh, M. Spindler, and C.W. Sullivan, 1992. Ecology of sea ice biota. 2 Global significance. Polar Biol. 12:429–444.

Matsuoka, A., Y. Huot, K. Shimada, and S.-I. Saitoh, 2007. Bio-optical characteristics of the western Arctic Ocean: Implications for ocean color algorithms. Can. J. Rem. Sens. 33:503-518.

Matsuoka, A., P. Larouche, M. Poulin, T. Hirawake, H. Hattori, and S.-I. Saitoh, 2008. Phytoplankton community adaptation to changing light levels in the southern Beaufort Sea and the Amundsen Gulf, Canadian Arctic. Est. Coast. Shelf. Sci. Submitted.

Michel, C., M. Gosselin, and C. Nozais, 2002. Preferential sinking export of biogenic silica during the spring and summer in the North Water Polynya (northern Baffin Bay): Temperature or biological control? J. Geophys. Res. 107, C7, doi:10.1029/2000JC000408.

Renaud, P.E., A. Riedel, C. Michel, M. Morata, M. Gosselin, T. Juul-Pedersen, and A. Chiuchiolo, 2007. Seasonal variation in benthic community oxygen demand: A response to an ice algal bloom in the Beaufort Sea, Canadian Arctic? J. Mar. Syst. 67:1–12.

Riedel, A.J., 2006. Influence des substances exopolymériques et des microorganismes hétérotrophes sur le cycle du carbone de la glace de mer, sur le plateau du Mackenzie, dans l'Arctique canadien. Thèse (Ph.D.), Université du Québec à Rimouski, 141 p.

Riedel, A., C. Michel, and M. Gosselin, 2006. Seasonal study of sea-ice exopolymeric substances (EPS) on the Mackenzie shelf: implications for the transport of sea-ice bacteria and algae. Aquat. Microb. Ecol. 45:195–206.

Riedel, A., C. Michel, M. Gosselin, and B. LeBlanc, 2007a. Enrichment of nutrients, exopolymeric substances and microorganisms in newly formed sea ice on the Mackenzie shelf. Mar. Ecol. Prog. Ser. 342:55–67.

Riedel, A., C. Michel, and M. Gosselin, 2007b. Grazing of large-sized bacteria by sea-ice heterotrophic protists on the Mackenzie shelf during the winter-spring transition. Aquat. Microb. Ecol. 50:25–38.

Riedel, A., C. Michel, M. Gosselin, and B. LeBlanc, 2008. Winter-spring dynamics in sea-ice carbon cycling in the coastal Arctic Ocean. J. Mar. Syst.74: 918-932

Różańska, M., M. Poulin, and M. Gosselin, 2008. Protist entrapment in newly formed sea ice in the coastal Arctic Ocean. J. Mar. Syst. 74:887-901.

Simpson, K.G., J.-É. Tremblay, Y. Gratton, and N.M. Price, 2008a. An annual study of inorganic and organic nitrogen and phosphorus and silicic acid in the southeastern Beaufort Sea. J. Geophys. Res. (Oceans) 113, C07016, doi:10.1029/2007JC004462.

Simpson, K.G., J.-É. Tremblay, and N.M. Price, 2008b. Nutrient dynamics in the Amundsen Gulf and Cape Bathurst Polynya: New production in spring inferred from nutrient draw-down. Mar. Ecol. Prog. Ser. Submitted.

Simpson, K.G., J.-É. Tremblay, and N.M. Price, 2008c. Nutrient dynamics in the Amundsen Gulf and Cape Bathurst Polynya: Stable isotope tracer estimates of new and regenerated production. In preparation.

Stein, R., and R.W. Macdonald, 2004. The organic carbon cycle in the Arctic Ocean. Springer-Verlag, New York.

Tremblay, J.-É., K. Simpson, J. Martin, L. Miller, Y. Gratton, D. Barber, and N.M. Price, 2008. Vertical stability and the annual dynamics of nutrients and chlorophyll fluorescence in the coastal, southeast Beaufort Sea. J. Geophys. Res. (Oceans) 113, C07S90, doi:10.1029/2007JC004547.

Xie, H., and M. Gosselin, 2005. Photoproduction of carbon monoxide in first-year sea ice in Franklin Bay, southeastern Beaufort Sea. Geophys. Res. Lett. 32, L12606, doi:10.1029/2005GL022803.

CHAPTER 5

Microbial Communities and Carbon Fluxes

Warwick F. Vincent[1]*, Carlos Pedrós-Alió[2], Curtis Suttle[3], Connie Lovejoy[1], Jody Deming[4], Chris Osburn[5], Lance Lesack[6], Huixang Xie[7], Marcel Babin[8], Annick Wilmotte[9]

5.1 Introduction and Rationale

"Imagine overlooking 50% of the total marine biomass and then suddenly discovering it!" (Proctor & Karl, 2007).

Marine microbiology has become a rapidly developing field of exciting discoveries. New technologies (including DNA and RNA based techniques) are being applied throughout the world's oceans in combination with assays of biomass stocks and biogeochemical fluxes. The emerging consensus is that microbes not only contribute much of the total biomass and biological production of marine ecosystems, but they also constitute a major proportion of their biodiversity. This diversity is expressed both in terms of species richness (i.e., the number of genetically distinct taxa) and functional richness. This latter dimension refers to their performance of key ecological and biogeochemical roles, including photosynthesis, grazing and parasitism, the breakdown of organic compounds, nutrient redox reactions and re-cycling, and the production and consumption of gases that regulate the atmosphere and biosphere (notably oxygen, carbon dioxide, methane, hydrogen, nitrous oxide and dimethyl sulphide).

The CASES program has provided a remarkable opportunity to examine the community structure and seasonal dynamics of microbial communities within coastal Arctic Ocean waters (Fig. 5.1). Through the microbiological component of this program, we were able to study the diversity and activities of many forms of microscopic life, including viruses, Archaea and Bacteria (known collectively as prokaryotes), and Eukarya (single-celled members, also known as protists). The CASES microbiological subprogram began

LEFT: Arctic community.
Photo: Marc Tawil / ArcticNet.

[1] Département de Biologie, Université Laval, Québec QC, G1V 0A6, Canada

[2] Dept Biologia Marina i Oceanografia, Institut de Ciències del Mar, Barcelona, Spain

[3] Department of Earth and Ocean Sciences, University of British Columbia, 6270 University Blvd., Vancouver, BC, V6T 1Z4, Canada

[4] School of Oceanography, University of Washington, Campus Box 357940, Seattle, WA 98195-7940, USA

[5] US Naval Research Laboratory, Washington DC 2075, USA

[6] Departments of Geography and Biological Sciences, Simon Fraser University, Burnaby, Canada V5A 1S6, BC

[7] Institut des Sciences de la Mer (ISMER), Université du Québec à Rimouski, Allée des Ursulines, Rimouski, QC, G5H 3Z4, Canada

[8] Laboratoire océanologique de Villefranche-sur-mer, BP 28, 06234 Villefranche-sur-mer Cedex, France

[9] Centre for Protein Engineering, Institute of Chemistry B6, Sart Tilman, University of Liège, Liège, Belgium

* Corresponding author: Warwick.Vincent@bio.ulaval.ca

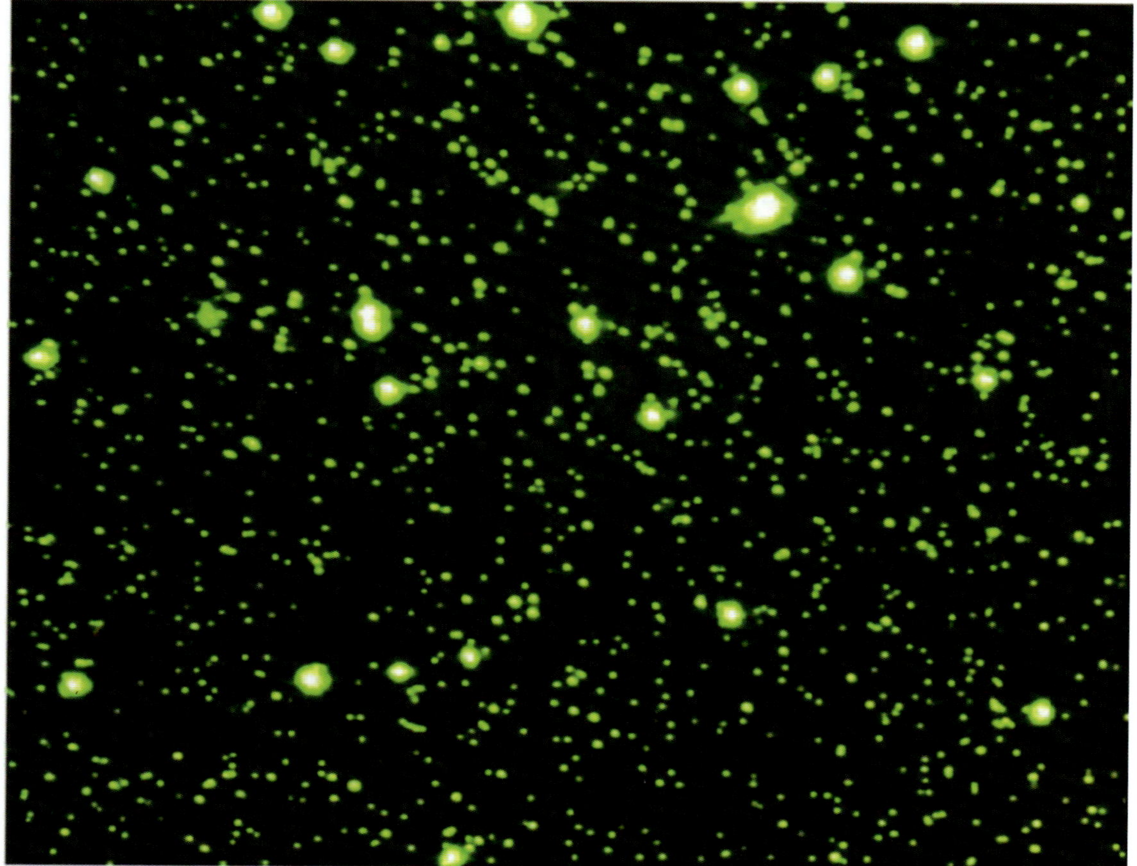

Figure 5.1

A galaxy of microbial populations in Arctic seawater. Here a sample from 3 m at CASES Station 200 (May 16, 2004) was stained with YoPro-I and then examined by epifluorescence microscopy. The smaller fluorescent dots are viruses (each about 20-200 nm in diameter) and the larger fluorescent particles are bacteria (about 200-500 nm in diameter). DNA analyses showed that these populations are composed of diverse taxa, many with different functional roles in the ocean. Micrograph from Jérôme Payet and Curtis Suttle.

with inaugural "Leg 0" aboard CCGS Pierre Radisson in September-October 2002, which allowed sampling of Franklin Bay and the eastern Beaufort Sea, as well as a transect from the Mackenzie River to the edge of the Arctic pack ice over the continental slope. The microbial research then continued from September 2003 to August 2004 aboard CCGS Amundsen, which was stationed in Franklin Bay from November 2003 to June 2004 for measurements during winter and spring (see Chapter 1 for the distribution of sampling sites).

The satellite program ARDEX (Arctic River Delta Experiment) was undertaken in July-August 2004, aboard a shallow draft research vessel, CCGS Nahidik, in order to sample a transect from Inuvik, 250 km upstream of the mouth the Mackenzie River, to a coastal marine station 50 km offshore. ARDEX provided observations and experiments that were complementary to CASES; specifically, riverine nutrients, organic carbon, and microbial processes that influence the shelf ecosystem. Given the central role of dissolved organic matter (DOM) in microbial processes within river and coastal ecosystems, both CASES and ARDEX were interested in the optical and biogeochemical properties of DOM, as well as its effects on primary production and its photochemical reactivity. Both programs also provided outreach opportunities to northern communities and the public at large, including school visits, the 'Schools on Board' program, websites, and an Inuvialuit high school internship aboard the ARDEX cruise (see http://jurban.es.umb.edu/ardex_home.aspx).

5.2 Overview of Results

5.2.1 Organic carbon composition, nutrients and carbon fluxes

Analyses of samples obtained from the Mackenzie River showed that dissolved organic carbon (DOC) and its coloured fraction (coloured dissolved organic matter, CDOM) were present in high concentrations and that these terrigenous materials exerted a strong influence on the surface properties of the coastal Beaufort Sea region (for example, in spectral light absorption). The highest CDOM absorbance per unit DOC for the river was observed in June, near its period of peak discharge. Shelf waters sampled in September-October appeared to retain the fluorescence signature of river discharge from the high flow period (Retamal et al., 2007). The ARDEX analyses showed that CDOM controlled the underwater UV-radiation regime in the river, the estuary and the shelf, and that the combination of suspended particulate material and CDOM controlled the availability of light for underwater photosynthesis (Fig. 5.2; Retamal et al., 2008). Optical and photosynthetic measurements indicated that phytoplankton within both the turbid river water and the deep photosynthetic maxima of sea water were limited by light availability, and that future changes in CDOM and particulate organic carbon loading (for example, via eroding permafrost in a warmer climate) could lead to reduced primary production in both portions (freshwater and marine) of the ecosystem.

CDOM is broken down by photochemical reactions to dissolved inorganic carbon (DIC), a process measured in the Beaufort Sea by Bélanger et al. (2006) during CASES. They used an optical-photochemical model that incorporated water column optics and experimental measurements of DIC photoproduction. Apparent quantum yields were determined on water samples from sites throughout the CASES region, and incident UV irradiances were estimated using a radiative transfer model equipped with satellite-backscattered and passive microwave radiance measurements. The resulting estimates of DIC photoproduction were equivalent to 8% of new primary production rates and 2.8 ± 0.6% of the 1.3 Tg of dissolved organic carbon

Figure 5.2

Controls on underwater light in the coastal Arctic Ocean. Light absorption by phytoplankton (a_{ph}), non algal particles (a_{NAP}), CDOM (a_{CDOM}) and water (a_W) is plotted as a function of wavelength for three sections of the Mackenzie estuarine system: the Mackenzie River (R1 and R2 at the surface); the estuarine transition zone (R5 at the surface); and offshore water (R9 in the surface and in the deep chlorophyll maximum). Redrawn from Retamal et al. (2008).

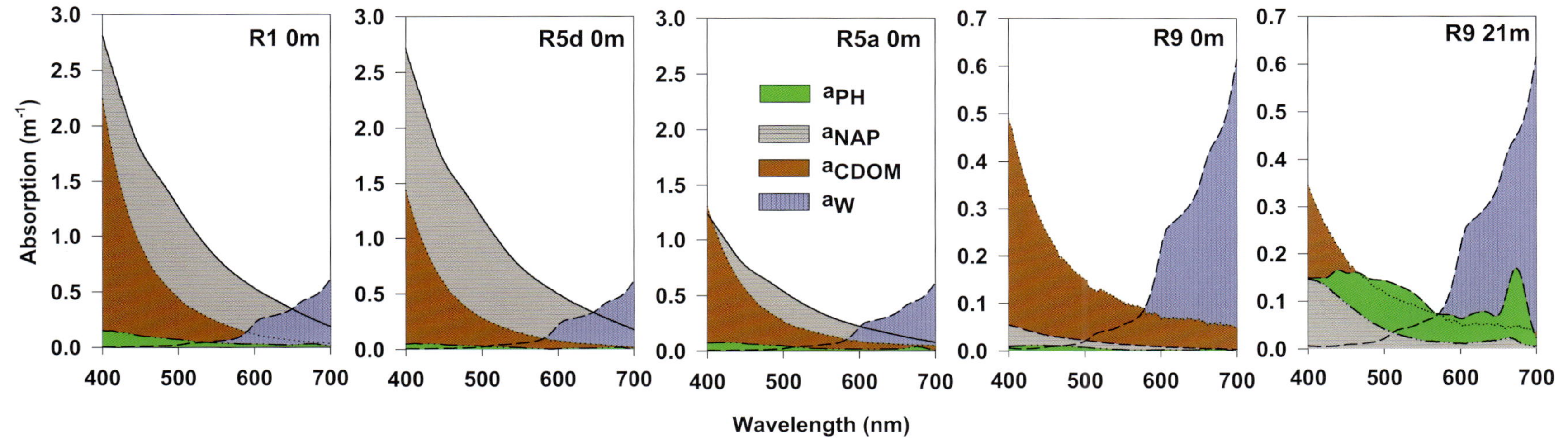

(DOC) discharged into the region each year by the Mackenzie River. The latter value was estimated to rise to 6.2% under ice-free conditions, indicating that ongoing contraction of sea ice will accelerate the photochemical breakdown of terrigenous DOM in Arctic coastal waters.

Photobleaching was examined in a surface water sample collected during 2002 over the 200 m isobath on the Beaufort Shelf (Station 66). This sample did not exhibit the terrigenous features of the Mackenzie River (determined by synchronous fluorescence) and had lower rates of photobleaching than surface waters collected from the Middle Atlantic Bight. Peroxide production was substantially less than that in surface waters of terrigenous nature and was not significantly related to CDOM photobleaching (O'Sullivan et al., 2005). However, the waters of the Mackenzie River were highly photoreactive, even during low-flow conditions in October (Osburn and Vincent, in preparation). Further analyses showed that stable carbon isotope ratios of DOM from Franklin Bay during winter ranged from –24 to –26 per mil, reflecting the influence of the Mackenzie River on surface waters of the Mackenzie Shelf (Osburn and St-Jean, 2007).

Riverine fluxes exert a strong influence on the Arctic ecosystem, and the ARDEX program therefore sought to obtain improved estimates of nutrient fluxes from the Mackenzie River to the Arctic Ocean. Of particular interest was the biogeochemical effect of the lake-rich Mackenzie Delta floodplain on riverine nutrient fluxes downstream of the gauging station where nutrients have historically been monitored (Fig. 5.3). Complete partitioning of the 13,000 km² delta into lakes, wetlands, channels, and non-wet floodplain through GIS

analysis (Emmerton et al., 2007) allowed the development of a nutrient mixing model based on two sources: the channel water and water stored off-channel within the delta. The predicted nutrient fluxes downstream from the delta were generally supported by observations; that is, river transport across the floodplain resulted in reduced concentrations of suspended particulates and dissolved inorganic nutrients, but enhanced concentrations of dissolved organic nutrients (Emmerton et al., 2008a).

Nutrient concentrations were also tracked between the estuarine transition zone and offshore marine stations on the adjacent Mackenzie Shelf (Emmerton et al., 2008b). Many constituents declined rapidly and non-conservatively across this transition zone, including particulates (suspended particulate matter, particulate organic carbon, particulate nitrogen and particulate phosphorus) and certain dissolved inorganic nutrients (nitrate, ammonium and silica). However, dissolved organic nitrogen (DON) and dissolved organic phosphorus (DOP) concentrations increased substantially from the river to the coastal shelf waters, reaching concentrations among the highest reported for the Arctic Ocean. Simultaneously, DOC concentrations declined from the river to the shelf and DOC : DON : DOP ratios changed significantly, from highly C-rich (terrestrial) values to near Redfield values. This apparent loss of C (but conservation of N and P) from the river to the shelf could have resulted from the microbial processing of organic carbon from a large river pulse (i.e., from the earlier freshet period). It was concluded that the nutrient stripping and processing of riverine DOC within this transition region was likely an important contributor to the low inorganic nitrogen and algal biomass state of the surface layer of the coastal ocean.

5.2.2 Microbial dynamics

The microbial processes addressed within CASES centered largely on the heterotrophic activities of bacteria (a term which generically refers to members of both the bacterial and archaeal domains of life) and of Eukaryotic protists (see also Chapter 4.8). An initial study by Garneau et al. (2006) showed that a large fraction of bacterial activity in turbid inshore waters was associated with particle-attached communities. This work was followed up by Vallières et al. (2008) during the ARDEX cruise. They measured size-fractionated bacterial production by ^{3}H-leucine uptake, and showed that the contribution of particle-attached bacteria (>3 μm) to total bacterial production was consistently high in the river (97% of total bacterial production in the river near Inuvik), lower at marine sites (minimum of 16% at an offshore station), and correlated with concentrations of particulate organic carbon (POC). In more recent analyses, Garneau et al. (2008a) confirmed a high presence of particle-attached bacterial communities in the inshore shelf, and found that the net abundance of free-living bacteria varied throughout the year in Franklin Bay.

The CASES over-winter deployment of a research platform (CCGS Amundsen) in Franklin Bay provided a unique opportunity to obtain a seasonal record of bacterial dynamics in the coastal Arctic Ocean (Garneau et al., 2008b). Integral bacterial production varied from 1 (winter) to 80 (summer) mg C m^{-2} d^{-1}, with a low overall annual rate consistent with a highly stratified, oligotrophic (low nutrient) environment. During winter-spring, the bacterial production was uncorrelated with chlorophyll a, implying a reliance on allochthonous carbon sources. However, these variables were significantly correlated during summer-autumn, suggesting a bottom-up control of carbon by autotrophs during that time of year. The annual bacterial CO$_2$ production

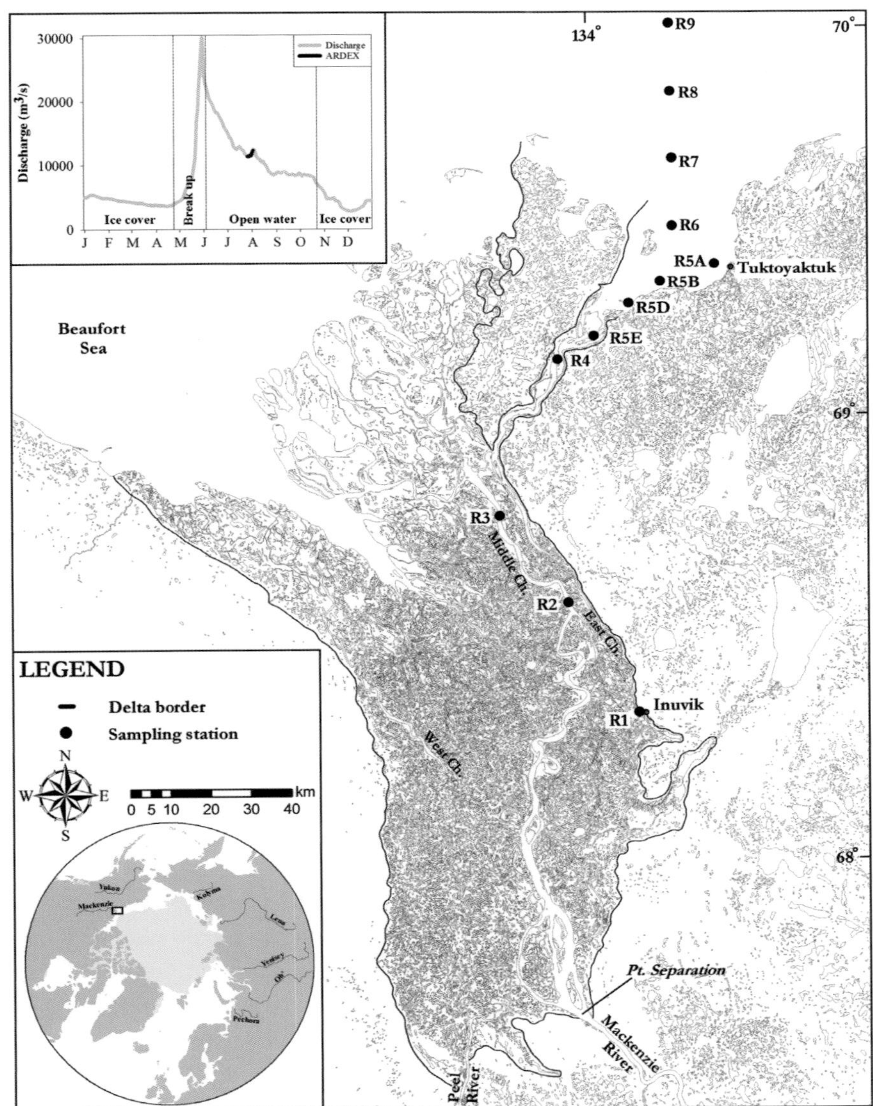

Figure 5.3

Sampling site locations and the annual hydrograph for the Mackenzie River during CASES and ARDEX. The data are from Arctic Red River WSC Station 10LC014. Modified from Emmerton et al. (2008a).

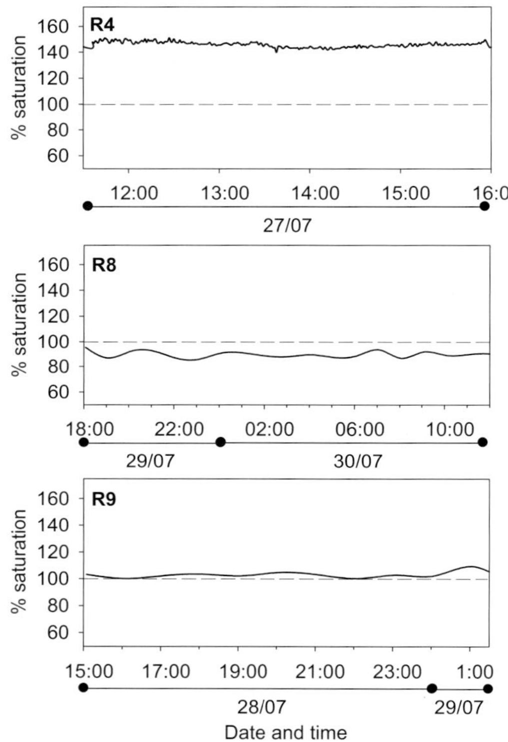

Figure 5.4

Percentage saturation of surface water CO_2 concentrations compared with mean values in the overlying atmosphere at three sites in the ARDEX cruise: R4 (Mackenzie River) and R8 and R9 (offshore marine sites). The results indicate a strong CO_2 efflux in the river, consistent with elevated partial pressures of CO_2 measured in the atmosphere over the Mackenzie River Delta during sampling. From Vallières et al. (2008).

rate for Franklin Bay was estimated as 1.7×10^{11} g C per year, a value equivalent to 13% of the terrestrial DOC flux from the Mackenzie River to the Beaufort Sea Shelf (and well above the photochemical DIC production rate). The results showed that the heterotrophic microbial community continued to be active throughout the year, including winter (see also Chapter 4.4), and constituted a major contributor to biological carbon fluxes in this coastal Arctic ecosystem.

The ARDEX cruise provided additional analyses of bacterial carbon dioxide fluxes relative to photosynthetic CO_2 fixation and measurements of partial pressure of CO_2 (Vallières et al., 2008). The results (Fig. 5.4) showed that, at the time of sampling, the Mackenzie River was heterotrophic (i.e., a net source of CO_2 to the atmosphere) while the Beaufort Sea tended towards equilibrium or autotrophy (net CO_2 sink). Results from enrichment experiments also showed that bacterial metabolism was organic C-limited in the Mackenzie River but not in the Beaufort Sea. Photochemical reactions increased the availability of organic carbon for bacterial metabolism in the Mackenzie River but decreased it in the Beaufort Sea.

5.2.3 Viruses

One of the many exciting research themes within environmental microbiology is the biodiversity and ecology of naturally occurring viruses. Viruses are the most abundant biological particles in aquatic ecosystems: each litre of lake or seawater will typically contain more than 1 billion viruses representing a myriad of different types (Figs 5.1 & 5.5). Characterised by a relatively simple yet efficient structure, viruses are one of the most elegantly successful forms of life in the biosphere. They play multiple roles in natural ecosystems, including the exchange of genetic material among microbes, the

control of host populations, and the biogeochemical cycling of carbon and other elements (Suttle, 2005).

During CASES, we found that virus concentrations in the water column of Franklin Bay varied from 1×10^5 to 2×10^7 ml^{-1} over the course of the year, with minimum concentrations occurring during winter (Payet and Suttle, 2008). Viral abundance, on average, was lower in Arctic Ocean waters than in Pacific Ocean waters or lakes; however, the relationship between viral abundance and prokaryotes was similar for both the Arctic and Pacific waters (Clasen et al., 2008). Two major groups of viruses were differentiated in the Arctic Ocean: a low SYBR-green fluorescence virus group which represented about 70% of the total viral abundance and whose presence correlated with that of high nucleic acid fluorescence (HNA) bacteria; and a high SYBR-green fluorescence viral group which especially prevalent during periods of high levels of the main phytoplankton pigment, chlorophyll a (chl*a*). Our findings suggested that the former group consisted of bacteriophages (viruses that infect bacteria) while the latter group were phycoviruses, which attack eukaryotic phytoplankton.

To examine the ecological dynamics of bacteriophages under strongly oligotrophic conditions, viral processes were measured in the bottom waters of Franklin Bay during winter (Wells and Deming, 2006a). In two experiments, the rates of virus-induced bacterial mortality (–0.006 to –0.015 h^{-1}) were similar to or exceeded the bacterial growth rate (0.010 h^{-1}), while flagellate grazing rates were low (-0.004 h^{-1}) or undetectable. The results showed that even during winter bacterial, communities remained active and subject to viral predation. In another set of winter experiments, samples of bacterial and viral assemblages from first year sea ice were incubated in brine for 8 days at a temperature (-12 °C) and salinity (160 psu) approximating the expected in situ values (Wells and Deming, 2006b). The results provided evidence of viral persistence and production as well as bacterial growth, even under extreme conditions.

5.2.4 Archaea

The archaeal domain of life has often been associated with extreme environments such as hot springs, acid waters and hypersaline ponds. However one archaeal group, the marine group I Crenarchaeota, appears to be common in the bacterioplankton of polar seas (Hollibaugh et al., 2007). Work prior to CASES by Wells et al. (2003) showed that Archaea were present throughout the water column of Amundsen Gulf. Their highest concentrations occurred in particle-rich 'nepheloid' layers, where they contributed up to 13% of the total prokaryotic cell count. This work was extended to Franklin Bay and the Mackenzie Shelf, where it was confirmed that archaeal abundance was highly correlated with particle concentrations, perhaps derived from the Mackenzie River (Wells et al., 2006). During the 2003-2004 over-winter period at Franklin Bay, the numbers of Archaea were followed by fluorescence in situ hybridization (FISH) using probes specific to the two main groups: Crenarchaeota and Euryarchaeota (Alonso-Sáez et al., 2008). Both groups were relatively abundant in winter, especially the Crenarchaeota (10% of the total prokaryotic count) and decreased to undetectable levels during the summer. FISH was combined with microautoradiography, a technique in which a radioactively labelled substrate is provided to cells and those active in its uptake show a dark halo under the microscope. Glucose, a mixture of amino acids and ATP were provided as substrates. The two groups of Archaea comprised a very small percentage of the cells that were labelled with any of

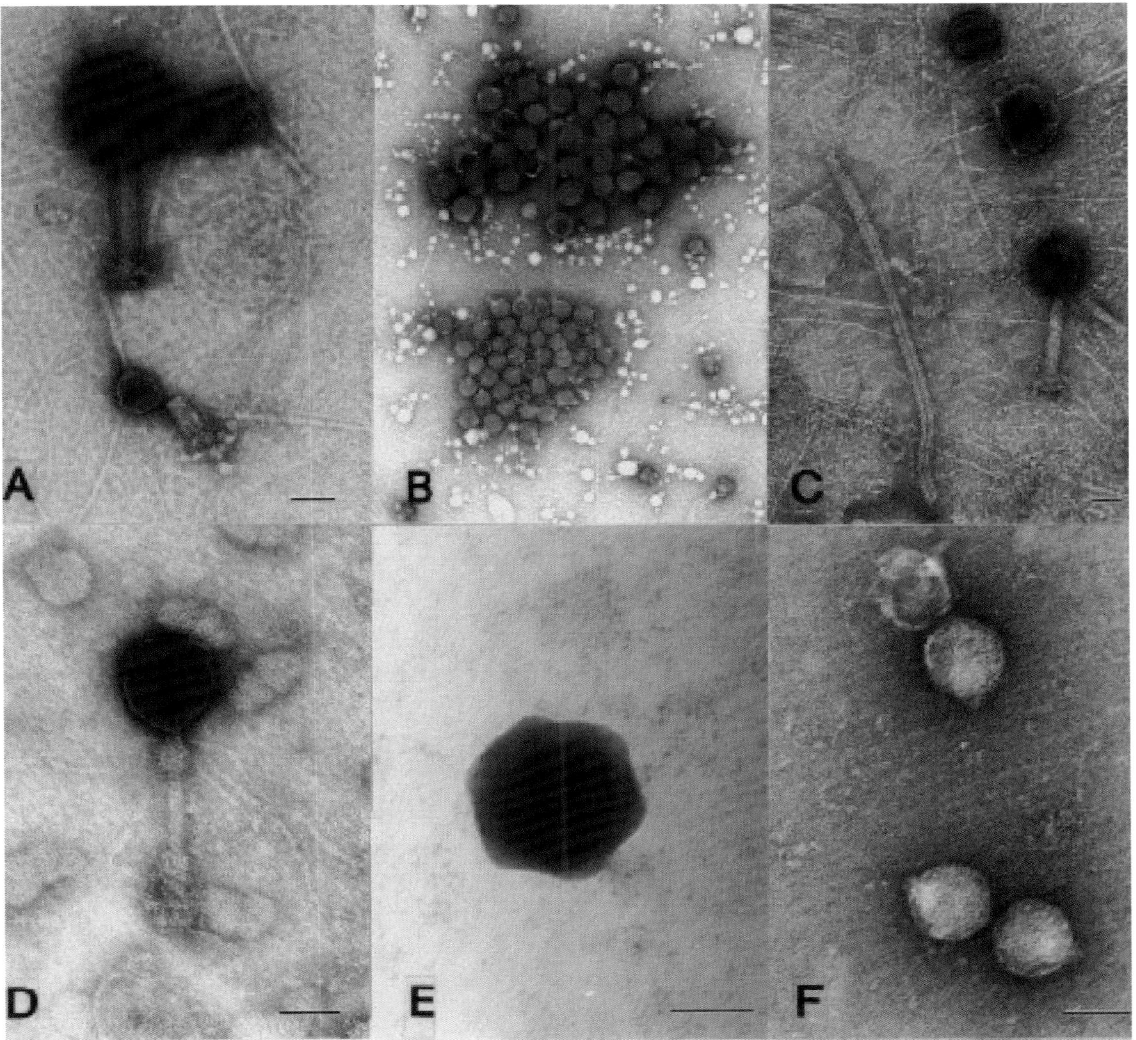

Figure 5.5

Electron micrographs of a variety of viruses found in the ocean. Reproduced from http://www. ocgy.ubc.ca/~suttle/genes.htm.

the substrates (a maximum of 8%). Thus, there was a chance that these microorganisms were not using organic matter for growth and were perhaps chemo-lithoautotrophs (i.e., CO_2 fixation using energy derived from reactions such as ammonium oxidation). This hypothesis will have to be tested in future cruises. If it is confirmed, then a significant portion of carbon input to the system during the winter may be due to this group of Archaea.

Molecular analysis of CASES archaeal samples (Galand et al., 2006), specifically 16S rRNA gene clone libraries, showed that coastal shelf assemblages were dominated by Group II Euryarchaeota, followed by Group I Crenarchaeota, and that they exhibited greater diversity (18 to 23 phylotypes) than had been previously reported for marine communities elsewhere. Extremely high diversity was observed in the river, with an estimated 286 archaeal phylotypes related to sequences previously obtained from flooded soils and sediments. The absence of overlap in taxa between the marine and river sites indicated that the coastal communities—at least those in shallow surface waters—were not derived from the Mackenzie River, contrary to earlier ideas. More detailed analysis showed that within particle-rich waters, the particle-associated Archaea had greater diversity than the free-living communities (Galand et al., 2008a). The results revealed a clear biogeographical separation of riverine, coastal and offshore marine waters, where each possessed a strikingly different archaeal community. Specifically, Euryarchaeota in the Arctic surface layer appeared to be associated with particle-rich waters, while Crenarchaeota appeared to be more characteristic of offshore Arctic Ocean waters.

5.2.5 Bacteria

Bacteria display pronounced genetic and functional diversity and are among the most abundant prokary-

otes in the world. A recent analysis suggested that the world's oceans may contain up to 10,000 taxa, most of which are extremely rare ("the rare biosphere"). However, the majority of core ecosysfunctions in the ocean might be performed by only a small, variable minority of bacterial taxa (Pedrós-Alió, 2007). Apart from the pathogenic role of some species, bacterial functions include recycling nutrients, decomposing waste, fixing and transforming nitrogen, producing and oxidizing methane, and providing carbon and nutrient sources to other microscopic organisms (especially protists).

CASES observations showed that there were strong inshore-offshore gradients in the abundance and community structure of Bacteria within the coastal Arctic Ocean, a finding that was consistent with the archaeal data. The dominant heterotrophic cell types found at river sites and offshore were β-Proteobacteria and a-Proteobacteria, respectively. Cells in the Bacteroidetes and γ-Proteobacteria groups were found in low concentrations at river sites, but accounted for more than 10% of the bacterial community in offshore samples (Garneau et al., 2006). The results implied broad functional diversity within heterotrophic microbial communities which break down terrestrial organic materials discharged into Arctic waters.

During the over-winter study in Franklin Bay (Alonso Sáez et al., 2008), Bacteria were detected using FISH equipped with a universal eubacterial probe. It was found that Bacteria represented 60% of the total hybridizable cells detected in the winter (a proxy for potentially active cells) and 99% during the summer. This suggested that the winter bacterial assemblage had a large fraction of inactive cells. Each bacterial group tested (α- and γ-Proteobacteria and Bacteroidetes) increased their contribution to the active cell pool at similar rates throughout spring and summer.

In comparison with temperate oceans, the percentage of active Bacteria found was high, even during the winter. Between 15 and 60% of the cells were active in the uptake of glucose, amino acids and ATP throughout the year. Cells of the Roseobacter group were the most active in taking up these small compounds, while SAR11 and Bacteroidetes showed very little activity. An analysis of 30 different organic compounds (Sala et al., 2008) indicated that polymers were the preferred carbon source for Bacteria during winter. Perhaps the Bacteroidetes, a group of Bacteria known to be specialized in the degradation of polymers, were responsible for this preference.

Microscopic analyses by epifluorescence (which does not distinguish Bacteria from Archaea; Fig. 5.1) showed that the total bacterial biomass within Franklin Bay fluctuated irregularly throughout the year. This fluctuation appeared consistent with the intermittent advection of particle-rich water into the bay (Garneau et al., 2008b). Peak biomass occurred during July, and was significantly correlated with phytoplankton stocks (as measured by chl*a*). Depth profiles for Bacteria were also measured by flow cytometry between December and May, during which two groups were distinguished: low-nucleic acid (LNA) and high nucleic acid (HNA) cells (Belzile et al., 2008). The HNA cells started to increase in concentration a few weeks after the spring onset of phytoplankton growth, while LNA grew more slowly (after a two-month lag). There was also a three-fold increase in bacterial abundance in spring—mostly due to HNA Bacteria—lending support to the idea that HNA cells were comparatively more active.

Bacterial concentrations within the sea ice of Franklin Bay were also determined throughout winter (Collins et al., 2008). Cell abundance was higher in the upper ice (likely because of rapid initial freezing which "captured" the cells) and lower in the underlying ice. Bacterial numbers also tended to decline in the coldest parts of the ice cover over the 3-month observation period. Particulate extracellular polymeric substances (pEPS) appeared to increase over time with decreasing ambient ice temperature. This was consistent with the notion that sea ice Bacteria produce EPS in situ as a survival (or cryoprotectant) mechanism.

CASES also provided an opportunity to make initial assessment of the microbiology of a stamukhi lake (Galand et al., 2008b). Stamukhi lakes occur throughout winter and spring along the Arctic coastline, forming behind barriers of thick, partially-grounded sea ice near large river inflows. Clone library analyses of the three domains of life (Archaea, Bacteria and Eukarya) in stamukhi Lake Mackenzie showed its microbial community was quite distinct from that of the Mackenzie River and offshore marine waters. Radioisotopic assays confirmed that it was a heterotrophically active ecosystem, which implied that this circumpolar ecosystem type plays a key functional role in processing riverine inputs to the Arctic Ocean.

In addition to bacterial protein and DNA synthesis measurements, experiments also were conducted to determine the ability of Bacteria to metabolize dimethylsulfoniopropionate (DMSP), an important intermediate in sulphur biogeochemistry (and precursor to the climate-influencing gas, dimethyl sulphide, DMS). By combining microautoradiography with FISH, it was shown that DMSP was taken up at slow rates by certain bacterial groups (Vila-Costa et al., 2008). With the exception of the Roseobacter group, DMSP uptake was generally less efficient than that of leucine. This differed substantially from the general situation found in temperate waters, where the uptake efficiency for DMSP and leucine are similar.

5.2.6 *Picocyanobacteria*

Cyanobacteria are an ancient group of photosynthetic microbes and include several genera with extremely small cells (i.e., picocyanobacteria) that are of widespread importance in the ocean, notably Synechococcus and Prochlorococcus. Cyanobacteria (including picocyanobacteria) are often the dominant cell types in polar lakes, ponds and rivers, yet are conspicuously absent or in low abundance in polar seas (Vincent, 2000). Initial CASES sampling showed that picocyanobacteria were present in the Mackenzie River but their concentrations declined precipitously in adjacent marine waters (Garneau et al., 2006). This observation was confirmed by the ARDEX transect: cell concentrations dropped from 51,000 cells ml^{-1} in the river to 30 cells ml^{-1} over the shelf (Vallières et al., 2008). Although picocyanobacteria populations were detected in low abundance at all CASES stations, molecular analysis showed that they did not contain the Prochlorococcus or Synechococcus clades that are typically found in temperate oceans (Waleron et al., 2007). The 16S rRNA gene sequences showed a close affinity to fresh water and brackish water phylotypes, which suggested that these declining coastal populations were derived from the surrounding rivers. In fact, 2 of the 6 phylotypes found in the Mackenzie River and estuary were also found in the ocean (up to several hundred km offshore near the edge of the pack ice). Although amplification of nifH gene sequences was attempted on the CASES 2002 samples using the appropriate primers (Zehr and McReynolds 1989), no amplicon was obtained. This implied a very low abundance or even absence of cyanobacterial nitrogen fixers, although bacterial N_2- fixers could be detected (Wilmotte et al., unpublished data).

5.2.7 *Protists*

Protists are single-celled members of the Eukarya domain, and include species that use sunlight (phototrophs), organic carbon (heterotrophs) or a mixture of both (mixotrophs) as their energy source. The CASES microbial program focused on picoeukaryotes (small-cell species, generally passing through a 3-μm filter) and found that Arctic seas contain a rich diversity of these organisms, including certain taxa that are highly adapted to low temperatures and that may be endemic to the region (Lovejoy et al., 2006 & 2007).

Pigment analyses made on Franklin Bay samples showed a continuous prevalence of picoeukaryotes in the green algal class Prasinophyceae throughout the year (Lovejoy et al., 2007). The most abundant photosynthetic cell types found were Micromonas-like picoprasinophytes, which persisted throughout winter darkness and then began to grow under the extreme low sub-ice irradiance conditions of January-February (Fig. 5.6). Genetic analyses made on samples from the CASES region and other sites within the Arctic Ocean showed that the psychrophilic Micromonas ecotype had a pan-Arctic distribution. The prevalence of such an obligate low-temperature, shade-adapted species in the phytoplankton indicated that the lower food web of the Arctic Ocean is likely to be vulnerable to ongoing climate change.

Detailed microscopic analysis of protists and other microbes in Franklin Bay confirmed the persistence of a photosynthetic inoculum throughout winter darkness, and the strong seasonal response of Arctic microbial food webs to sub-ice irradiance in early spring (Terrado et al., 2008). The growth of the picoprasinophytes was positively correlated with surface irradiance, and despite the continuing presence of sea ice, phytoplankton biomass rose by more than an order of magnitude (to 244 mg C m^{-2}) in the upper mixed layer by May. A shipboard experiment in April showed that this photo-trophic increase was not responsive to pulsed nutrient enrichment. That is, all treatments showed a strong growth response to increasing irradiance conditions. Additional experiments (Estrada et al., in preparation) showed that the early spring community of small flagellates was inhibited by irradiances higher than 50 µmol m^{-2} s^{-1} and could grow under irradiances as low as 10 µmol m^{-2} s^{-1}. The assemblage at the end of incubation was dominated by Pyramimonas (Fig. 5.7). The same experiments repeated in summer showed a much higher light optimum and growth of diatoms. While the spring flagellate assemblage did not respond to inorganic nutrient additions, the summer diatom assemblage grew better in the presence of nutrients than in their absence. It seemed, therefore, that phy-toplankton communities in the winter and spring had completely different strategies than those of the better known summer assemblages. The experiments also indicated that organic nitrogen may play an important role in the growth of algae in these polar waters.

An exciting discovery from molecular analyses of northern marine waters concerns a new group of pho-tosynthetic picoeukaryotes (Not et al., 2007). These microbes have very little evolutionary affinity to other Eukarya, and have been placed in a newly established group, named the 'picobiliphytes'. These micro-or-ganisms appear to be abundant in cold Arctic waters, and are commonly encountered in the Beaufort Sea and elsewhere in the Canadian Arctic (Lovejoy et al., 2006). A diverse range of heterotrophic flagellates was also observed throughout the CASES region, though marine Stramenopiles group 4 (a group which is gener-ally widespread and abundant throughout most of the world's oceans) was conspicuously absent (Massana et al., 2006).

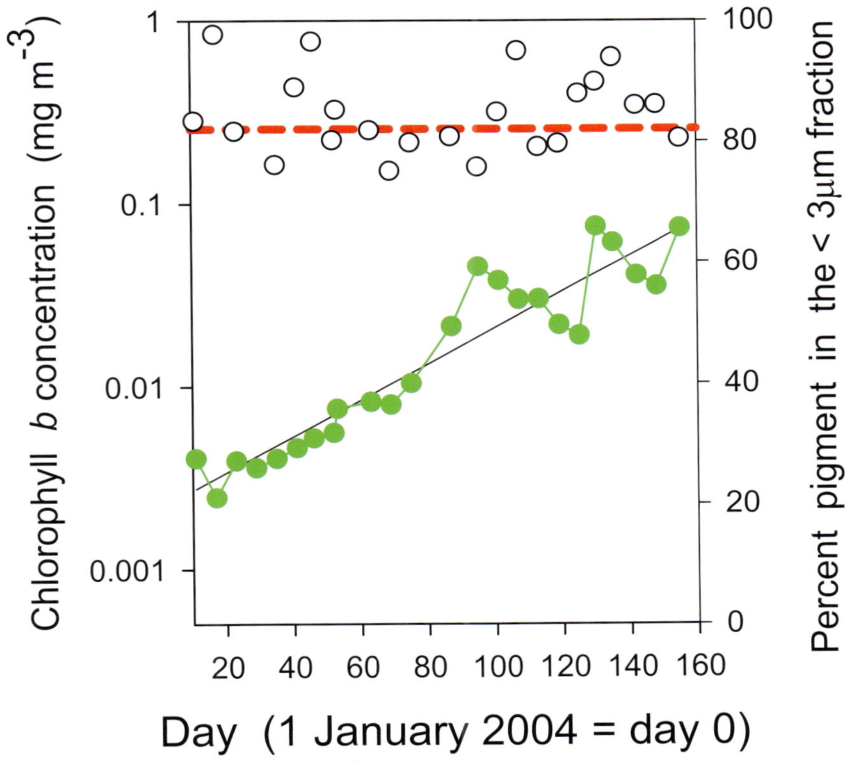

Figure 5.6

The remarkable growth of minute phytoplankton (picoprasinophytes) in the waters beneath the ice in Franklin Bay from first light onwards. The green circles show the concentration of chlorophyll b (a characteristic pigment of prasiniophytes), which increased exponentially through time. These concentrations correlated strongly with the population density of Micromonas-like cells. The upper circles show the percentage of chlorophyll b in the <3 µm fraction. On average (dashed red line), more than 80% of the pigment was associated with very small cells (picophytoplankton). Redrawn from Lovejoy et al. (2007).

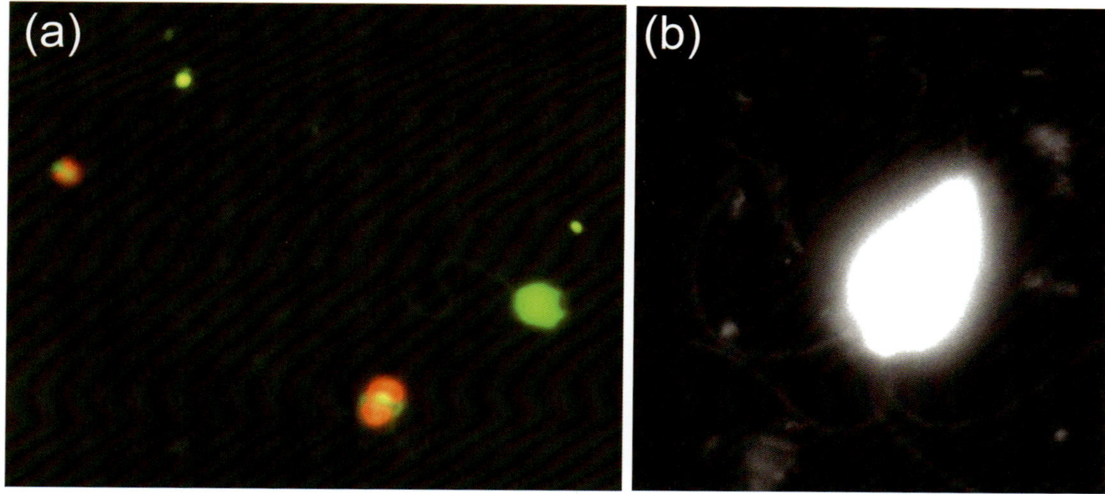

Figure 5.7

Epifluorescence micrographs of winter communities of microbes in Franklin Bay. Left: Natural community. The small green dots are fluorescently labeled bacteria added to estimate bacterivory rates. The green cell with a flagellum is a heterotrophic flagellate. The two other cells are autotrophic, as indicated by the red autofluorescence due to chlorophyll a. The smaller cell to the left is a Micromonas; the larger one is likely a Phaeocystis. Right: The assemblage after incubating a winter community for 20 days under low light, dominated by the flagellate Pyramimonas. Photomicrographs by D. Vaqué.

Heterotrophic protists were also studied by epifluorescence microscopy during the Franklin Bay over-winter study (Vaqué et al., submitted). An assemblage of relatively large flagellates developed in February, apparently in response to an increase in autotrophic flagellates (mostly Micromonas-like). Smaller flagellates increased in abundance later in the spring, when total bacterial numbers were increasing. Bacterivory rates were estimated by adding fluorescently labelled bacteria to natural samples (Fig. 5.7). Bacterial production rates were lower than or similar to winter bacterivory rates (from 0.011 ± 0.01 to 0.33 ± 0.47 µg C L^{-1} d^{-1}) and higher than spring bacterivory rates (from 0.90 ± 0.15 to 1.17 ± 0.16 µg C L^{-1} d^{-1}). Bacterial production

increased immediately in conjunction with chlorophyll a; however, this increase in prey did not translate into an increase in bacterial grazing rates. Possible explanations for this might include that the Bacteria were controlled by substrate supply or that the protist predators were feeding on different prey, such as other protists.

5.3 Implications of this work

CASES and its satellite program ARDEX have provided an unprecedented opportunity to explore the microbiology of the Arctic Ocean and to evaluate a broad suite of environmental processes and variables. These observations revealed a rich diversity of Bacteria, Archaea, protists and viruses. Abundant populations of each of these groups were found throughout the year. Contrary to expectation, winter in the Arctic was not found to be a period of biological quiescence, but instead microbial heterotrophic processes continued throughout the period of winter darkness at rates well above background. Even photosynthetic processes began much earlier in the year than expected, thanks to minute eukaryotic cells that were highly adapted to low water temperatures and to the capture of low irradiance levels beneath the ice during early spring.

Obviously, winter ecology within Arctic waters is very different from summer ecology. Such difference, and its implications, can only be studied through projects such as CASES, in which a suitable research platform is available on site throughout all seasons. As a result of CASES, we have learned that small cells dominate carbon fluxes throughout most of the year, and that much of the carbon-processing is associated with microbial consortia attached to particles. We also identified several aspects of this arctic shelf ecosystem that may be vulnerable to climate change, including the presence of obligate cold-adapted species within the plankton community, the sensitivity of deep chlorophyll abundance to changes in underwater irradiance, and the dependence of the stamukhi lake ecosystem on ice integrity, especially during the period of peak discharge from the Mackenzie River.

Several features of the studied region may lead to an increased net production of CO_2 in a warmer climate. These include an increased change in the photochemical production of CO_2 as a function of reduced sea-ice cover, an increased change in the transport and microbial degradation of terrigenous carbon, a loss of the stamukhi lake as a decantation system for particulate organic carbon during peak discharge, and more severe light limitation on photosynthetic CO_2-fixation due to increased runoff of CDOM and non-algal particulates.

5.4 Recommendations

The outstanding success of the CASES/ARDEX microbial program can be attributed to the diverse expertise of researchers who worked together and combined traditional oceanographic measurements with advanced analytical and molecular biological techniques. The resulting ensemble of data provides one of the first seasonal records (of any ocean) of microbial community dynamics and associated carbon fluxes.

CASES/ARDEX unveiled the striking genetic diversity of microscopic life in the coastal Arctic Ocean, and many new questions have now emerged. While analyses to date have focused on the surface layer of the ocean, the depth dependence of community structure and processes (including those associated with marine sediments) still needs to be defined. ARDEX provided a valuable snapshot of the Mackenzie River microbial ecosystem during late summer, and similar measurements are now required for the remainder of the year, including the logistically difficult period of peak discharge. The functional diversity

and biogeochemical characteristics of the microbial consortia on particles will require close attention in the future. Finally, further information needs to be collected concerning the microbial control of gas fluxes, specifically oxygen, carbon dioxide, nitrous oxide and dimethyl sulfide.

CASES has been only a first step in defining "the rare biosphere" of the coastal Arctic Ocean (see Pedrós-Alió, 2007) and more detailed analyses are required to understand the temporal and spatial complexity of Arctic microbiota. Emerging genomic tools offer the potential to capture the entire 'metagenome' of environmental samples (e.g. Tyson et al., 2004; Sogin et al., 2006) and provide new insights into the microbial diversity and biogeography of arctic lakes, rivers and seas. Some of these tools, such as qPCR and microarrays, are also providing new ways to address questions of microbial gene expression in the environment. All this has the potential to yield exciting new perspectives on the functional diversity of Arctic microbial communities and their responses to climate change.

Acknowledgements

We would like to thank Louis Fortier for his leadership of CASES; Martin Fortier, Marc Ringuette and Josée Michaud for their valuable coordination roles; the officers and crew of CCGS Pierre Radisson, CCGS Amundsen and CCGS Nahidik; Leah Braithwaite (Fisheries and Oceans Canada) for her advice and support for CASES data archiving; our many students, technicians and postdoctoral researchers; and the Natural Sciences and Engineering Research Council (NSERC) and the Canada Research Chair program for funding support. International contributions were also made through the Generalitat de Catalunya, the Spanish Ministerio de Educación y Ciencia, the National Science Foundation (USA), the European Science Foundation program CYANOFIX, and the National Fund for Scientific Research (FNRS) Belgium.

References

Alonso-Sáez, L., O. Sánchez, J.M. Gasol, V. Balagué, and C. Pedrós-Alió. 2008. Winter-to-summer changes in the composition and single-cell activity of near-surface Arctic prokaryotes. Environ. Microbiol. 10: 2444-2454.

Bélanger, S., H. Xie, N.Krotkov, P. Larouche, W.F. Vincent, and M. Babin, 2006. Photomineralization of terrigenous dissolved organic matter in Arctic coastal waters from 1979 to 2003: Interannual variability and implications of climate change. Global Biogeochemical Cycles 20, GB4005, doi: 10.1029/2006GB002708.

Belzile, C., S. Brugel, C.Nozais, Y. Gratton, and S. Demers. 2008. Seasonal variations of the abundance and nucleic acid content of heterotrophic bacteria in Beaufort Shelf waters during winter and spring: Dependence on autochthonous carbon and top-down regulation. J. Mar. Syst. doi:10.1016/j.jmarsys.2007.12.010.

Clasen, J.L., S.M. Brigden, J.P. Payet, and C.A. Suttle, 2008. Viral abundance across marine and freshwater systems is driven by different biological factors. Fresh. Biol. In press.

Collins, E., S. Carpenter, and J. Deming. 2008. Spatial heterogeneity and temporal dynamics of particles, bacteria, and pEPS in Arctic winter sea ice. J. Mar. Syst. doi:10.1016/j.jmarsys.2007.09.00.

Emmerton, C.A., L.F.W. Lesack, and P. Marsh, 2007. Lake abundance, potential water storage, and habitat distribution in the Mackenzie River Delta, western Canadian Arctic. Water Resour. Res. 43: W5419, doi:10.1029/2006WR005139.

Emmerton C.A., L. Lesack, and W.F. Vincent. 2008. Mackenzie River nutrient fluxes to the Arctic Ocean and effects of the Mackenzie River delta. Global Biogeochem. Cycles. 22, GB1024, doi:10.1029/2006GB002856..

Emmerton C.A., L. Lesack, and W.F. Vincent. Inorganic and organic nutrient gradients in a coastal arctic system during sea-ice recession: The Mackenzie River, estuary and shelf. J. Mar. Syst. doi:10.1016/j.jmarsys.2007.10.001.

Estrada, M., M. Bayer, J. Felipe, C. Marrasé, M.M. Sala, and M. Vidal. Growth of phytoplankton collected from an ice-covered water column in the Coastal Arctic Ocean. In preparation.

Fortier, L., and J.K. Cochran 2008. Introduction to special section on Annual Cycles on the Arctic Ocean Shelf, J. Geophys. Res. 113, C03S00, doi:10.1029/2007JC004457.

Galand, P.E., C. Lovejoy, and W.F. Vincent, 2006. Remarkably diverse and contrasting archaeal communities in a large arctic river and the coastal Arctic Ocean. Aquat. Microb. Ecol. 44: 115-126.

Galand., P.E., C. Lovejoy, J. Pouliot, and W.F. Vincent. 2008a. Archaeal diversity in the particle-rich environment of an arctic shelf ecosystem. J. Mar. Syst. doi:10.1016/j.jmarsys.2007.12.001.

Galand, PE, C. Lovejoy, J. Pouliot, M-E. Garneau, and W.F. Vincent. 2008b. Microbial commudiversity and heterotrophic production in a coastal arctic ecosystem: A stamukhi lake and its source waters. Limnol. Oceanogr. 53: 813-823.

Garneau, M-È., W.F. Vincent, L.Alonso-Sáez, Y.Gratton, and C. Lovejoy, 2006. Prokaryotic community structure and heterotrophic production in a river-influenced coastal arctic ecosystem. Aquat. Microb. Ecol. 42: 27-40.

Garneau, M-È., W.F. Vincent, R. Terrado, and C. Lovejoy 2008a. Importance of particle-associated bacterial heterotrophy in a coastal Arctic ecosystem. J. Mar. Syst. doi:10.1016/j.jmarsys.2008.09.002.

Garneau, M-È., S. Roy, C. Lovejoy, Y. Gratton, and W.F. Vincent. 2008b. Seasonal dynamics of bacterial biomass and production in a coastal arctic ecosystem: Franklin Bay, western Canadian Arctic. J. Geophy. Res. Oceans 113, C07S91, doi:10.1029/2007JC004281.

Hollibaugh, J., C. Lovejoy, and A. Murray, 2007. Microbiology in polar oceans. Oceanogr. 20 (special issue): 138-143.

Lovejoy, C., R. Massana, and C. Pedrós-Alió, 2006. Diversity and distribution of marine microbial eukaryotes in the changing Arctic. Appl. Environ. Microbiol. 72: 3085-3095.

Lovejoy, C., W.F. Vincent, S. Bonilla, S. Roy, M-J. Martineau, R. Terrado, M. Potvin, R.Massana, and C. Pedrós-Alió, 2007. Distribution, phylogeny and growth of cold-adapted picoprasinophytes in arctic seas. J. Phycol. 43: 78-89. doi: 10.1111/j.1529-8817.2006.310.x.

Massana, R., R. Terrado, I. Forn, C. Lovejoy, and C. Pedrós-Alió, 2006. Distribution and abundance of uncultured heterotrophic flagellates in the world oceans. Environ. Microbiol. doi: 10.1111/j.1462-2920.2006.01042.

Not, F., K. Valentin, K. Romari, C. Lovejoy, R. Massana, K. Töbe, D. Vaulot, and L. Medlin, 2007. Picobiliphytes: A Marine picoplanktonic algal group with unknown affinities to other eukaryotes. Science 315: 253-255.

Osburn, C. L. and G. St-Jean, 2007. The use of wet chemical oxidation with high-amplification isotope ratio mass spectrometry (WCO-IRMS) to measure stable isotope values of dissolved organic carbon in seawater. Limnol. Oceanogr. Meth. 5: 296-308.

Osburn, C. L. and W. F. Vincent. Photochemical reactivity of riverine DOM transported to the Western Canadian Arctic. in preparation.

O'Sullivan, D.W., P.J. Neale, R.B. Coffin, T.J. Boyd, and C.L. Osburn, 2005. Photochemical production of hydrogen peroxide and methylhydroperoxide in coastal waters. Mar. Chem. 97: 14-33.

Payet, J., and C. Suttle. 2008. Physical and biological drivers of virus dynamics in the southern Beaufort Sea and Amundsen Gulf. J. Mar. Syst. doi:10.1016/j.jmarsys.2007.11.002.

Pedrós-Alió, C., 2007. Dipping into the rare biosphere. Science 315: 192-193.

Proctor, L.M. and D.M.Karl (eds), 2007. A sea of microbes. Oceanogr. (Special Issue) 20 (2): 14-171.

Retamal, L., W.F. Vincent, C. Martineau, and C.L. Osburn, 2007. Comparison of the optical properties of dissolved organic matter in two river-influenced coastal regions of the Canadian Arctic. Est. Coast. Shelf Sci. 72: 261-272, doi:10.1016/j.ecss.2006.10.022.

Retamal, L., S. Bonilla, and W.F. Vincent, 2008. Optical gradients and phytoplankton production in the Mackenzie River and coastal Beaufort Sea. Polar Biol., 31:363–379, doi : 10.1007/s00300-007-0365-0.

Sala, M.M., R. Terrados, C. Lovejoy, F. Unrein, and C. Pedrós-Alió, 2008. Large winter bacterio-plankton metabolic diversity in the ice-covered Coastal Arctic Ocean. Environ. Microbiol. doi:10.1111/j.1462-2920.2007.01513.x.

Sogin, M.L., H.G. Morrison, J. A. Huber, D. M. Welch, S. M. Huse, P. R. Neal, J. M. Arrieta, and J. Herndl, 2006. Microbial diversity in the deep sea and the underexplored "rare biosphere". Proceeding of the National Academy of Sciences (USA) 103: 12115–12120. doi: 10.1073_pnas.0605127103.

Suttle, C.A., 2005. Viruses in the sea. Nature 437: 356-361.

Terrado, R., C. Lovejoy, R. Massana, and W.F. Vincent. 2008. Microbial responses light and nutrients beneath arctic sea ice during the winter spring transition. J. Mar. Syst. doi:10.1016/j.jmarsys.2007.11.001.

Tyson, G .W., J. Chapman, P. Hugenholtz, E. E. Allen, R. J. Ram, P. M. Richardson, V. V. Solovyev, E. M. Rubin, D. S. Rokhsar, and J. F. Banfield 2004. Community structure and metabolism through reconstruction of microbial genomes from the environment. Nature 428: 37-43.

Vallières, C., L. Retamal, C. Osburn, and W.F. Vincent. 2008. Bacterial production and microbial food web structure in a large arctic river and the coastal Arctic Ocean. J. Mar. Syst. doi:10.1016/j.jmarsys.2007.12.002.

Vaqué, D., O. Guadayol, F. Peters, J. Felipe, L. Angel-Ripoll, R. Terrado, C. Lovejoy, and C. Pedrós-Alió. Seasonal changes in planktonic bacterivory rates under the ice in Franklin Bay. Coastal Arctic Ocean. Submitted.

Vila-Costa, M., R. Simó, L. Alonso-Sáez, and C. Pedrós-Alió. 2008. Number and phylogenetic affiliation of bacteria assimilating dimethylsulfoniopropionate and leucine in the ice-covered coastal Arctic Ocean. J. Mar. Syst. doi:10.1016/j.jmarsys.2007.10.006.

Vincent, W.F. 2000. Cyanobacterial dominance in the polar regions. In B.Whitton & M. Potts. Ecology of the Cyanobacteria: their Diversity in Space and Time. Kluwers Academic Press, The Netherlands, p.321-340.

Vincent, W.F. and C. Pedrós-Alió. 2008. Sea ice and life in a river-influenced Arctic shelf ecosystem. J. Mar. Syst. doi:10.1016/j.jmarsys.2008.03.003.

Waleron, M., K. Waleron, W.F. Vincent, and A. Wilmotte, 2007. Allochthonous inputs of riverine picocyanobacteria to coastal waters in the Arctic Ocean FEMS. Microb. Ecol. 59: 356-365; doi: 10.1111/j.1574-6941.2006.00236.x.

Wells, L.E. and J.W. Deming, 2003. Abundance of Bacteria, the Cytophaga-Flavobacterium cluster and Archaea in cold oligotrophic waters and nepheloid layers of the Northwest Passage, Canadian Archipelago. Aquat. Microb. Ecol. 31: 19–31.

Wells, L.E. and J.W. Deming, 2006a. Significance of bacterivory and viral lysis in bottom waters of Franklin Bay, Canadian Arctic, during winter. Aquat. Microb. Ecol. 43: 209-221.

Wells, L.E. and J.W. Deming, 2006b. Modelled and measured dynamics of viruses in Arctic winter sea-ice brines. Environ. Microbiol. 8: 1115–1121.

Wells, L.E., M. Cordray, S. Bowerman, L.A. Miller, W.F. Vincent and J.W. Deming, 2006c. Archaea in particle-rich waters of the Beaufort Shelf and Franklin Bay, Canadian Arctic: Clues to an allochthonous origin? Limnol. Oceanogr. 51: 47-59.

Zehr, J.P. and L.A. McReynolds, 1989. Use of degenerate oligonucleotides for amplification of the nifH gene from the marine cyanobacterium Trichodesmium thiebautii. App. Environ. Microbiol. 55: 2522-2526.

CHAPTER 6

The Pelagic Food Web: Structure, Function and Contaminants

Don Deibel[1]*, Louis Fortier[2], Gary Stern[3], Lisa Loseto[4], Gérald Darnis[2], Delphine Benoit[2], Tara Connelly[1], Pascale Lafrance[2], Lena Seuthe[5], Yvan Simard[6], Piotr Trela[1]

6.1 Introduction and Rationale

Within an oceanic context the term *food web* refers to the vast array of phytoplankton and marine animals which depend upon each other for survival. The predator-prey connections formed among species determine a food web's *structure*, while the rates at which species eat one another and affect energy and carbon flows determine its *function*. Food web structure and function are crucially important to the long-term sustainability of ocean productivity (including the yield of top predators, such as fish, seals, whales and bears) and are related to many physical and chemical environmental conditions, such as weather, climate, rainfall, temperature, salinity and sea ice. The primary objective of the CASES Pelagic Food Web Group was to determine how present-day changes in climate and sea ice cover are affecting the food webs of the Beaufort Sea Shelf, including the Gulf of Amundsen and Franklin Bay. Achieving such an understanding is key to predicting

how further climatic changes within the area will affect local dominant food webs.

The specific objectives of our group were to determine 1) the abundance, depth distribution and vertical migration of zooplankton, and juvenile and adult fish, 2) the predation rates of zooplankton herbivores and carnivores on their prey, 3) the respiration and egg production rates of copepods and appendicularians (two dominant zooplankton groups on the Beaufort Sea Shelf), 4) the predation, growth and survival rates of the early life stages of Arctic cod, 5) the detailed chemical composition of zooplankton and fish (i.e., the fats, fatty acids and stable isotopes which reflect information on their prey), and 6) the bioaccumulation and biomagnification of contaminants such as mercury and polychlorinated biphenyls (PCB's) within food webs. In order to achieve these objectives and fully comprehend the structure and function of food webs within the Beaufort Sea Shelf—an area which

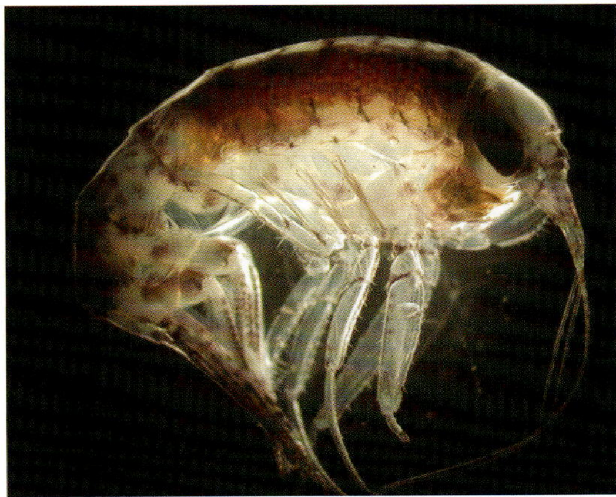

Themisto libellula. Photo: *Gérald Darnis.*

LEFT: Scientists deploying a mooring. Photo: *Ramon Terrado.*

[1] *Ocean Sciences Centre, Memorial University, St. John's, NL, A1C 5S7, Canada*

[2] *Québec-Océan, Departement de Biologie, Université Laval, Québec, QC, G1V 0A6, Canada*

[3] *Freshwater Institute/Fisheries and Oceans Canada, 501 University Cres., Winnipeg MB, R3T 2N6, Canada and Department of Environment & Geography, University of Manitoba, 500 University Cres., Winnipeg MB, R3T 2N2, Canada*

[4] *Department of Zoology, University of Manitoba, 500 University Cres., Winnipeg MB, R3T 2N2, Canada*

[5] *Norwegian College of Fishery Science, University of Tromsø, 9037 Tromsø, Norway*

[6] *Institut des sciences de la Mer (ISMER), Université du Québec à Rimouski, Allée des Ursulines, Rimouski, QC, G5L 3A1, Canada*

* Corresponding author: ddeibel@mun.ca

Figure 6.1a

*Collecting fish with a large gillnet during the over-wintering study in Franklin Bay.
Photo: Gérald Darnis.*

experiences pronounced seasonal changes in sea ice cover and river outflow—it was important to study biological dynamics over an entire year. Such an extended study period would further enable us to gather information on how zooplankton adapt to the long winter night, when there is little light and little food.

The Beaufort Sea Shelf is a large study area, and its physical and chemical environments vary a great deal from place-to-place and over time. Fresh water input is dominated by the Mackenzie River to the west, while outflow from smaller rivers has an impact on salinity and circulation in Franklin Bay to the east. Early opening of sea ice within the Amundsen Gulf Polynya is likely to be important for local food webs in the spring, while the dynamic boundary between land-fast ice and the drifting polar pack may also be an area of potentially significant biological productivity. The challenge presented to the CASES Research Network was to design a field program which included all space and time-scales relevant to the region's unique environmental makeup; from the Mackenzie Delta to Amundsen Gulf, and from far upriver to the shelf break, several hundred kilometres offshore. The resulting 12-month program has since provided a wealth of data on the potential links between the physical and chemical environments of the Beaufort Sea Shelf and the structure and function of its food webs.

We used a wide variety of sampling devices and strategies to study zooplankton and fish on the Beaufort Sea Shelf, including gillnets (Fig. 6.1a), vertically-towed nets (HYDROBIOS, Fig. 6.1b), a BIONESS multi-net system, a rectangular mid-water trawl (RMT), a multi-frequency echosounder, two video plankton recorders (VPR), an epibenthic sled and net system, bottom-tripping Niskin bottles to collect water samples and under-ice zooplankton nets pulled by snowmobiles.

Figure 6.1b

The HYDROBIOS multiple zooplankton net sampling system positioned above the moonpool inside the ship. Photo: Gérald Darnis.

Our sampling strategy included wide spatial coverage during summer and fall (when open water permitted) and continuous temporal coverage during winter by freezing our ship into the ice in Franklin Bay. Here, samples were taken through the moonpool of the ship and through ice holes from two fixed stations, one beside the ship and the second 20 km away. In the next section, we present highlights from our results and discuss their potential to further our understanding of the food webs of the Beaufort Sea Shelf. We also discuss how humans might better manage their interactions with shelf ecosystems in the face of global climate change.

6.2 Overview of Results

Prior to CASES, there existed only one scientific paper pertaining to the food webs of the Beaufort Sea Shelf (Parsons et al., 1989). It indicated that there were two separate food webs on the shelf: one based upon nutrients flowing onto the shelf from the Mackenzie River (the 'terrestrial-based' food web) and one based upon nutrients supplied by the shelf itself (the 'marine-based' food web). The terrestrial-based food web included a group of zooplankton known as appendicularian tunicates and its top predator was the rainbow smelt. The marine-based food web included marine phytoplankton and a different group of zooplankton known as copepods, and its top predator was the Arctic cod (*Boreogadus saida*). Unfortunately, the Parsons et al. (1989) study reflected results from a very short time span in August during a single year on a single transect of stations off Tuktoyaktuk. In contrast, scientists in the CASES program were able to study food webs over the entire shelf for a 12 month period. Thus, we were able to see if the Parsons et al. (1989) food web model could be extended to a larger area and an annual time scale. This kind of research is important to fishers and managers if they are to effectively sustain the shelf ecosystem, including both its marine and terrestrial components.

6.2.1 Surprising mid-winter food web activity

We discovered an active food web under the ice during mid-winter and early spring in Franklin Bay (Seuthe et al., 2007; Garneau et al., 2008). Some of the copepod species within this web became active much earlier than was expected (Seuthe et al., 2007), and there were remarkably high abundances of adult Arctic cod (Benoit et al., 2008); enough to account for the mystery

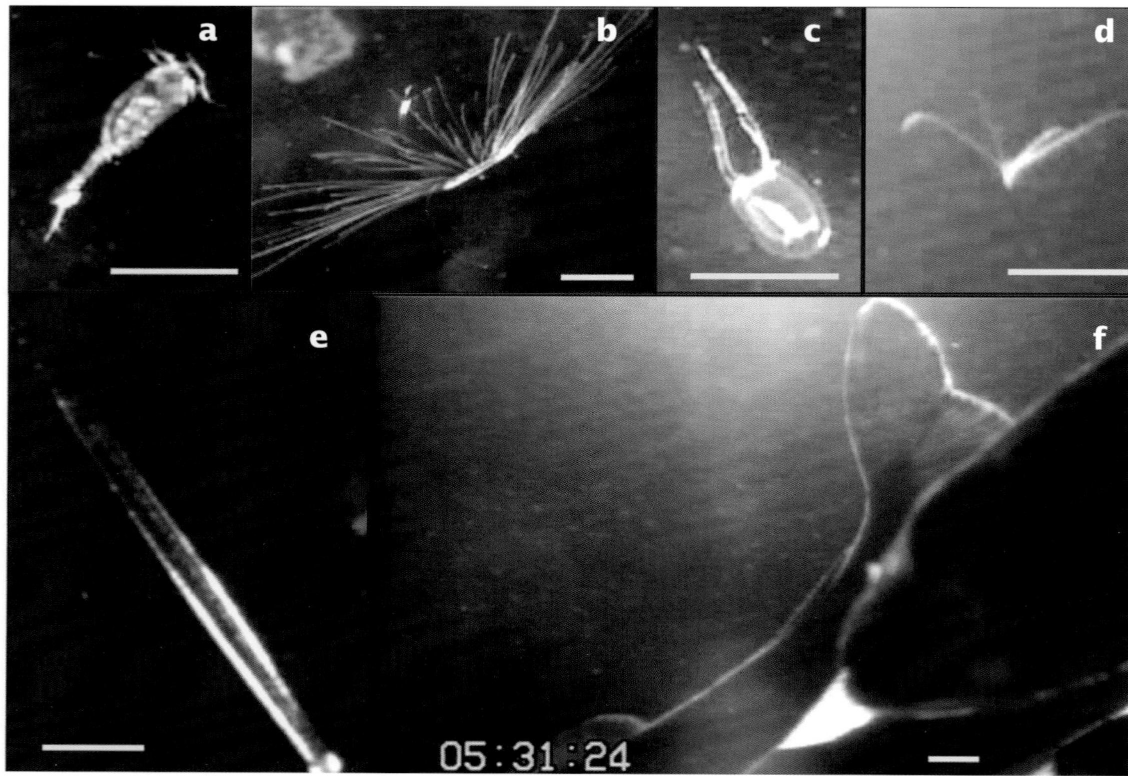

Figure 6.2

Dominant taxa present in the underwater video profiles from the over-wintering station in Franklin Bay, late April 2004: a) copepods; b) the hydromedusa Aglantha sp.; c) other hydromedusae; d) unidentified organism (probably a meroplanktonic larval stage of an echinoderm); e) chaetognaths; f) fish (Arctic cod). The scale bar in each panel represents 5 mm.

of the 'missing cod' on the Beaufort Sea Shelf (Welch et al, 1992). Our video plankton recorder revealed pronounced layers of particles and zooplankton throughout the water column during winter (Figs. 6.2 & 6.3). Some of the organisms, like *Aglantha* sp. and other hydromedusae, stayed in the upper half of the water column, while copepods and Arctic cod stayed in the warmer lower half of the water column. Net tows revealed the copepod community to be primarily composed of three large species: *Calanus hyperboreus, C. glacialis* and *Metridia longa* (Darnis et al., 2008). Unexpectedly, the faecal pellet production rate (an indirect measure of feeding rate) increased in all three copepod species under the land-fast ice during April and May, even though the concentration of phytoplankton (one of their major prey) remained low and most of the copepods were still deep in the water column (Seuthe et al., 2007). Fatty acid analyses indicated that the proportion of both terrestrial-based and marine-based prey increased within copepod communities over this time. This was perhaps a result of the uptake of river-borne nutrients by bacteria, which were eaten by flagellates and ciliates and then consumed by the copepods (Businski et al., in preparation). This scenario was supported by a 20-fold increase in flagellate and ciliate abundance between March and May and an increase in the growth rate of bacteria even though the water was still cold (Garneau et al., 2008). Finally, the large copepod *Calanus hyperboreus* had its annual spawning maximum in March, long before the seasonal increase in phytoplankton abundance (Darnis et al, unpublished). However, the Horton River delivered significant amounts of freshwater to Franklin Bay over winter (R. Macdonald, personal communication), suggesting that the spring-spawning *C. hyperboreus* might depend upon river-borne nutrients for egg maturation.

Net tow, echosounder (Benoit et al., 2008, Fig. 6.4) and in situ video information (Fig. 6.3) indicated that the active copepod communities under the ice in Franklin Bay were accompanied by an abundance of adult Arctic cod. Echosounder results indicated that the maximum abundance of Arctic cod was between 140 m depth and the bottom. This agreed with in situ video records, which showed the depth of maximum abundance in April around 190 m. The depth distribution of Arctic cod corresponded closely to that of copepods in the video records and included two peaks in population abundance, at around 130 and 180 m depth (Fig. 6.3). Quantitative estimates derived from echosounder and trawl data indicated that these deep-living cod gradually increased in abundance by a factor of about 100 during late winter and spring (Fig. 6.4). This may account for the 'missing' cod on the Beaufort Sea Shelf required to support the yield of higher predators (Welch et al., 1992).

6.2.2 Strong spatial variability in Beaufort Sea Shelf food webs

We documented strong spatial variability in the productivity of food webs on the Beaufort Sea Shelf which appeared to depend primarily upon the timing of sea ice disappearance. For example, in the Amundsen Gulf Polynya the production rates of copepods were more than twice as high as those under the ice in Franklin Bay at the end of May, probably due to higher concentrations of phytoplankton in the open waters of the polynya (Darnis et al., 2008). The carbon content of copepod faecal pellets was also higher within the polynya than anywhere else under the ice. Fatty acid analyses indicated that the highly productive copepod community in Amundsen Gulf had a different diet than those living in the rest of the study area, i.e., it was relatively low in river-based components and relatively

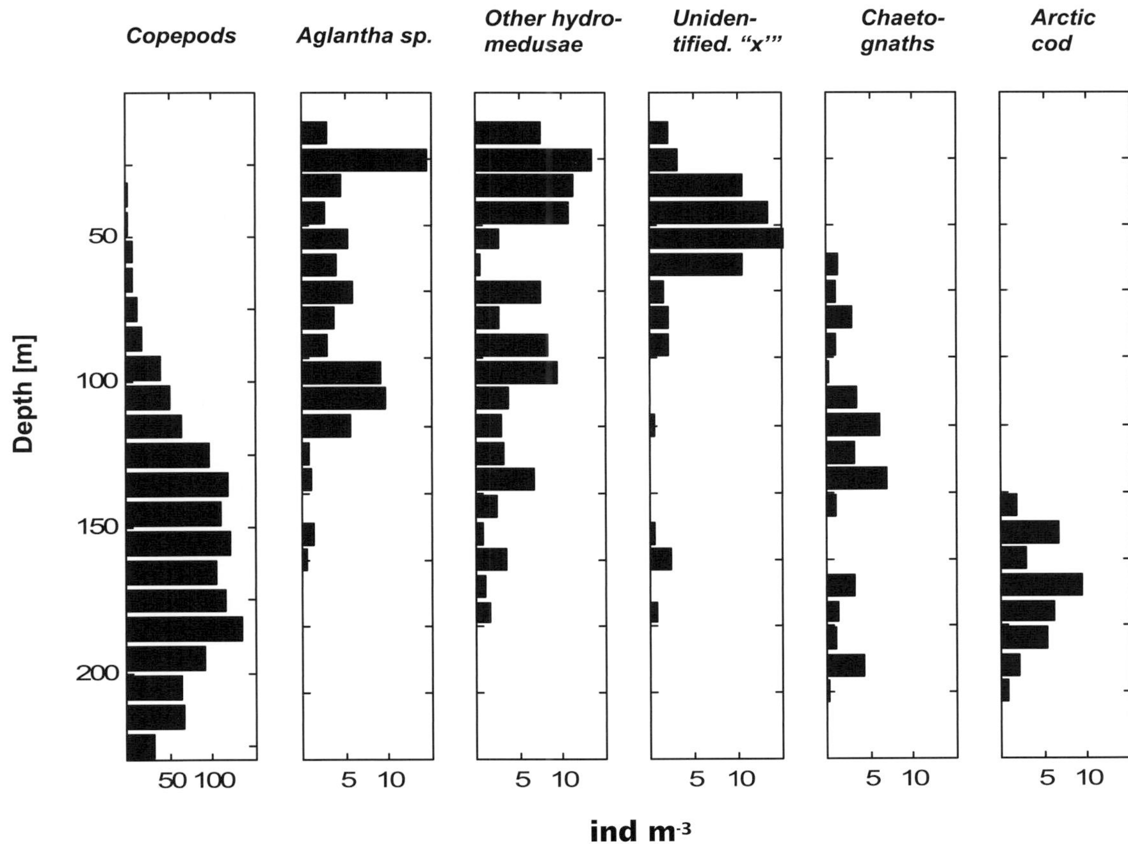

Figure 6.3

Vertical distribution of the dominant taxa in the underwater video profiles from the over-wintering station in Franklin Bay, late April 2004. The taxa are those shown in Fig. 2.

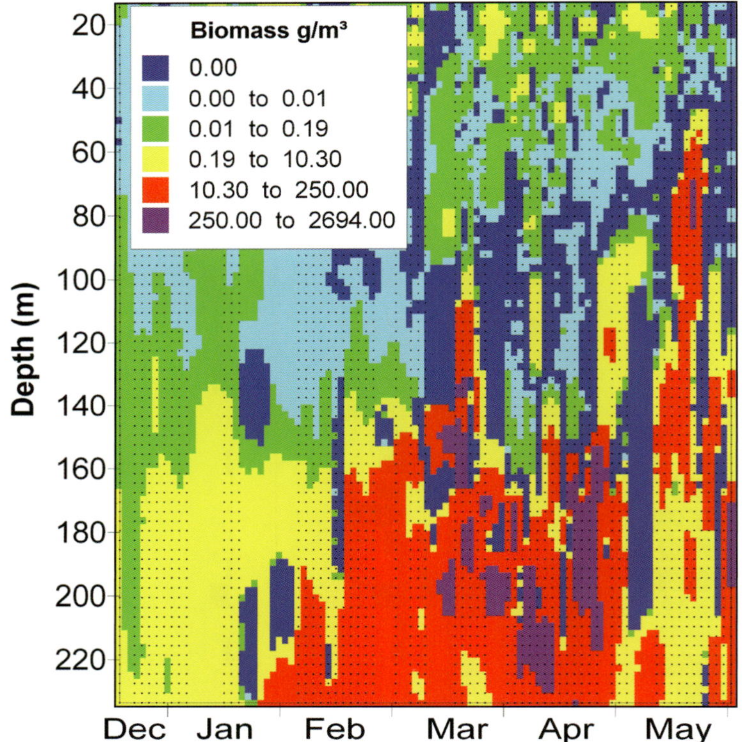

Figure 6.4

Acoustic determination of the abundance of Arctic cod at the Franklin Bay over-wintering station as a function of depth (m) and time. Adapted from Benoit et al. 2008.

high in nutritious dinoflagellates (Businski et al., in preparation). We suspect that the higher productivity of copepods within Amundsen Gulf was due to the dietary contribution of marine-based dinoflagellates rather than lower quality material from the river-based food web.

Analysis of the species making up the zooplankton community over the study region indicated three different assemblages which were roughly associated with three distinct areas of depth and ice cover. The coastal community was negatively related to ice cover and was dominated by the small copepod *Pseudocalanus* spp., which fed herbivorously on phytoplankton (Darnis et al., 2008; Fig. 6.5). *Pseudocalanus* was the most abundant prey item found in the guts of young-of-the-year Arctic cod during summer and fall (Lafrance et al., 2005). The shelf break community was dominated by two copepod species, *Cyclopina* sp. and *Microcalanus* sp. The third community, associated with Amundsen Gulf and Cape Bathurst, was dominated by *Oithona* sp., *Oncaea* sp. and the large copepods *Metridia longa* and *Calanus hyperboreus* (Darnis et al., 2008, Fig. 6.5). This latter community of large copepods fed primarily omnivorously (on plant and animal prey) or carnivorously, and fatty acid analyses indicated that they fed on a high proportion of river-borne material in the vicinity of Cape Bathurst, but on much less river-borne material in the Amundsen Gulf (Businski et al., in preparation). These results demonstrate that the Beaufort Sea Shelf is a diverse region with a variety of food webs, and that it should not be managed as a single ecosystem of uniform food web structure and function.

6.2.3 Early and long hatching season of Arctic cod

We discovered that Arctic cod hatched throughout late January to mid-July and exhibited maximum hatching rates between March and May. Otolith analysis and age-body area keys (used to determine the age of all individuals captured) demonstrated that much of the hatching period occurred well before seasonal ice break-up and the subsequent warming of the surface layer and phytoplankton bloom. In the marine-based food web of the Arctic Ocean, Arctic cod plays a central role in mediating energy flow from plankton to vertebrate predators such as marine mammals and seabirds. Previous studies elsewhere in the Arctic Ocean have demonstrated the importance of the hatch-date of spring and summer cohorts in determining the survival of young-of-the-year Arctic cod (Fortier et al., 2006; Ringuette et al., unpublished). In fact, it is believed that the pre-winter size achieved by pelagic juveniles plays a major role in first-winter survival and subsequent year class strength. Our study suggested that Arctic cod from the Beaufort Sea must hatch early in order to achieve the minimum size required to assure survival during the first winter. A comparison of the hatch date frequency distributions (HFD) of larvae sampled in polynyas and non-polynya regions suggested that Arctic cod hatch earlier in ice-covered, low-temperature regions, where more time is required to achieve a minimum over-wintering size (Lafrance et al., in preparation). An interannual comparison of hatch-date distributions for Arctic cod juveniles collected during the fall also indicated that survival and growth dynamics of early stages of this key species were closely related to environmental factors such as surface temperature and ice concentration (Lafrance et al., in preparation).

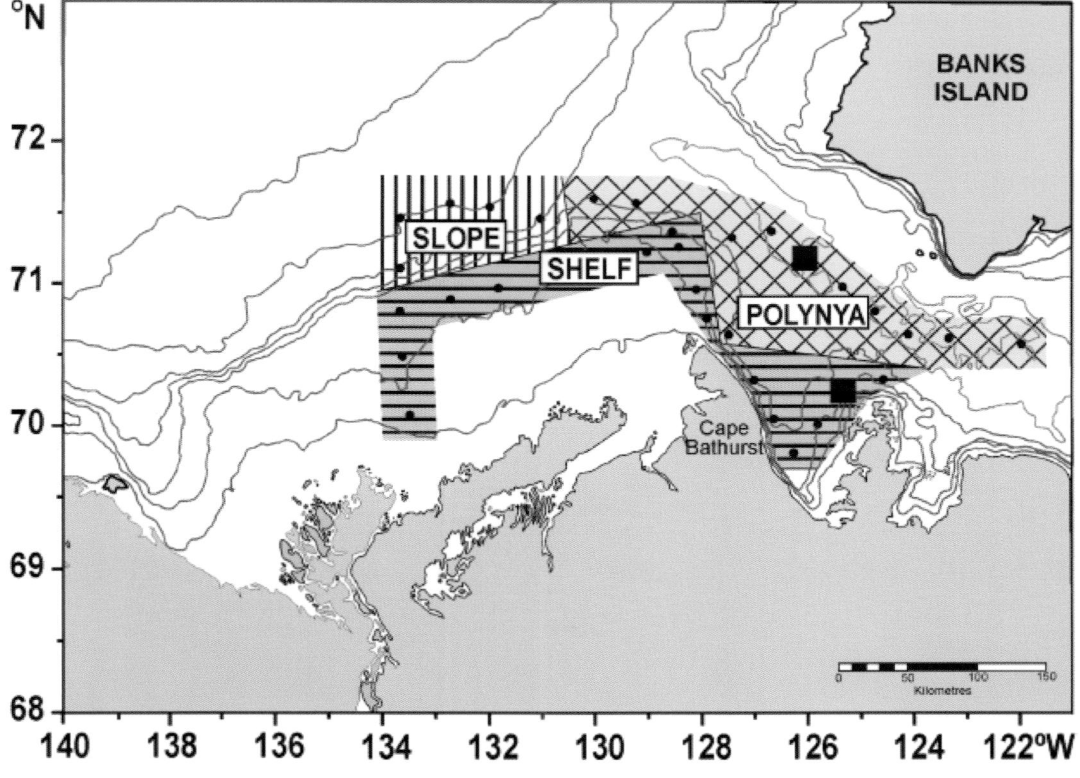

Figure 6.5

Zooplankton communities within the Beaufort Sea Shelf region, including those associated with the continental slope, the Mackenzie Shelf and the Amundsen Gulf Polynya (see text for the species making up these communities). Black dots indicate stations at which samples were taken and the two black squares indicate outlier stations with distinct zooplankton assemblages. Adapted from Darnis et al. 2008.

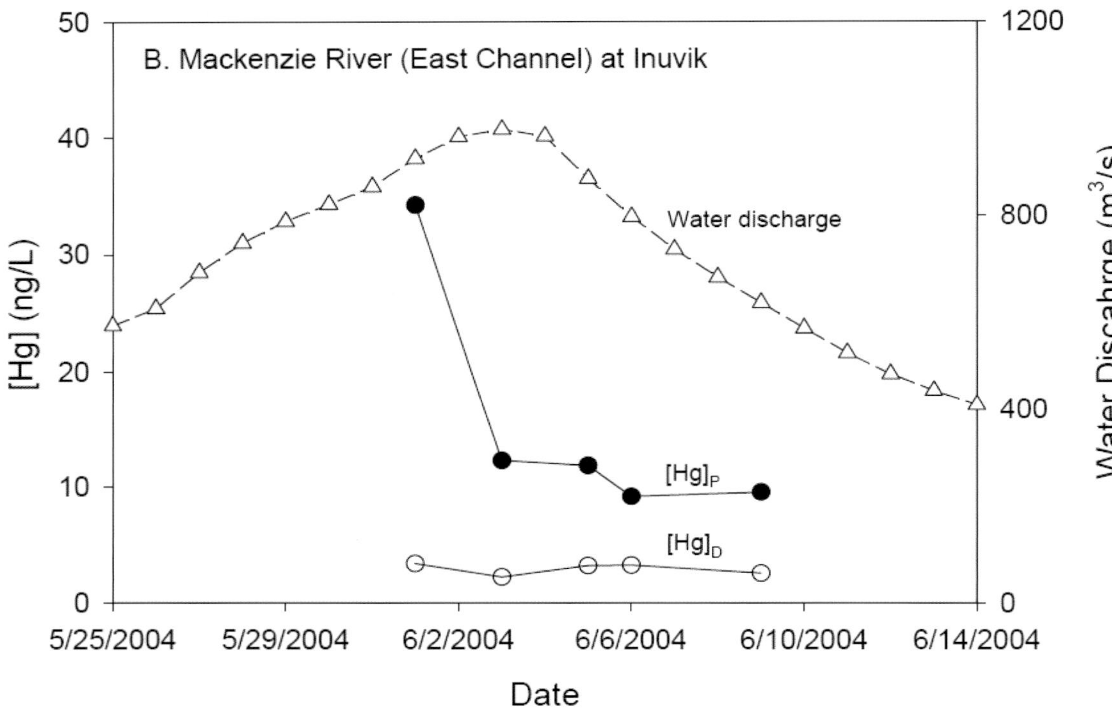

Figure 6.6

Total mercury flux and volume discharge from the Mackenzie River between 25 May and 14 June, 2004. Hg$_P$ = particle bound mercury. Hg$_D$ = dissolved mercury. Adapted from Leitch et al., 2007.

6.2.4 Beluga habitat selection

We found that beluga habitat selection differed as a function of length, sex, and reproductive status. Females with calves and smaller males selected open water habitats near the mainland, while large males selected high ice concentrations within or near the Arctic Archipelago and smaller males and two females with calves selected habitats near the ice edge (Loseto et al., 2006). Sex, age, and reproductive segregation of habitat may have been related to the different resources required for the various beluga life stages, or

represented characteristics of beluga social structure. Differences in habitat selection may have also resulted in differences in dietary contaminant sources. These differences are presently being investigated.

6.2.5 Mercury flux to the Beaufort Sea Shelf and to beluga

About 55% of the 2004 input of mercury (Hg) from the Mackenzie River to the Beaufort Sea Shelf occurred over a period of a few weeks in June during the spring freshet. 75% of this input of Hg was associated with small particles (Leitch et al., 2007; Fig. 6.6). This riverine input of contaminants was reflected in high mercury levels in Beaufort Sea *Calanus* spp. and Arctic cod (Stern and Macdonald, 2005). However, there were low mercury levels in the estuarine fish inhabiting the Mackenzie Delta, suggesting either that the food web associated with these fish did not depend upon riverine source materials or that the mercury was not yet bioavailable in the delta (Loseto et al., 2008).

Mercury levels in beluga from the eastern area of the Beaufort Sea were among the highest reported in the Arctic Ocean (Lockhart et al., 2005). During the 1990's, mercury concentration in Beaufort Sea beluga tripled relative to levels in the 1980's. However, over the last decade concentrations have decreased to levels comparable to those of beluga populations in the eastern Arctic. Because mercury levels in belugas are directly related to those of their prey, achieving a better understanding of the beluga diet becomes a crucial objective. During CASES, zooplankton and fish samples were collected for fatty acid, stable isotope and contaminants analyses, providing a basis for understanding mercury concentrations in beluga.

It is generally known that mercury is retained by organisms, resulting in its 'biomagnification' within food webs by a factor of 10 to 1,000 at each trophic level (Morel et al., 1998). Thus, mercury levels in top predators, such as some fish species and marine mammals, can be many orders of magnitude higher than in the phytoplankton at the base of the food web. A food web diagram from the Beaufort Sea Shelf indicates that mercury concentrations increased about 10-fold between each trophic level, from level 2 (herbivorous copepods), through level 3 (carnivorous invertebrates and planktivorous fish), to level 4 (including piscivorous fish such as sculpin and the top predator, beluga; Fig. 6.7). In addition, we found that Arctic cod were at the same trophic level and had a similar mercury load as several species of near-bottom dwelling invertebrates and fish (such as shrimp and flounder). Mercury levels in Arctic cod and near-bottom invertebrates were higher than those of all of the larger estuarine fish species (e.g. Pacific herring and Arctic cisco), suggesting that Arctic cod may spend a great deal of time feeding deep within the water column (along with near-bottom invertebrates), where they may be exposed to food sources high in mercury.

6.2.6 *Terrestrial energy sources are important*

Fatty acid analyses made on samples from the freshwater plume of the Mackenzie River confirmed the 'plume food web' model of Parsons et al. (1989), in which large copepods prey predominantly on marine-based plankton and the gelatinous appendicularian tunicates prey both on marine-based and river-based organisms (Businski et al., in preparation; Fig. 6.8). However, our results showed that this food web structure was not typical of the entire CASES study region. We concluded that marine-based diatoms constitute a major component of the copepod zooplankton diet within the Beaufort Sea Shelf and that river-borne

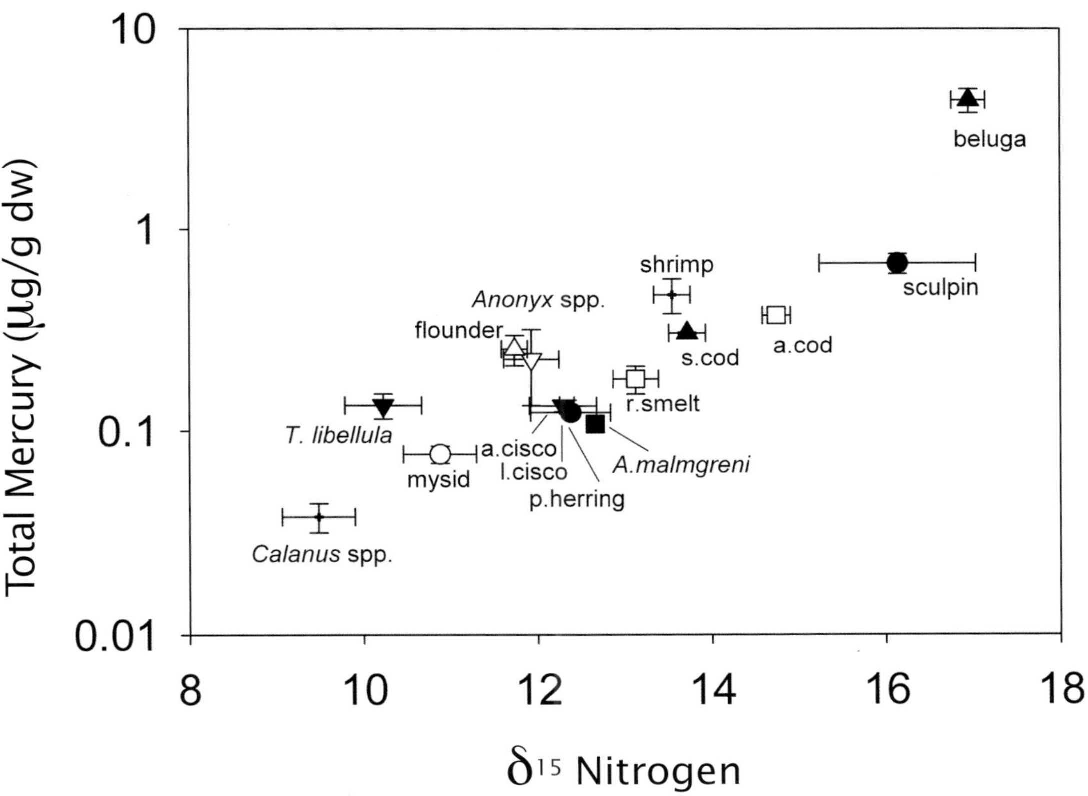

Figure 6.7

Trophic relationships among species, depicted by average (± standard error) total mercury concentrations (µg/g dry weight) versus δ15N. Species include Calanus spp., Pacific herring, Arctic cisco, least cisco, rainbow smelt, saffron cod, Themisto libellula, Arctic cod, mysids, Acanthostephia malmgreni, flounder, Anonyx spp., shrimp, sculpin and beluga. Adapted from Loseto et al. 2007.

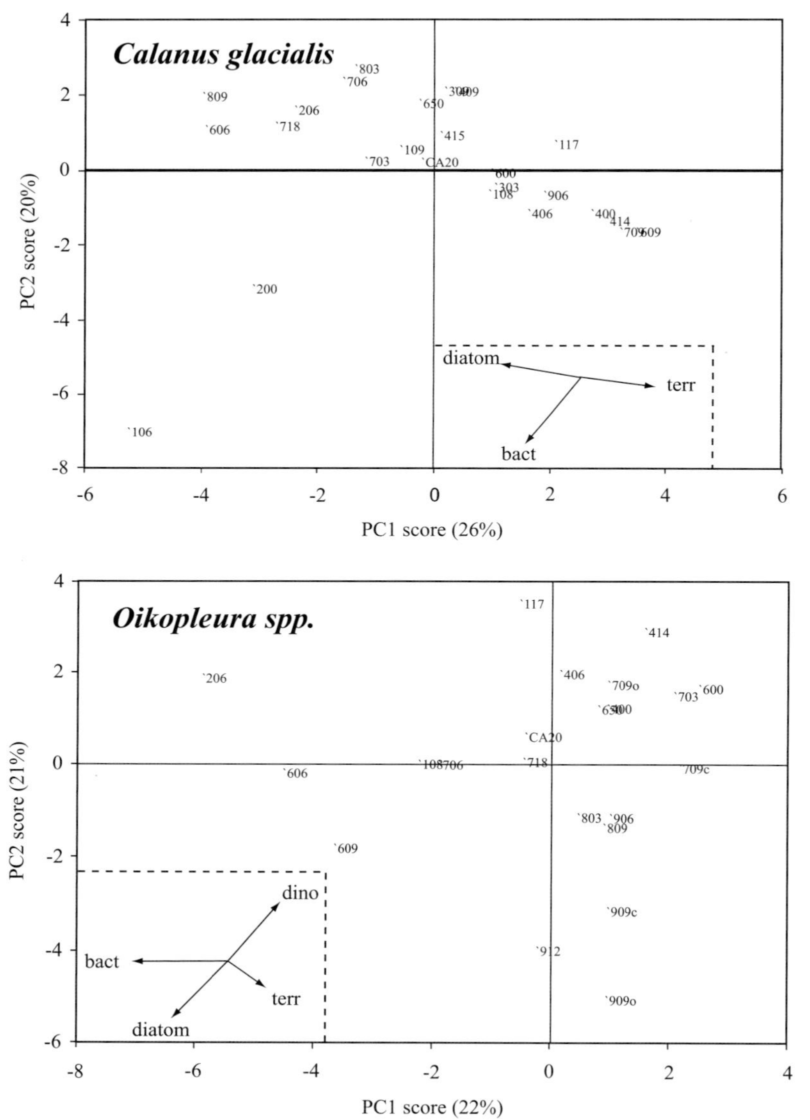

Figure 6.8

Fatty acid analyses of the diets of copepod Calanus glacialis (upper) and appendicularian tunicate Oikopleura spp. (lower) on the Beaufort Sea Shelf. The vector inserts show the diet composition in quadrants from the origin. 'bact' = bacterial food sources, 'terr' = terrestrial food sources, ' dino' = dinoflagellate food sources. The numbers in the principle components plots correspond to the stations where the zooplankton were collected.

nutrients may also enter the food web, not only through appendicularian tunicates (as concluded by Parsons et al., 1989), but also through copepod feeding. This means that terrestrial energy sources delivered to the shelf by the Mackenzie River can be very important to the productivity of the surrounding region.

6.3 Implications of this Work

Adult Arctic cod are dominant predators within food webs on the Beaufort Sea Shelf. However, more research is needed to establish the importance of this keystone fish species in the Beaufort Sea Shelf ecosystem (including the Amundsen Gulf Polynya), especially in relation to its early life stages.

The Beaufort Sea Shelf can be subdivided into three areas characterized by three fundamentally different food web structures. These are the Mackenzie River outflow, characterized predominantly by river-based production; the Beaufort Sea Shelf and Cape Bathurst, characterized by both river-based and marine production; and the Amundsen Gulf, a predominantly marine-based food web. In light of these fundamentally different food webs, the Beaufort Sea Shelf should not be managed as a single ecosystem. Finally, more research is required into the impact of other rivers (for example, the Horton River) on local biological production.

The Mackenzie River is a major source of elemental mercury for the Beaufort Sea Shelf, and its impact on local food webs and top predators like the beluga warrants additional research. Specifically, the effect of sex-based habitat selection on the diet and contaminant load of male, female and juvenile beluga remains poorly understood. It is also of importance to determine why the mercury load of the Beaufort Sea beluga is among the highest of beluga populations in the Canadian Arctic.

6.4 Recommendations

Many important questions arise from this work and recommend both research directions for the future and caution to environmental managers.

a) What causes the winter aggregation of Arctic cod under the sea ice of Franklin Bay? Is it possibly a response to the early feeding and abundance of deep-living copepods which may serve as prey? Or perhaps it reflects an avoidance of seal predation. Are there other possible explanations?

b) How are river-based materials transported from near shore to Cape Bathurst and the shelf break? What are the roles of physical and biological processes in causing river-borne particles to sink to the bottom of the Beaufort Sea Shelf?

c) What is the relative importance of the Mackenzie River vs. other rivers in transporting nutrients, sediment and contaminants to the Beaufort Sea Shelf?

d) What is the effect of sex-based habitat selection on the diet and contaminant load of male, female and juvenile beluga? Why is the mercury load of the Beaufort Sea beluga among the highest of other beluga populations in the Canadian Arctic?

References

Benoit, D., Y. Simard, and L. Fortier, 2008. Hydro-acoustic detection of large winter aggregations of Arctic cod (*Boreogadus saida*) at depth in ice-covered Franklin Bay (Beaufort Sea). J. Geophys. Res. 113, C06S90, doi:10.1029/2007JC004276.

Businski, T. Seasonal changes in zooplankton food sources in Franklin Bay. Ph.D. Thesis, Memorial University, St. John's. In preparation.

Dalsgaard, J., M. St. John, G. Kattner, D. Müller-Navarra, and W. Hagen, 2003. Fatty acid trophic markers in the pelagic marine environment. Adv. Mar. Biol. 46: 225-340.

Darnis, G., D.G. Barber, and L. Fortier, 2008 Sea ice and the onshore-offshore gradient in pre-winter zooplankton assemblages in the southeastern Beaufort Sea. J. Mar. Syst. 74(3-4): 994-1011

Fortier, L., P. Sirois, J. Michaud, and D.G. Barber, 2006. Survival of Arctic cod larvae (*Boreogadus saida*) in relation to sea ice and temperature in the Northeast Water Polynya (Greenland Sea). Can. J. Fish. Aquat. Sci. 63: 1608-1617.

Garneau, M-È., S. Roy, C. Lovejoy, Y. Gratton, and W.F. Vincent, 2008. Seasonal dynamics of bacterial biomass and production in a coastal arctic ecosystem: Franklin Bay, western Canadian Arctic. J. Geophy. Res. Oceans 113, C07S91, doi:10.1029/2007JC004281.

Lafrance, P., L. Fortier, and J.A. Gagné, 2005. Interannual variations in the diet of young-of-the-year Arctic cod (*Boreogadus saida*) in the Mackenzie Shelf and Amundsen Gulf, Canadian Arctic, during fall. Proceedings of the ArcticNet Annual Scientific Meeting, Banff, Canada, 2005.

Leitch, D.R., J. Carrie, D. Lean, R.W. Macdonald, G.A. Stern, and F. Wang, 2007. The delivery of mercury to the Beaufort Sea of the Arctic Ocean by the Mackenzie River. Sci. Total Environ. 373: 178-195.

Loseto, L.L., P. Richard, G.A. Stern, J. Orr, and S.H. Ferguson, 2006. Segregation of Beaufort Sea beluga whales during the open-water season. Can. J. Zool. 84:1743-1751.

Loseto, L.L., G.A. Stern, D. Deibel, T.L. Connelly, A. Prokopowicz, D.R.S. Lean, L. Fortier, and S.H. Ferguson, Linking mercury exposure to habitat and feeding behaviour in Beaufort Sea beluga whales J. Mar. Syst. 74 (3-4): 1012-1024

Lockhart, W.L., G.A. Stern, R. Wagemann, R.V. Hunt, D.A. Metner, J. DeLaronde, B. Dunn, R.E.A. Stewart, C.K. Hyatt, L. Harwood, and others, 2005. Concentrations of mercury in tissues of beluga whales (*Delphinapterus leucas*) from several communities in the Canadian Arctic from 1981 to 2002. Sci. Total Env. 351: 391-412.

Morel, F.M.M., A.M.L. Kraepiel, and M. Amyot, 1998. The chemical cycle and bioaccumulation of mercury. Ann. Rev. Ecol. Syst. 29: 543-566.

Parsons, T.R., D.G. Webb, B.E. Rokeby, M. Lawrence, G.E. Hopkey, and D.B. Chiperzak, 1989. Autotrophic and heterotrophic production in the Mackenzie River/Beaufort Sea estuary. Polar Biol. 9: 261-266.

Seuthe, L., G. Darnis, C.W. Riser, P. Wassman, and L. Fortier, 2007. Winter-spring feeding and metabolism of Arctic copepods: insights from faecal pellet production and respiration measurements in the southeastern Beaufort Sea. Polar Biol. 30: 427-436.

Stern, G.A. and R.W. Macdonald, 2005. Biogeographical provinces of total and methyl mercury in zooplankton and fish from the Beaufort and Chukchi seas: Results from the SHEBA drift. Env. Sci. and Tech. 39: 4707-4713.

Welch, H.E., M.A. Bergmann, T.D. Siferd, K.A. Martin, M.F. Curtis, R.E. Crawford, R.J. Conover, and H. Hop, 1992. Energy flow through the marine ecosystem of the Lancaster Sound region, Arctic Canada. Arctic. 45: 343-357.

CHAPTER 7

Organic and Inorganic Fluxes

Alfonso Mucci[1*], Alexandre Forest[2], Louis Fortier[2], Mitsuo Fukuchi[3], John Grant[4], Hiroshi Hattori[5], Philip Hill[6], Gwyn Lintern[6], Ryosuke Makabe[7], Cédric Magen[1], Lisa Miller[8], Makoto Sampei[2], Hiroshi Sasaki[7], Bjorn Sundby[1], Tony Walker[4], Paul Wassmann[9]

7.1 Introduction and Rationale

Models predict that global climate change will accelerate under current CO_2 emission scenarios and that its greatest effects will be felt in the Polar Regions (ACIA, 2004). Satellite and field observations over the Arctic Ocean reveal that sea ice cover is steadily decreasing and that the most recent summers have set new records in minimum ice concentration (Serreze et al., 2003; Mueller et al., 2003; Stroeve et al., 2005). Similar trends have also been reported in sea ice thickness during the winter months (Rothrock et al., 1999; Wadhams and Davis, 2000; Laxon et al., 2003; Yu et al., 2004). One climate change scenario (Holland et al., 2006) predicts that Septembers will be nearly ice-free by 2040. The potential implications of such changes to the Arctic ecosystem are numerous. Decreased ice cover would most likely translate into more efficient gas exchanges at the air-sea interface, along with increased sediment resuspension (by wind stress), wave action, and shoreline erosion. It would also lead to increased primary production due to greater light penetration and organic matter fluxes through the water column. Increased river discharge might lead to greater primary production as more nutrients would be delivered to coastal waters. Conversely, increased suspended particulate matter via sediment resuspension, erosion, and river discharge might impede primary production by limiting light penetration. Increased delivery of terrestrial organic matter via permafrost melting (Payette et al., 2004) might lead to increased organic carbon respiration, and consequently, an increased release of CO_2 from coastal waters to the atmosphere. Finally, increased stratification due to higher

Cloud reflection on water. Photo: Hiroshi Hattori.

LEFT: Arctic fox. Photo: Alexandre Forest.

[1] Department of Earth and Planetary Sciences, McGill University, 3450 University St., Montréal, QC, H3A 2A7, Canada

[2] Québec-Océan, Département de Biologie, Université Laval, Québec, QC, G1V 0A6, Canada

[3] National Institute of Polar research (NIPR), 1-9-10 Kaga, Itabashi-ku, Tokyo 173-8515, Japan

[4] Department of Oceanography, Dalhousie University, Halifax, NS, B3H 4J1, Canada

[5] Hokkaido Tokai University, 5-1-1-1 Minami-sawa, Minami-ku, Sapporo 005-8601, Japan

[6] Natural Resources Canada, Geological Survey of Canada - Pacific, 9860 West Saanich Road, P.O. Box 6000, Sidney, BC, V8L 4B2, Canada

[7] Senshu University of Ishinomaki, Shinmito, Minamisakai, Ishinomaki, Miyagi 986-8580, Japan

[8] Department of Fisheries and Oceans, Institute of Ocean Sciences, P.O. box 6000, Sidney, BC, V8L 4B2, Canada

[9] Norwegian College of Fishery Science, University of Tromsø, N-9037 Tromsø, Norway

*corresponding author: alm@eps.mcgill.ca

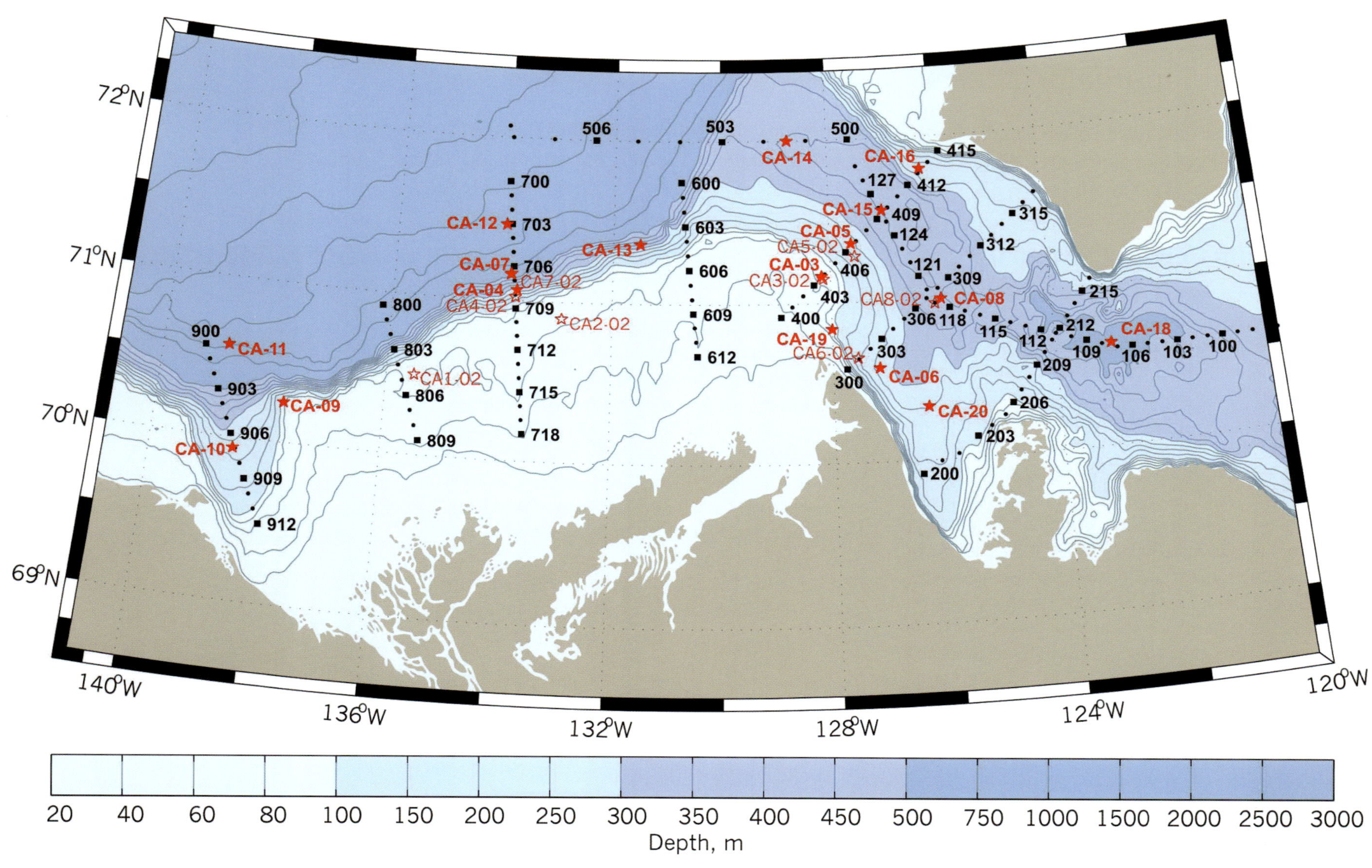

Figure 7.1

CASES study area, bathymetry, and location of sampling stations.

river outflow might cause a decrease in atmospheric carbon drawdown (as the thin freshwater lens would rapidly come into equilibrium with the atmosphere). Understanding these potential feedbacks is critical in determining the direction and intensity of the carbon cycle in Polar Regions and its response to further climate variations.

The western Canadian Arctic—specifically, the southern Beaufort Sea (Fig. 7.1)—has experienced alarming increases in surface temperatures over the last few decades (IPCC, 2001; Comiso, 2003), as well as steady decreases in sea ice concentration (Barber and Hanesiak, 2004). The Mackenzie River basin covers 1.8×10^6 km^2 and flows northwards into the Beaufort Sea and Arctic Ocean (Hill et al., 2001). It discharges large amounts of fresh water (330 km^3 annually) and is the largest single source of sediment to the Canadian Arctic Ocean (Carson et al., 1998; Macdonald et al., 1998; O'Brien et al., 2006). The shelf is currently dominated by seasonal ice from October until August, with continuous freshwater input throughout the year. Most of the annual discharge of water and sediments occurs between May and September and is associated with the break up of land fast ice (Hill et al., 2001). The nearshore zone is dominated by the river plume, creating strong gradients of turbidity, salinity, and temperature (Lintern et al., 2005). During the winter ice-covered period, a 'winter plume' still manages to discharge substantial quantities of sediment and freshwater (Carmack et al., 2004).

Of the 250×10^6 metric tons per year (t/y) of suspended particulate matter delivered by all rivers to the Arctic Ocean, the Mackenzie River discharges nearly half—including approximately 3×10^6 t/y of organic carbon (Telang et al., 1991; Droppo et al., 1998; Macdonald et al., 1998). Consequently, land-derived organic carbon accounts for the majority of the organic matter in Beaufort Shelf sediments (Yunker et al., 1995; Goñi et al., 2000; Yunker et al., 2005). Although coastal erosion provides an additional source for terrestrial organic carbon to the Mackenzie Shelf, its contribution is minimal (~7%; Macdonald et al., 1998). Horizontal and vertical fluxes of organic carbon and suspended particulate matter (SPM) in the surface, intermediate, and bottom layers of the Mackenzie Shelf and Beaufort Sea show significant seasonal variations, depending on ice cover, winds (and the resulting currents), river flow, and primary productivity. The Mackenzie River significantly impacts the total carbon budget of the Arctic through the flow of terrestrial organic carbon (via food webs, horizontal fluxes of SPM and the delivery of organic matter to benthic communities; Macdonald et al., 2003). During brief open water conditions in the shallow Mackenzie Delta region, gales cause strong storm surges. These result in major resuspension of bottom sediments and have important consequences for organic carbon fluxes (Lintern et al., 2006a; Walker et al., in press). Sediments reworked during storm events are transported across the outer shelf during open water periods (via surface plumes and near bottom suspensions) by wind driven currents (O'Brien et al., 2006). During sea ice growth, thermohaline convection can also resuspend and transport bottom material offshore by spreading the benthic nepheloid layer beyond the shelfbreak along isopycnals (Forest et al., 2007). Finally, the shelf is traversed by several troughs—features which are key to understanding the overarching extent of erosion and sediment transport in the Arctic Ocean (Yunker et al., 2005).

*The CCGS Amundsen in a storm.
Photo: Alfonso Mucci.*

Figure 7.2 AT RIGHT

a) Contour plots of average pCO₂ (µatm) in the surface mixed layer (0-15 m) calculated from pH and At or DIC measurements carried out on Legs 1 and 2 of the CASES program (late September through early December, 2003); b) Contour plots of average pCO₂ (µatm) in the surface mixed layer (0-50 m) calculated from pH and At or DIC measurements carried out on Legs 7, 8, and 9 of the CASES program (June-August, 2004).

7.2 Overview of results

7.2.1 CO_2 fluxes at the air-sea interface

In this section, we report the partial pressure of carbon dioxide in the surface mixed layer of the southern Beaufort Sea during two ice-free seasons of the CASES program: September-November 2003, and June-August 2004. This quantity (pCO_2) represents the potential for CO_2 to move between the surface ocean and the atmosphere, and allows us to estimate carbon fluxes at the air-sea interface. The flux of CO_2 across the air-sea interface is driven by the difference in CO_2 partial pressure between the surface ocean and the overlying atmosphere. If the difference in partial pressure ($\Delta pCO_2 = pCO_2(SW) - pCO_2(air)$) is negative, the surface ocean becomes a net *sink* of CO_2 (i.e., the surface ocean absorbs CO_2 from the atmosphere). The data presented here served to establish whether seasonal variations in surface water pCO_2 could be associated with the hydrological cycle or the onset of plankton blooms. It also allowed us to estimate the strength of the CO_2 sink in this region of the Arctic Ocean.

We collected samples at 76 stations. These samples represented conditions at the surface (1-3 m) and throughout the water column; under open ice conditions in the Beaufort Sea and on the Mackenzie Shelf (Fig. 7.1). Most of these samples were analyzed onboard for total dissolved inorganic carbon (DIC), titration alkalinity (At), and pH. These measurements allowed us to calculate the pCO_2 (as averages in the surface mixed layer) at each station. Our data spanned over two time periods: the Fall of 2003 (late September to early mid-November), and the Summer of 2004 (June to August). Most of the freshwater discharge from the Mackenzie River occurs within these periods,

i.e., May to October (Carmack and Macdonald, 2002). Results from our calculations (Fig. 7.2 a) revealed that the relatively fresh (21 < S < 28) and cold (-1.52 °C < t < -0.87 °C) surface waters of the fall were undersaturated (249 µatm < pCO_2 < 325 µatm) with respect to the overlying atmosphere (~381 µatm) and served as a sink for atmospheric CO_2. In contrast, the fresh (7.65 < S < 31.8) and warm (-1.36 °C < t < 8.07 °C) waters sampled soon after ice breakup and throughout the summer (Fig. 7.2 b) were closer to equilibrium (257 µatm < pCO_2 < 397 µatm) with respect to the atmosphere.

7.2.2 Sediment resuspension in the Beaufort Sea and along the Mackenzie Shelf

Reduced sea ice thickness in response to climate change is expected to increase the turbidity of surface waters during spring and summer and result in an increased flux of sediments (due to erosion and resuspension) to the outer shelf (Carmack et al., 2004). Obtaining measurements of sediment and carbon fluxes on the Mackenzie Shelf (both under open water and under sea ice) therefore becomes critical in understanding physical and biogenic carbon cycles across the shelf and their dependence on river discharge, ice cover and storms (Macdonald, 2000). The interaction between marine and estuarine conditions can be a useful analog to climate-related changes in sediment transport, coastal erosion, and sediment resuspension in the Arctic. Determining sediment erosion thresholds by currents and waves is therefore an important step in predicting sediment resuspension as well as fluxes of particulate organic and inorganic matter.

Several methods were used to investigate sediment resuspension in the Beaufort Sea. Walker et al. (in press) used a sediment erosion device called the Benthic Environmental Assessment Sediment Tool

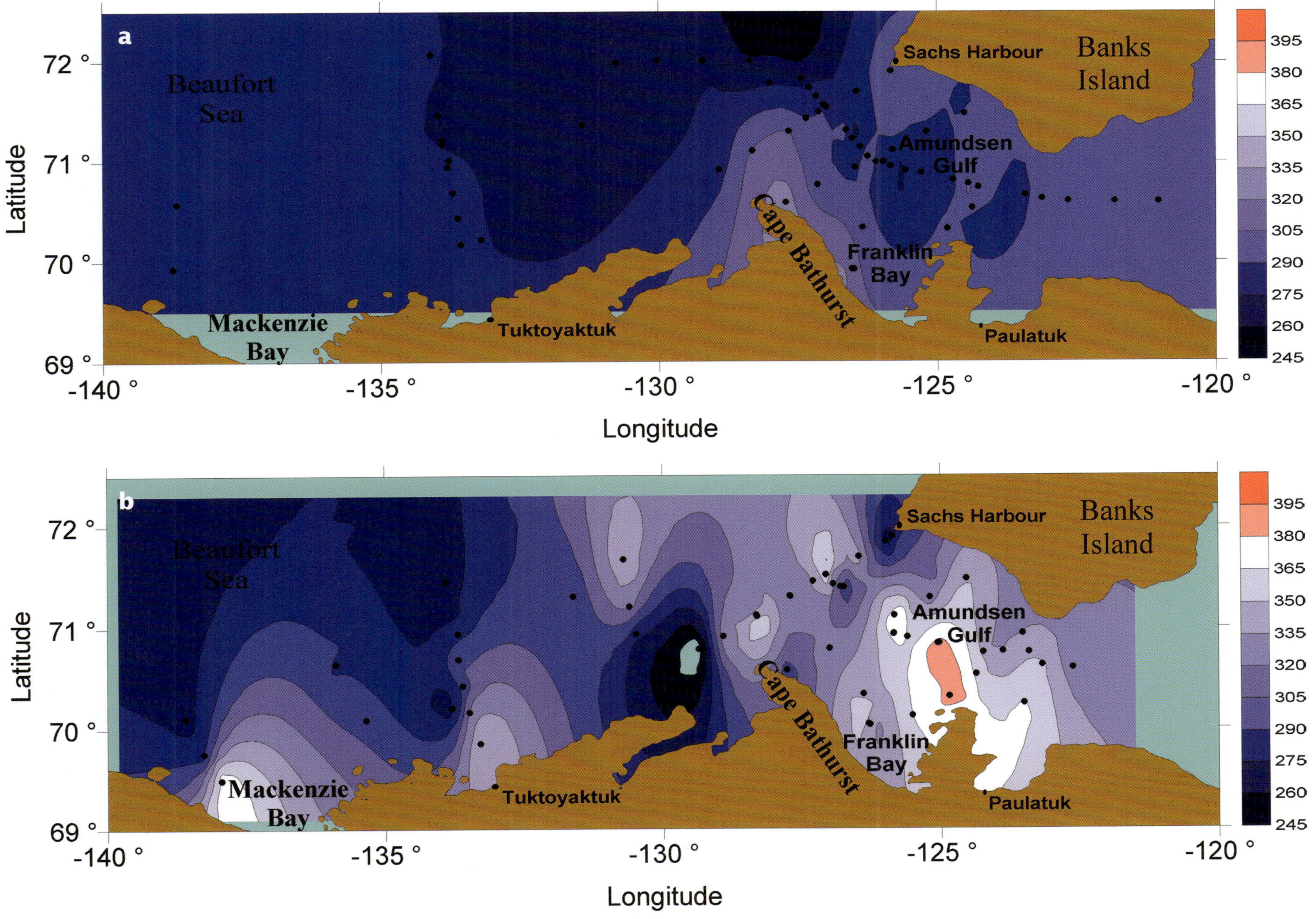
a
Beaufort
Sea
Sachs Harbour
Banks
Island
Amundsen
Gulf
Cape Bathurst
Franklin
Bay
Tuktoyaktuk
Paulatuk
Mackenzie
Bay
Latitude
Longitude
72 °
71 °
70 °
69 °
-140 °
-135 °
-130 °
-125 °
-120 °
395
380
365
350
335
320
305
290
275
260
245
b
Beaufort
Sea
Sachs Harbour
Banks
Island
Amundsen
Gulf
Cape Bathurst
Franklin
Bay
Tuktoyaktuk
Paulatuk
Mackenzie
Bay
Latitude
Longitude
72 °
71 °
70 °
69 °
-140 °
-135 °
-130 °
-125 °
-120 °
395
380
365
350
335
320
305
290
275
260
245

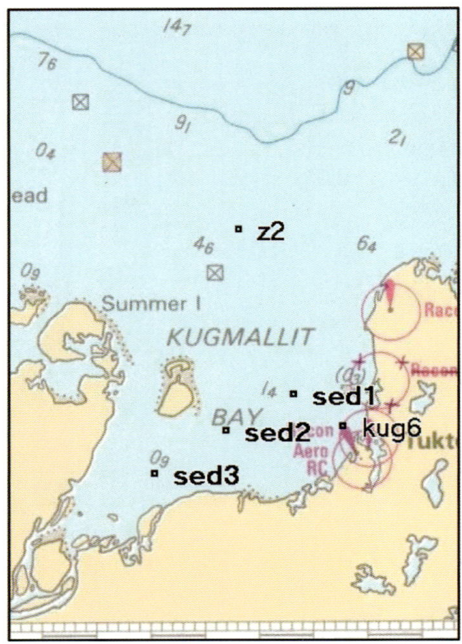

Figure 7.3

The study area, Beaufort Sea and Kugmallit Bay.

(BEAST) to quantify sediment erosion. Lintern et al. (2005) made measurements of sediment concentrations during moderate storms in Kugmallit Bay, as well as observations of bed properties in the area. Lintern et al. (2006a) re-examined data taken during an earlier deployment of pore pressure and backscatter sensors, and used a state of the art model (SedTrans) to compare measured data to model outputs. Finally, Lintern et al. (2006b) examined the relationship between the extent of the Mackenzie River plume and local weather (i.e. storm) conditions.

In the first of the studies listed above, sediment erosion thresholds were measured using BEAST on undisturbed fresh sediment cores collected from stations in the Beaufort Sea (Walker et al., in press). Erosion thresholds were also measured using the *in-situ* instrument Miniflume (Lintern et al., unpublished). These resuspension tests were then compared to observed sedimentation rates, sediment characteristics, suspended particulate matter (SPM) fluxes and the timing of increased primary production during spring, and also compared with data collected from the same site during the previous summer.

During July and August 2004, several shallow water (0-5 m depth) sampling stations were selected in Kugmallit Bay at the mouth of the East Channel of the Mackenzie River (Fig. 7.3). Efforts were concentrated on the transect formed by stations SED3, SED2 and SED1. Lintern et al. (2005) reported grain size measurements, vane shear strengths, and storm-driven resuspended concentrations at these stations, whereas Walker et al. (in press) reported results from 1-2 day current meter and sediment trap deployments.

According to Lintern et al. (2005), sediments in the area consisted mainly of sandy silts or silts (Fig. 7.4). Median

grain sizes decreased from 25 to 50 µm at the river mouth (station SED3) to around 10 µm in the middle of Kugmallit Bay (SED1). A third sample, near Tuktoyaktuk harbour (station KUG6) showed intermediate grain sizes between 10 and 20 µm. Samples from 9 cm depth exhibited smaller median diameter (D50) than those at shallower depths (with the exception of station SED3, which had a surface layer of finer sediment). Figure 7.5 shows the vane shear results for all of the samples from the river mouth to SED1. Overall, decreasing shear strength was observed from the mouth of the East Channel to more seaward stations. This was true for both peak vane and residual strengths. Just below the surface (3 cm depth), the shear strengths were less than at the surface. Nevertheless, the relationship of decreasing strength to distance from the river mouth held true for peak strengths at both depths. The vane strengths appeared to correlate with grain size, and indicated a slightly weakened (and possibly mobile) layer a few cm below the surface (Lintern et al, 2005). Potential implications of this on sediment erosion and resuspension are currently under investigation.

Under open-water conditions, turbidity is clearly wind-driven. A plot of SPM concentration versus wind speed for a period of low-to-moderate wave heights (July 2004) is shown in Figure 7.6 (Walker et al., in press). Between August 1 and August 3, a typical storm for this area occurred, with wind velocities reaching over 40 km/hr from the northwest. This caused a significant storm surge with over a 1 m rise in sea level as well as moderate-to-high seas, with breaking waves estimated at 1.5 to 2 m (Lintern et al., 2005). During this particular storm, it was not possible to deploy instruments in Kugmallit Bay. Rather, the instruments were deployed in the relatively sheltered harbour behind Tuktoyaktuk Spit. The spit normally protects the harbour but

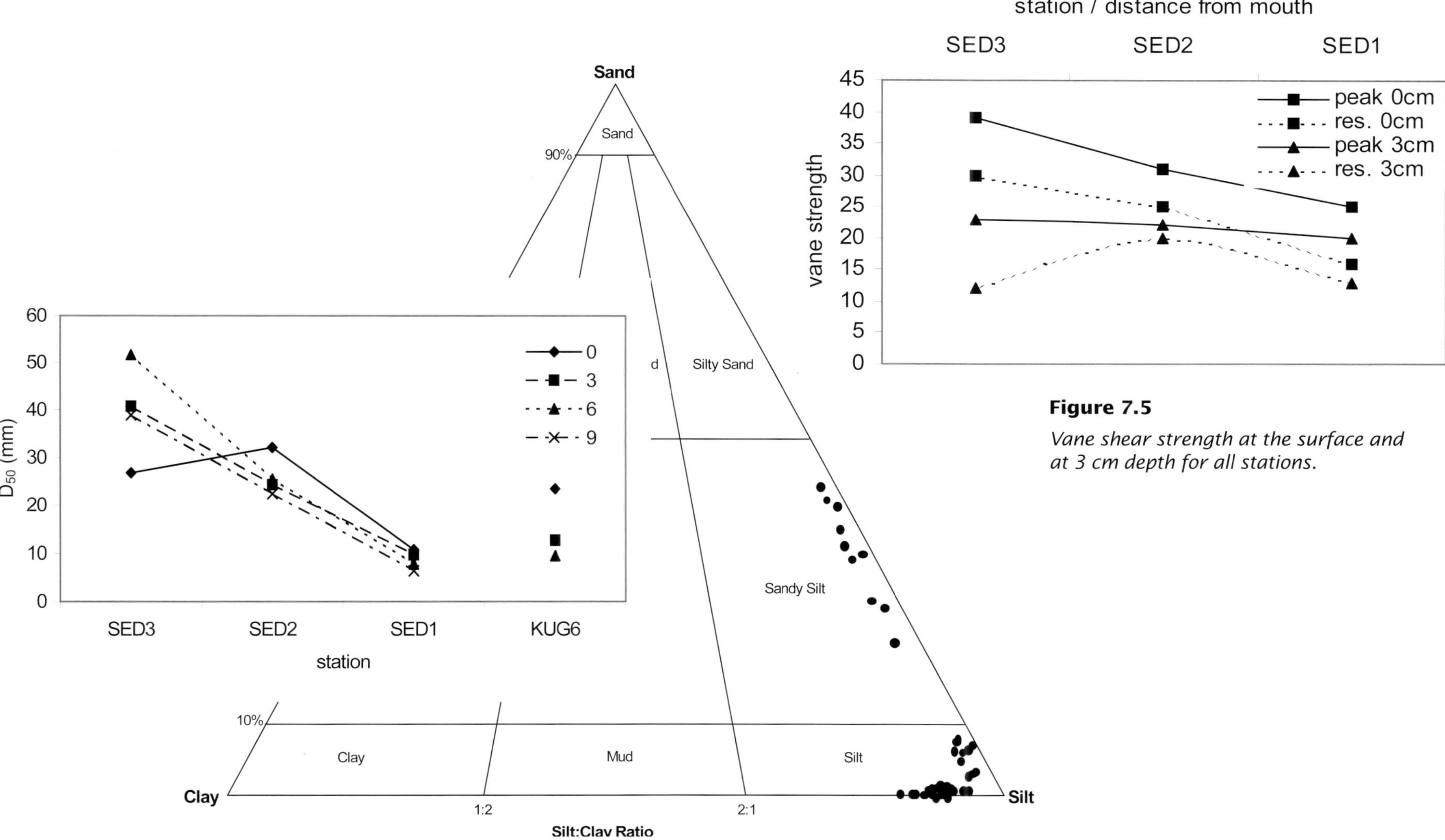

Figure 7.5

Vane shear strength at the surface and at 3 cm depth for all stations.

Figure 7.4

Results of sediment grain size analysis. All samples are sandy silt or silt. Inset: D50 at multiple depths (0, 3, 6 and 9 cm) for three stations along a seaward transect (stations SED3 to SED1) and at station KUG6.

during this storm it was overtopped by sediment laden waves. Concentrations of SPM in the harbour are shown in Figure 7.7 alongside wind speed. Four filter samples taken during this storm period (August 1-3) revealed that SPM concentrations increased dramatically to 200mg/L. This contrasted against measurements taken under similar (but not sustained) wind velocities during the period of July 29-31.

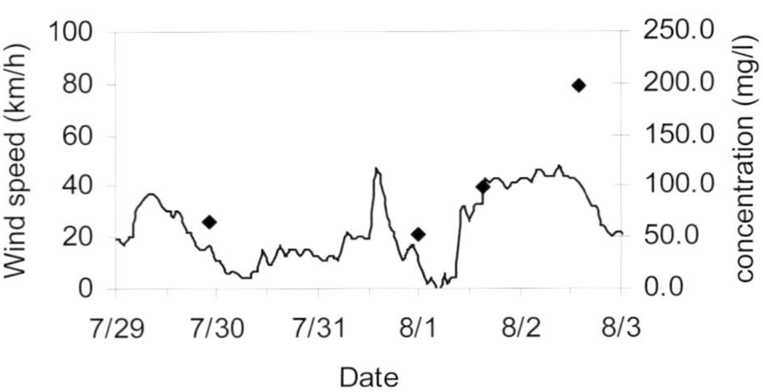

Figure 7.7

Wind speed and suspended particulate matter concentration measured using filtered water samples.

Walker et al. (in press) reported observations of sediment captured in traps at stations SED1, SED2 and SED3 during a period in late July 2004. Mean organic matter (OM) content was fairly high in SPM residues (11%) and surface sediments (8%). SPM concentrations were strongly correlated with wind speed ($R^2 = 0.71$) in inner Kugmallit Bay (Fig. 7.6). Sediment traps deployed at the same stations yielded sedimentation rates between 3700-5400 g m^{-2} d^{-1}. The large input of fresh, warm turbid water from the Mackenzie River into Kugmallit Bay produced a freshwater lens, and increased salinity was strongly related to lower water temperatures at the surface ($R^2 = 0.96$). This resulted in stratification characterized by pronounced horizontal gradients in salinity, temperature and SPM concentration.

Particle sizes and their settling rates were measured before and after storm events using a Perspex settling chamber and DV camera. A strong relationship between equivalent spherical diameter (ESD) and particle settling rate was found after prolonged high winds ($R^2 = 0.91$). Mean particle settling rates were 0.72 cm s^{-1}, with corresponding mean ESD values of 0.9 mm when wind speeds were >11 km h^{-1}.

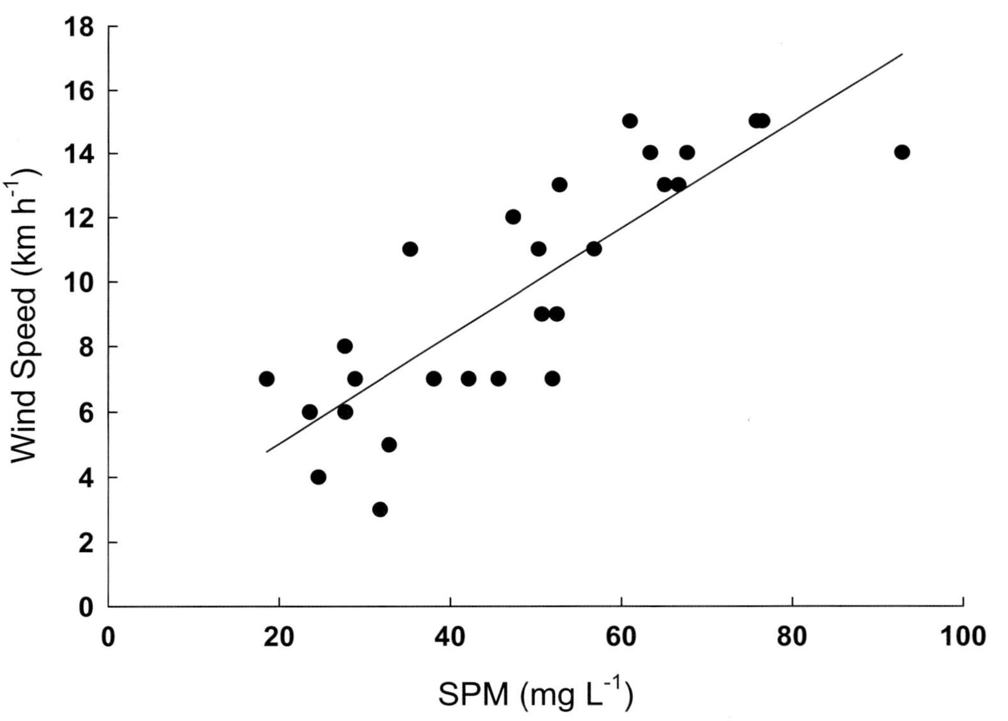

Figure 7.6

Relationship between wind speed and SPM concentration in inner Kugmallit Bay. (r 2 = 0.71, n = 27). Adapted from Walker at al. 2008b.

During a sampling period through sea ice in May, 2005, a steady increase in chlorophyll a concentration was observed (R^2 = 0.81, P<0.001). This indicated the first signs of a spring bloom (which may have begun further south in the drainage basin), though the mean value (2.2 µg L^{-1}) was low compared to typical summer values (202 µg L^{-1} during July, 2004). There was also an apparent increase in current velocity over the mooring period, from 13-20 cm s^{-1} between May 5 and 9 (R^2 = 0.75, P<0.001). Simultaneously, under-ice measurements showed that current velocity fluctuations were due to semi-diurnal tidal periods. Normally, under open water conditions, current circulation is primarily wind-driven. Sediment fluxes estimated from sediment trap deployments in Kugmallit Bay during the spring produced appreciable sedimentation rates of 51 g m^{-2} d^{-1}. These were comparable to winter rates reported by Macdonald and Thomas (1991).

Significant differences were observed in temperature, chlorophyll a, SPM concentration and sedimentation rates between spring and summer at the same site (P < 0.001 in all cases). A strong relationship was observed between water column depth and total organic carbon (TOC) content in the surface sediments of Kugmallit Bay and the Beaufort Sea (R^2 = 0.91) (Fig. 7.8). There were also significant differences in OM and TOC content in surface sediments between spring and summer (P < 0.001). Differences in sediment C:N ratios were small, but were significantly different (P < 0.05). Specifically, the spring ratios were slightly higher, probably due to lower OM and TOC contents during winter. Sediments in Kugmallit Bay were dominated by sandy silt or silt. Grain size homogeneity in Kugmallit Bay corresponded well to the Mackenzie plume in terms of its fine components and associated carbon content. TOC and OM values in sediments at this site were heavily influenced by the fine sediment

fractions of the Mackenzie River. However, the values were less than half of those from the previous summer. This is consistent with C:N ratios which indicated that there was less carbon but more nitrogen in the surface sediments during the winter (although both winter and summer C:N ratios remained high). The high sediment C:N ratios in Kugmallit Bay were similar to those reported along the Siberian Arctic coastline by Guo et al. (2004). At sites <10 m, Hill et al. (2001) reported mean TOC values of 5% during summer and concluded that organic carbon principally consisted of terrigenous components derived from the Mackenzie River (Yunker et al., 2005). Our TOC values reported here (1% in the spring; 4% in the summer) compare well with these values.

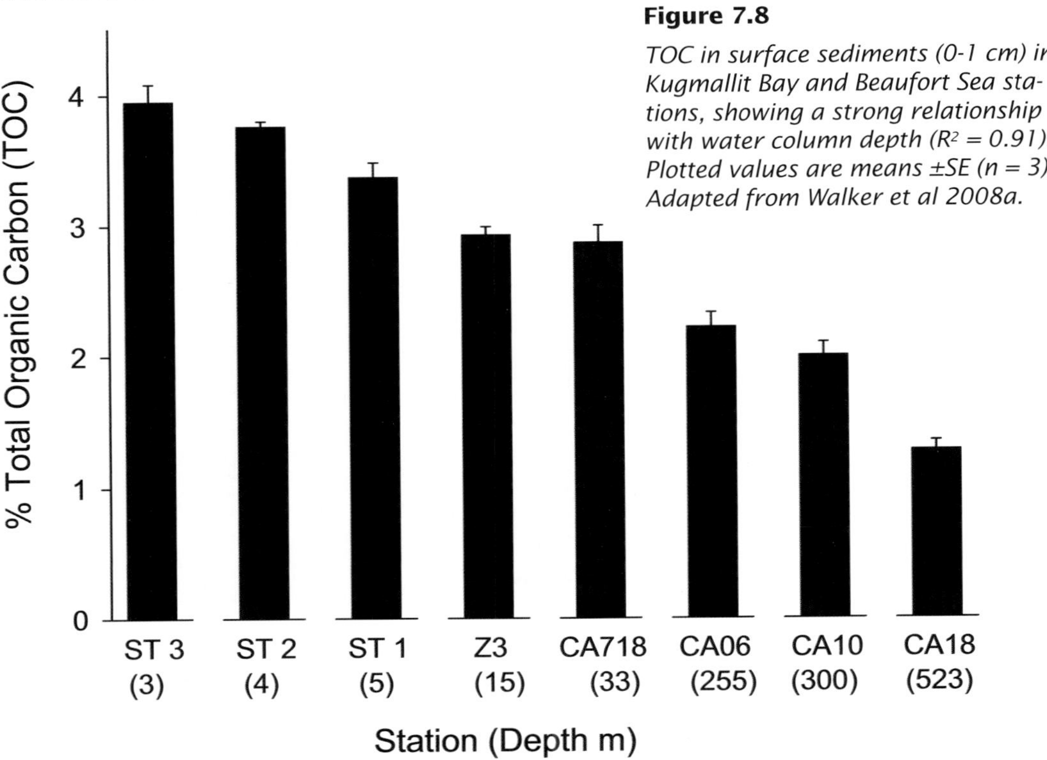

Figure 7.8

TOC in surface sediments (0-1 cm) in Kugmallit Bay and Beaufort Sea stations, showing a strong relationship with water column depth (R^2 = 0.91). Plotted values are means ±SE (n = 3). Adapted from Walker et al 2008a.

Comparison of organic *quality* in seston and trapped material demonstrated substantial seasonal differences (Table 7.1). The seston was nine times richer in total pigments per unit mass under sea ice than during summer. Such phytoplankton-rich seston appeared to result from a combination of reduced river-borne detritus and a developing spring bloom. Correspondingly, trap contents in spring were five times richer in pigment content under ice compared to summer. The TOC content of trapped particles was similar between seasons but the organic content of seston in the spring samples was twice as high as that in the summer samples. These results suggested that pigment content had a greater differential between traps and seston than organic content. The effect was intensified in spring: the seston was 47 times richer in pigments than in TOC content, reflecting the relatively greater contribution of phytoplankton to seston in the former season.

Measurements of erosion rates, critical shear stresses, particle size distributions and resuspension thresholds for bottom sediments were examined at four regionally contrasting stations (all between 33-523 m depth) between September and October, 2003, using a new method for assessing sediment erosion. Values for critical erosion thresholds (u*) varied between 1.3-1.9 cm s^{-1}, with the deepest station in Amundsen Gulf (CA-18) showing the lowest erosion thresholds. Based on the limited number of cores available for erosion assessment, there appeared to be as much variation in erosion thresholds between cores from the same sites as those from different sites. Sediments in the Mackenzie Trough (CA10) and Amundsen Gulf yielded the highest erosion rates (22-54 g m^{-2} min^{-1}). They were dominated by sediments composed of silt and clay and were furthest from the Mackenzie River plume. Sediments in the Kugmallit Trough (CA718) contained more sand and required higher erosion thresholds before sediment resuspension, thereby resulting in overall lower erosion rates. Particle sizes, expressed as estimated spherical diameter (ESD), were measured by videography of resuspended sediment at different erosion thresholds. Particle sizes ranged from 100 µm (the visible limit of the camera and analysis system) to 930 µm for all stations. Amundsen Gulf contained particles in the smaller size classes, whereas the largest particles were recorded in the Kugmallit Trough (reflecting grain size gradients and SPM concentrations of the Mackenzie plume). There was no correlation in critical erosion thresholds or sediment erosion rates between

LEFT: Removing an ice block.
Photo: Alexandre Forest.

Table 7.1
Summary of water column environmental data (1 m) and physical characteristics of surface sediments (0-1 cm) in Kugmallit Bay during spring and summer. Values represent means (±SE, *n* = 3-1955).

	Water column (1 m)						Sediments (0-1 cm)					
Season	Temp. (°C)	Salinity (PSU)	Chl *a* (µg L^{-1})	SPM (mg L^{-1})	SPM OM (%)	Sedimentation rate (g m^{-2} d^{-1})	Mean density (g mL^{-1})	Mean porosity	OM (%)	TOC (%)	C:N ratio (µm)	Grain-size characteristics
Spring (May)	-0.431 (0.0001)	0.118 (0.00004)	2.213 (0.008)	8.514 (0.024)	1.17 (0.13)	51	1.99 (0.33)	0.48 (0.01)	2.87 (0.15)	1.15 (0.08)	22.12 (0.49)	Sandy silts or silts (D$_{50}$ = 25-50 µm)
Summer (July)[a]	12.360 (0.552)	0.100 (0.01)	202.23 (13.52)	47.98 (2.970)	10.86 (0.77)	5386	2.01 (0.21)	0.49 (0.01)	7.88 (0.05)	3.95 (0.13)	19.94 (0.39)	Sandy silts or silts (D$_{50}$ = 25-50 µm)

[a]Data from Walker et al. (unpublished data).

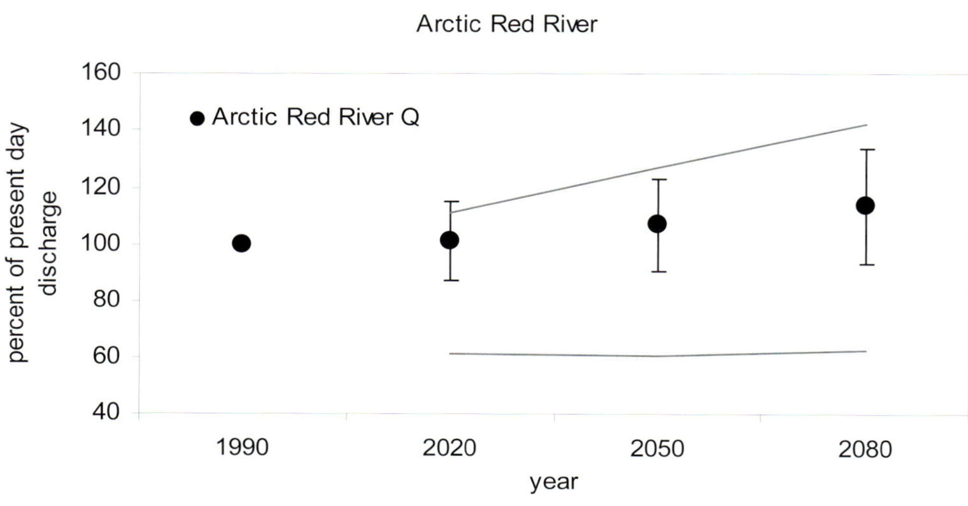

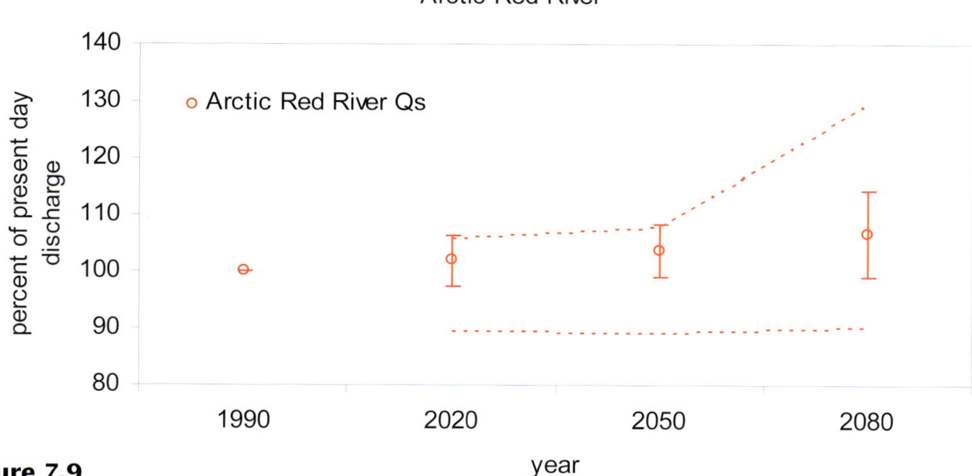

Figure 7.9

Discharge of a) water and b) sediment under future climate change scenarios. The symbols represent the mean and standard deviation of 20 different simulated scenarios, and the lines represent minima and maxima.

sediment chlorophyll *a* or chloroplastic pigment equivalents (CPE), often used as indicators of sediment stability (P >0.05) but chlorophyll *a*, was correlated with sediment OM and density (P = 0.007 and 0.032, respectively). Relative sediment stability in the Kugmallit Trough was most likely influenced by a combination of environmental variables, such as sediment grain size, OM content, density, porosity, and the current velocity of the Mackenzie River itself. The downward flux of low organic, low density sediment transported by the Mackenzie plume may have contributed to increased sediment resuspension rates in these deeper shelf stations.

7.2.3 Modelling water and sediment discharge from the Mackenzie River and sediment transport in the shallow Beaufort Sea

The immense size of the Mackenzie River system presents a challenge to modellers. Climate varies widely between the relatively temperate region of northern British Columbia and the high Arctic. Topography also varies greatly between the central Canadian plains and the Rocky Mountains. To add to these modelling difficulties, the Mackenzie is almost completely frozen for more than 6 months of the year (causing sediment-laden water to be released abruptly during breaches in ice dams) and thousands of lakes contained within the Mackenzie Delta act as sediment traps during ice free seasons. Some of these difficulties have been overcome (using a combination of software like Rivertools, Hydrotrend, Matlab, etc.) to arrive at realistic discharge estimates for the river under current and potential future climate scenarios (Haaf and Lintern, submitted; Lintern et al., 2006d). Further models (ST-Wave, Delft3D with SWAN, SedTrans) apply wind, wave and current forcing to resuspend plume-deposited material, and transport sediment along the shore (Jakes and Lintern, submitted; Lintern et al., 2006d).

Figure 7.9 shows a summary of potential future discharge scenarios. This is based on a culmination of over 20 SRES and IS92 predictions. The location represented is at Arctic Red River, where all of the sub-basins of the Mackenzie discharge through a single point in the Lower Mackenzie (except the Peel River which joins in at the delta). The figure shows the mean of all scenarios, the minimum and maximum of all scenarios, and their standard deviation. Figure 7.9a shows that the mean result produces a water discharge up to 114 % of present day discharge by the year 2080. Sediment discharge is less impacted and, according to the mean, is increased by 107 % of present day values. Minimum and maximum calculated discharges show much larger changes. Finally, the model reports an earlier melting of the Mackenzie River sub-basins, predicting that the peak melt for several of the sub-basins will occur in April-May, rather than May-June, in 2080.

Results from the Beaufort Sea Storm Modelling Study indicate that storm events may cause large resuspensions along the Beaufort Coast at the 5 m depth contour (Fig. 7.10). These events may be focused on the eastern side of Mackenzie Bay and the western side of Liverpool Bay. Substantial resuspensions during storm events may extend from the 1 m isobath out to the 15 m isobath along the Tuktoyaktuk Peninsula. The most significant dissipation of wave energy is predicted to occur along the Tuktoyaktuk Peninsula, particularly along its north east tip. Noticeable wave dissipation is also forecasted for Mackenzie Bay.

In order to better understand the potential impact of diminishing ice cover, a simulation similar to that of the 1982 storm was conducted wherein wind-generated waves were allowed to propagate across the computational boundary (in essence, extending the fetch). The time-dependant wave heights were based on the wave

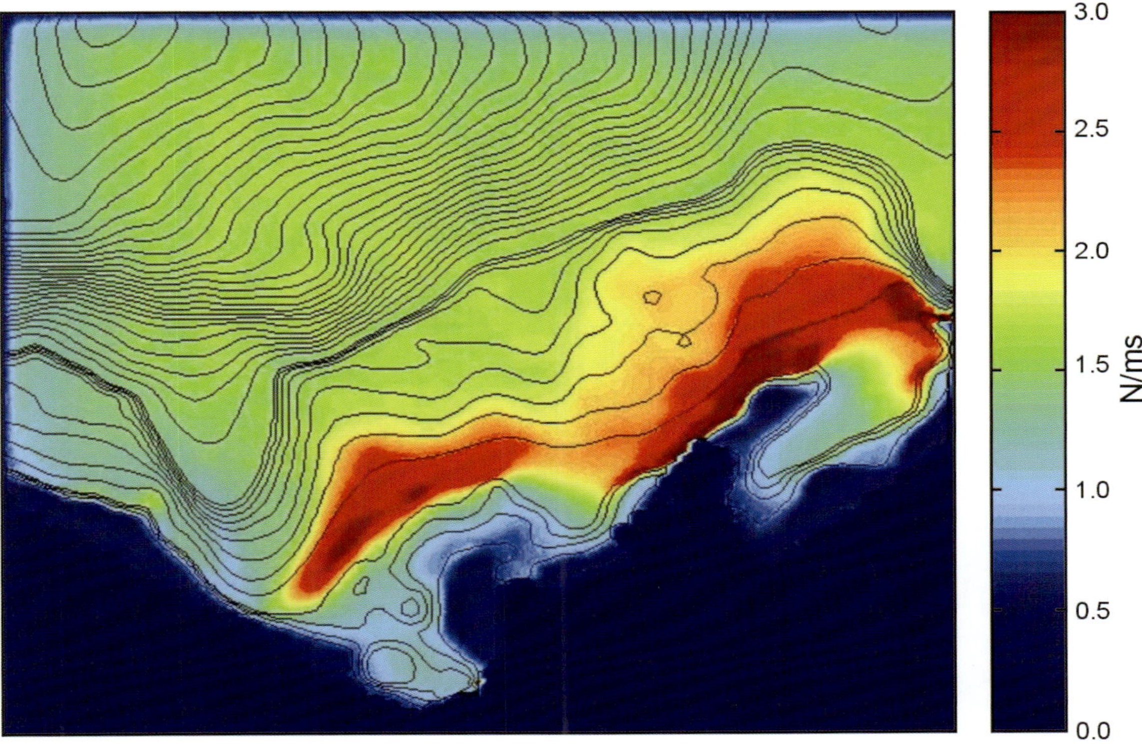

Figure 7.10

Modeled wave energy dissipation along the Beaufort coastline.

growth equations of the Shore Protection Manual from the US Corp of Engineers (1984). The wave heights and directions were based on hourly wind data from Tuktoyaktuk and the extended ice retreat was assumed to be 100 km.

Figure 7.11 shows significant wave heights throughout the storm. Here, the solid thin line represents measured wave heights, the thick dotted line represents calculated wave heights under actual ice conditions for the period in 1982, and the thick solid line is the

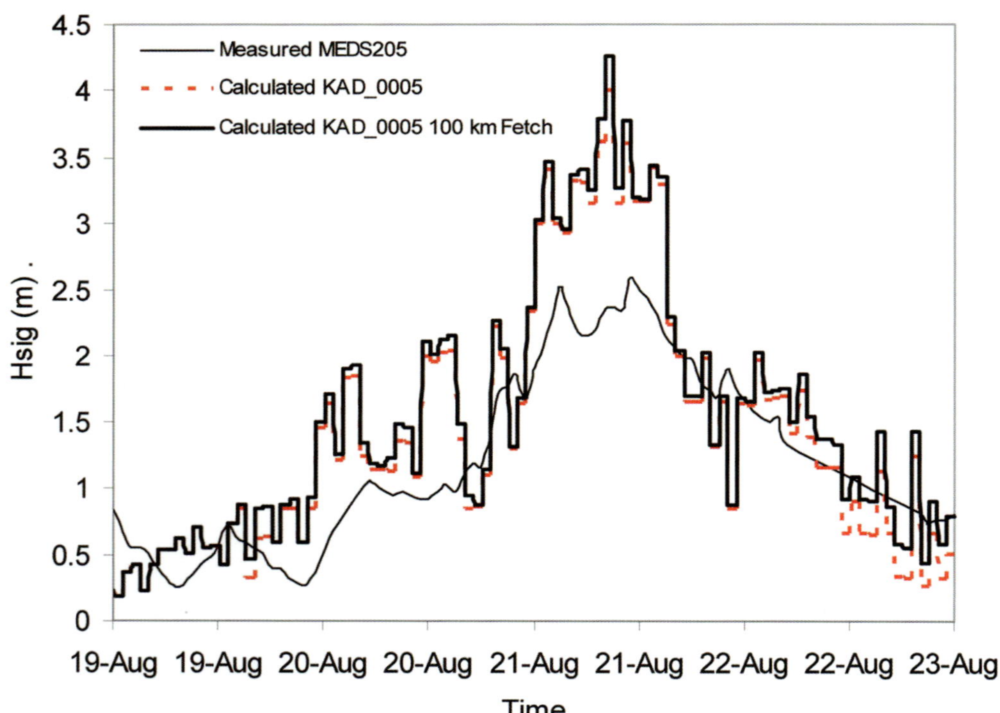

Figure 7.11

Measured and simulated wave heights throughout the storm period.

modelled wave heights with increased fetch. The build up and dissipation of the storm were well captured by the model, though the model generally led to higher wave heights. The results suggested that extending the wave fetch by 100 km might increase wave heights by as much as 20 cm near the shore.

7.2.4 Off-shelf transport of particulate matter during the fall and winter of 2003-2004

A major pathway to export material from the continental shelf to the deep ocean is the detachment of the benthic nepheloid layer (BNL) at the shelf break or at the upper slope (e.g. Cortesogno et al., 1998; Pickart,

2000; McPhee-Shaw, 2006). The consequent dispersal of bottom material across the margin produces an intermediate nepheloid layer (INL) that intrudes the upper water column along isopycnals (e.g. van Weering et al., 2001; Oliveira et al., 2002; Linder et al., 2004). The resuspended particulate matter transported within a BNL/INL can then settle down offshore (on the lower part of the slope or over a deeper area). The physical processes that generate the lateral transport of particles within the BNL/INL include coastal windstorms, convective flows, internal wave interactions with sloping topography, current surges, shelfbreak eddies, and bottom gravity currents (e.g. McPhee-Shaw and Kunze, 2002; McPhee-Shaw et al., 2004; Puig et al., 2004). Accordingly, the final burial location of particulate organic carbon (POC) carried within the horizontal particle plume can be displaced from the initial biological production area (e.g. Inthorn et al., 2006). Discrepancies between low primary production rates and high sediment fluxes on continental slopes are often explained by such mechanisms (e.g. Bonnin et al., 2002). In the Arctic Ocean, most of the primary production takes place on continental shelves (~85%; Stein & Macdonald, 2004). Moreover, Arctic shelves are largely influenced by terrigenous inputs from river discharge and coastal erosion (Rachold et al., 2004). Hence, knowing the vertical fluxes generated by BNL detachments and INL intrusions in Arctic seas is crucial to understand the ultimate fate of particulate organic matter and its implications to the pelago-benthic food web (Puig et al., 2001; Wegner et al., 2003; Witbaard et al., 2005).

In a recent paper, Forest et al. (2007) proposed that vertical POC fluxes on the slope of the Mackenzie Shelf (from October 2003 to April 2004) were fully accounted for by the offshore dispersal of shelf bottom particles. They therefore concluded that thermohaline convection was the major driving force behind sediment resuspension and shelf-slope transport during that period. Complementary to this study, we present here a summary of the spatial variability (over the fall/winter period of 2003-2004) of the vertical particulate matter flux associated with the off-shelf export of resuspended material via BNL/INLs (specifically, particle mass and POC content).

Vertical particle fluxes were measured from October 2003 to April 2004 by way of automated sequential sediment traps (Nichiyu Gyken Kogyo©) attached at ~200 m depth on CASES moorings CA-04, CA-07, CA-12, CA-15, CA-18, and CA-20 (Fig. 7.1). Before their deployment, each sediment trap was prepared following the JGOFS protocol (Knap et al., 1996). Trap sample cups were filled with filtered seawater (GFF 0.7 µm) adjusted to 35 PSU with NaCl. Formalin was added for preservation (5% v/v, sodium borate buffered). After retrieval, sample cups were checked for accurate salinity and set aside for 24 hours to allow particles to settle. Quantitative splitting was completed onboard using an automated McLane Wet Samples Divider©. Zooplankton swimmers collected in the traps were removed either with a 1 mm sieve or by hand-picking

under a stereomicroscope (see section 7.2.5 for more details). Sediment trap sub-samples were filtered in triplicate trough pre-weighed Whatman glass fiber filters (GFF 0.7 µm, 25 mm, combusted 4 h at 450 °C) in order to determine POC fluxes. Filters were dried for 12 hours at 60 °C and weighted again for dry weight (DW). After a 12-hour exposure to concentrated HCl fumes (to remove the inorganic carbon fraction), samples were analyzed with a Perkin Elmer CHNS 2600 Series II elemental analyzer.

Starting in October 2003, all recovered samples exhibited a vertical mass flux increase until a first maximum in November, and then a decline to a minimum in December (Fig. 7.12). Weaker particle fluxes persisted during winter at all stations. The vertical mass flux was lowest at CA-12 in the Canada Basin (Fig. 7.12c), and was particularly high at CA-20 at the outer Franklin Bay station (Fig. 7.12f). The percentage of POC in dry weight oscillated between 1.5 and 7.5% during fall and winter, and was generally lower at locations ≤ 300 m depth (CA-04 and CA-20; Figs. 7.12a and f). Cumulated particle mass and POC fluxes from October to April are presented in Table 7.2.

TABLE 7.2

Cumulated vertical fluxes of particle mass (dry weight) and POC recorded at 200 m depth from October 2003 to April 2004 in the CASES study area.

Mooring	**CA-04**	**CA-07**	**CA-12**	**CA-15**	**CA-18**	**CA-20**
Total mass flux [g DW m^{-2}]	50.1	19.5	7.5	9.4	23.9	107.5
Total POC flux [g C m^{-2}]	1.0	0.5	0.3	0.4	1.2	2.3
Percent of POC in mass flux [%]	2.0%	2.6%	4.2%	4.7%	4.9%	2.1%

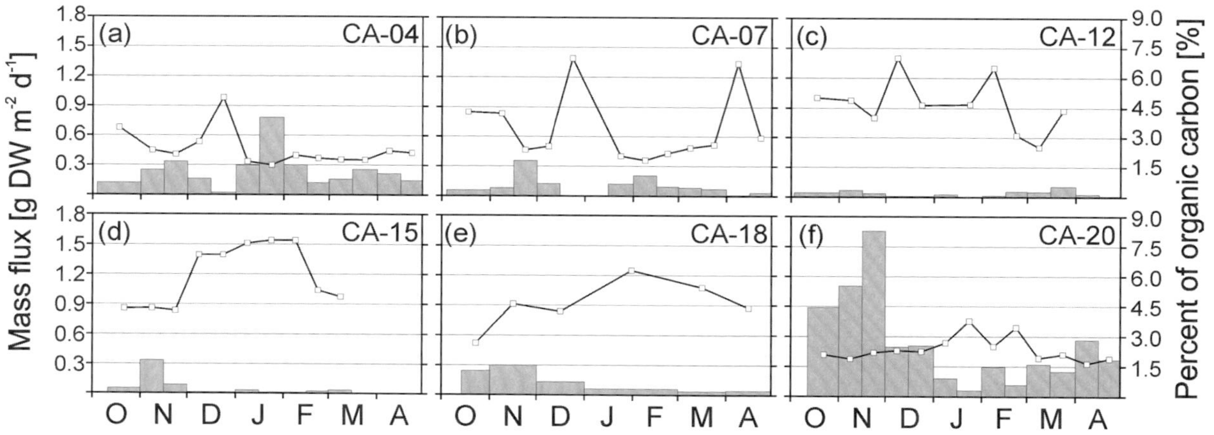

Figure 7.12

Magnitude of the vertical mass flux (dry weight) and organic carbon content (%) recorded at 200 m depth from October 2003 to April 2004. Locations of moorings are mapped on Figure 7.1.

During the Fall of 2003, two hydrographic cross-shelf transects were completed by the research icebreaker *CCGS Amundsen* on the Mackenzie Shelf (transects 700 and 400, on 21-23 October and 5-7 November, respectively). A caged rosette profiler carrying a CTD (Seabird SBE-911©) and a transmissometer (WETlabs©) was deployed at each station (at intervals of 10 km or less) along the sections. Validated data from all casts were averaged over 1-m bins. In both cross-shelf sections, the transmissivity signal indicated a well defined BNL extending vertically over the bottom of the Mackenzie Shelf (Fig. 7.13). The suspended particulate matter load was highest inshore. The BNL extended horizontally over the slope along isopycnals, forming a mid-depth INL beyond the shelf break. The suspended particulate load became less dense as the INL spread over the slope. Discrete and weaker INLs were also detected further off-shelf along the same isopycnals, suggesting that the production and/or offshore dispersal of particles at the shelf break were

discontinuous processes. In transect 400 (November 2003), the BNL advanced farther beyond the shelf break compared to transect 700 (October 2003), and a distinct INL originating from the upper slope was observed. This likely indicated that the particle load which was feeding BNL/INL development was increasing with the progression of Fall 2003. This BNL/INL event coincided with an increase in vertical particle fluxes which culminated at all mooring locations in November (Fig. 7.12). This maximum in vertical particle flux followed a fast onset of ice formation on the Mackenzie Shelf (Barber et al., pers. com.), and supported the hypothesis of sediment resuspension and shelf-slope transport by brine rejection and thermohaline convection. In addition, the increase in vertical particle flux during the fall coincided with coastal windstorms that likely combined with the convective mixing flow to resuspend and advect the BNL in November 2003 (Forest et al., 2007).

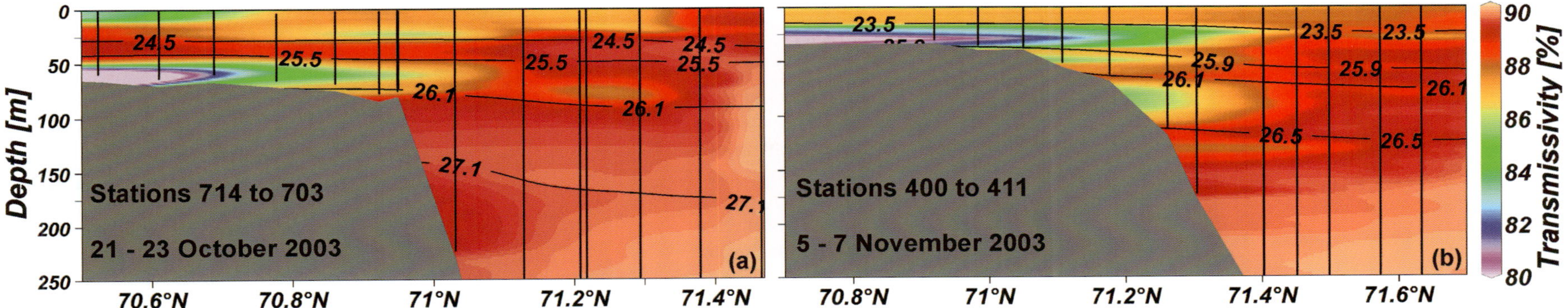

Figure 7.13

The distribution of transmissivity (a, b) along the cross-shelf sections conducted in Fall 2003 on the Mackenzie Shelf. The vertical lines represent the position of profiles along the sections. The overlaid isopycnals are labeled in σt units (kg m⁻³).

Due to logistical constraints, hydrographic cross-shelf sections could not be obtained during winter to confirm the persistence of BNL/INL processes beyond the fall. Nevertheless, the transmissivity time-depth series recorded at the over-winter site of *CCGS Amundsen* (December 2003 to May 2004) in Franklin Bay (70° 2.73′ N, 126° 18.1′ W) indicated the presence of strong turbidity patches which may have been linked to BNL detachments (see Fig. 6b in Forest et al., 2008). This implies that the discontinuous formation of particle-rich INLs along the Mackenzie Shelf might be related to the persistent vertical particle fluxes recorded at the mooring stations during the winter. Mechanisms that could generate such INLs during wintertime include convective mixing caused by ice growth (thermohaline convection), and turbulent diffusion caused by Pacific-derived shelf break eddies (Forest et al., 2007; Forest et al., 2008).

7.2.5 Zooplankton swimmers collected in sediment traps on the slope of the Mackenzie Shelf

Zooplankton swimmers are defined as living organisms that enter sediment traps actively rather than through passive sinking (e.g. Ota et al., in press; Sampei et al, in prep.). These organisms are generally removed from trap samples before vertical flux estimates are made. Taxonomic studies made on the sediment trap collected-zooplankton (TCZ) allow us to track the evolution of the zooplankton community over a long period. Here, for example, we investigated the composition, abundance and the seasonal variation of mesozooplankton on the slope of the Mackenzie Shelf during 2003-2004.

Two moorings, equipped with time-series sediment traps, were deployed respectively on the 300- and 500 m isobaths of the slope of the Mackenzie Shelf from October 2003 to August 2004 (site CA-04 at 71° 05.16′ N, 133° 43.39′ W and site CA-07 at 71° 8.99′ N, 133° 53.88′ W, respectively). The conical sediment traps (Nichiyu Gyken Kogyo©) were placed at ~200 m depth on each mooring line. TCZ of sizes > 1 mm were collected from the sample cups before splitting. TCZ of sizes <1 mm were isolated in the laboratory from non-living particles using a 1-mm sieve and by hand sorting under a stereomicroscope. The abundance of TCZ, in terms of flux (individuals m⁻² d⁻¹), was highest in autumn (October-November) at CA-04 and CA-07, and decreased in mid-winter (January-February; Figs. 7.14 & 7.15). The TCZ flux increased again in late

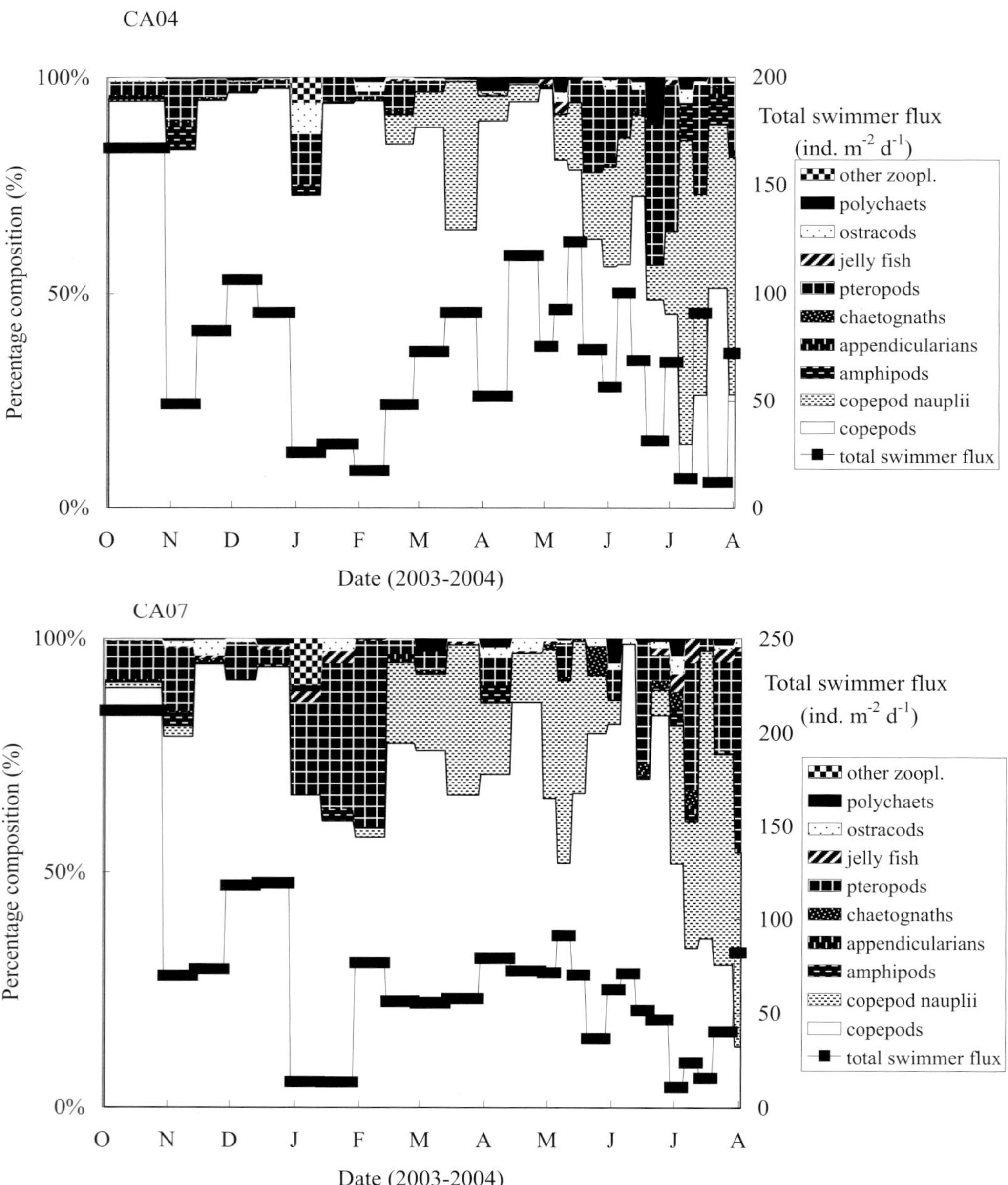

winter-spring (March-June) at both sites. The taxa of TCZ were typical of communities found in Arctic Seas and were primarily dominated by large copepods, such as *Calanus hyperboreus*, *C. glacialis* and *Metridia longa*, *Gaetanus tenuispinus*, *Heterorhabdus norvegicus*, *Aetideopsis spp.*. Among these, *M. longa* was dominant at both sites, irrespective of season (17-84% of large calanoid copepods at CA04; 20-86% at CA07). Fluxes of *C. hyperboreus* (AF: adult female) varied with season, and their relative abundance was high in autumn-late winter. The highest contribution relative to the total flux was 21% at CA-04 and 31% at CA-07 in October. Less marked seasonality was found for *C. glacialis*, and the contribution to total calanoid copepods was highest at CA04 in April (29%) and at CA07 in March (50%). The TCZ increase in autumn was probably due to entrapments of descending migrants such as AF *C. hyperboreus*. Small calanoids, cyclopoids (*Oithona similis*, *Oncaea borealis*, *Cyclopina sp.*) and copepod nauplii were also abundant in the traps. Copepod nauplii dominated in late winter and derived mostly from *C. hyperboreus*. In addition to copepods, many pteropods (*Limacina sp.*) were present in winter (CA-07) and spring-summer (CA-04). Pteropod fluxes observed in this study were less than those reported at the same locations in 1987-1988 (Forbes et al., 1992).

Figure 7.14

Seasonal variability of zooplankton swimmer flux in terms of absolute abundance and relative composition at CA-04 in 2003-2004.

Figure 7.15

Seasonal variability of zooplankton swimmer flux in terms of absolute abundance and relative composition at CA-07 in 2003-2004.

7.2.6 The origin and remineralization of organic matter in sediments of the Mackenzie Shelf and Amundsen Gulf

The Mackenzie River is the principal source of sediment to the Arctic Ocean (Macdonald et al., 1998). Some of the sediment carried in suspension is exported directly to the deep ocean; the remainder settles out on the Beaufort Sea shelf, including the Mackenzie Delta (Hill et al., 1991). Much of this latter portion of sediment is reworked, resuspended, and ultimately exported under the action of ice scour and tidal currents but a fraction remains buried and is subjected to *diagenesis*. Primary production in the Beaufort Sea adds reactive organic matter (OM) of marine origin to the less reactive terrestrial OM carried by Mackenzie River sediment. Both terrestrial and marine OM can fuel diagenesis, although to different degrees because of their contrasting reactivities. Using the stable isotope composition of organic carbon and nitrogen (C_{org} and N_{org}), we determined the relative contribution of terrestrial and marine OM to the sediments of the Mackenzie Shelf and the adjacent Amundsen Gulf. Pathways by which OM is remineralized/oxidized within the sediments were examined through the distribution of oxidants which are consumed during OM remineralisation.

Undisturbed sediment cores were recovered from the Mackenzie Shelf/Slope and Amundsen Gulf using a 0.12 m² Ocean Instruments Mark II box corer. The 30 to 50 cm long cores were divided into horizontal layers (0.5 cm at the surface, and up to 5 cm at depth). They were then placed in a glove-box purged by a continuous flow of nitrogen in order to minimize the oxidation of sediment components (Edenborn et al., 1986). Porewaters were extracted using Reeburgh-type squeezers (Reeburgh, 1967). Solid samples were then transferred to pre-weighed scintillation vials and stored at -20 °C on-board.

Upon return to McGill University, the vials were weighed, freeze-dried and re-weighed in order to determine their sediment water content and porosity. The freeze-dried sediments were homogenized by grinding for subsequent solid phase analysis. The total carbon (TC) and nitrogen (TN) contents were determined with an elemental analyzer. The total inorganic carbon (TIC) content was measured independently by coulometry (following the acidification of the samples and CO_2 extraction). The Total Organic Carbon (TOC) content was obtained by subtracting TIC from TC. The easily reducible Fe and Mn oxide content of the sediment was extracted using a buffered ascorbate reagent according to the method described by Kostka and Luther (1994). Porewater NO_3^- concentrations were measured on board using standard colorimetric methods (Grasshoff, 1999) adapted for use on a Bran+Luebbe® Auto-Analyzer 3.

7.2.6.1 Organic matter origin

The source of the sedimentary organic matter can be determined from its stable isotopic composition (e.g. Hayes, 1993; Meyers, 1994; Meyers, 1997). Assuming that bulk sedimentary organic matter is a mixture of allochtonous/terrigenous and autochtonous/marine organic matter with distinct $\delta^{13}C_{org}$ and $\delta^{15}N_{org}$ signatures, the relative contribution of each end-member component can be estimated. $\delta^{13}C_{org}$ values typically fall between -22 and -20‰ for autochtonous/marine organic matter (Meyers, 1994) and -27 ‰ for allochtonous/terrigenous C3 land plants (most of the Arctic vegetation) (Bickert, 2000). The $\delta^{15}N_{org}$ values for terrigenous organic matter are highly variable and range from -5 to +18‰ (with an average of +3‰). For C3 plants, the $\delta^{15}N_{org}$ is similar to the atmospheric nitrogen signature, with values close to 0.4‰ (Peterson and Howarth, 1987). The $\delta^{15}N_{org}$ of sedimentary marine organic matter usually falls between +7 and +10‰ (Meyers, 1997; Muller and Voss, 1999).

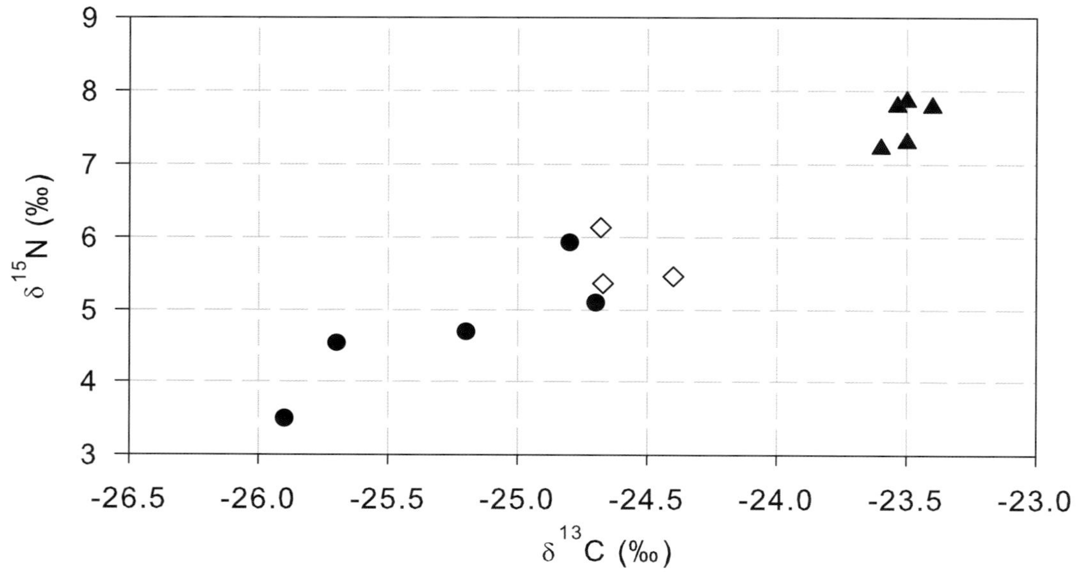

Figure 7.16

Stable isotope composition of the organic matter in surface sediments (0 - 0.5 cm) of the Mackenzie Shelf (●), slope of the Mackenzie Shelf (◇) and Amundsen Gulf (▲).

Our results showed that, in the Mackenzie Shelf/Amundsen Gulf area, the $\delta^{13}C_{org}$ signature of organic matter in surface sediments ranged from -25.9 to -23.4‰, and that $\delta^{15}N_{org}$ ranged from 2.54 to 7.86‰ (Fig. 7.16). These results were consistent with values reported in previous work (Naidu et al., 2000). Mean values in Amundsen Gulf for $\delta^{13}C_{org}$ and $\delta^{15}N_{org}$ (-23.5‰ and 7.6‰, respectively) suggested that the organic matter accumulating in these sediments was predominantly marine. In contrast, $\delta^{13}C_{org}$ and $\delta^{15}N_{org}$ values in Mackenzie Shelf sediments were generally lower (i.e. lighter). The lowest of these values was observed at the mouth of the Mackenzie River (Station 912) and indicated a predominantly land-derived OM. The terrigenous end-member for the Beaufort Sea has been repeatedly observed to be around -27‰ (e.g. Schell, 1983; Ruttenberg and Goni, 1997; Naidu et al., 2000). Our results for Station 912 were consistent

with this (Fig. 7.16). The marine end-member was not as well constrained. In the Chukchi Sea, Naidu et al. (1993) found a $\delta^{13}C_{org}$ value of -21‰ for the marine end-member, a result which was significantly different from the marine end-member in the Beaufort Sea (-24‰; Naidu et al., 2000). The $\delta^{13}C_{org}$ values in the Amundsen Gulf were all between -23.6‰ and -23.4‰ (Fig. 7.16), showing that, in general, the marine end-member for this area was slightly heavier than that of the Beaufort Sea.

7.2.6.2 Organic carbon remineralization

Although the sources of organic matter to the sediments of Amundsen Gulf and the Mackenzie Shelf are clearly different, the concentration of total organic carbon (TOC) in their surface sediments is quite similar (averaging 1.26% and 1.38%, respectively). In Mackenzie Shelf sediments, TOC concentrations varied little throughout the cores, suggesting that little TOC was lost by oxidation after burial (Fig. 7.17). In contrast, loss of organic carbon by oxidation appeared to be significant in Amundsen Gulf sediments (indicated by a large decrease in TOC over the top 30-40 cm of the cores). This difference could have been partly due to different sedimentation rates in the two environments. Acoustic measurements revealed that only a very thin layer of sediment has accumulated in Amundsen Gulf since the last deglaciation (Steve Blasco, pers. comm.), suggesting very low sedimentation rates (low sedimentation rates allow more time for carbon oxidation). This was confirmed by our own observations on a box core recovered from the inner part of Amundsen Gulf: It contained fine-grained brown-coloured mud down to about 40 cm, below which the sediment consisted of coarse pink gravel indicative of a glacial deposit. The observation implied a local sediment accumulation rate of ~0.04 mm yr^{-1}, a value several orders of magnitude less than those observed on the Mackenzie Shelf.

The microbial oxidation of organic matter uses available electron acceptors according to a sequence determined by the free energy yield of the participatory reactions (Froelich et al., 1979). Thus oxygen is used first, followed by nitrate, and the oxides of manganese and iron. The distribution of solid phase metals and metabolic by-products (NO_3^-, Mn(II) and Fe(II)) in the porewaters recoverd from two stations (one in Amundsen Gulf, the other in the Mackenzie Shelf region) are presented in Figure 7.17. In the shelf sediments, we observed: rapid depletion of NO_3^-; low abundance of reactive manganese; dissolved Mn(II) near the water/sediment interface; and depletion of reactive Fe immediately below the water/sediment interface. Such characteristics revealed that the demand for oxidants (electron acceptors) was high in these sediments. No significant decrease in organic carbon was observed with depth in these cores, which suggested that the reactive carbon driving diagenesis was consumed near the sediment-water interface well before it could be buried. Accordingly, the vertical profiles of organic carbon reflected background levels of low-reactivity refractory carbon (likely terrestrial in view of its isotopic signature). The early diagenetic reactions which involved NO_3^- and metal oxide reduction would have also taken place near the water/sediment interface. This would have allowed dissolved metals to escape the sediment and accounted for the weak accumulation of authigenic Mn-oxides observed in the shelf sediment. In contrast, low sedimentation rates in Amundsen Gulf allowed the diffusion of dissolved O_2 and NO_3^- deeper within the sediment and reactive Mn and Fe to be cycled and trapped within the sediment column.

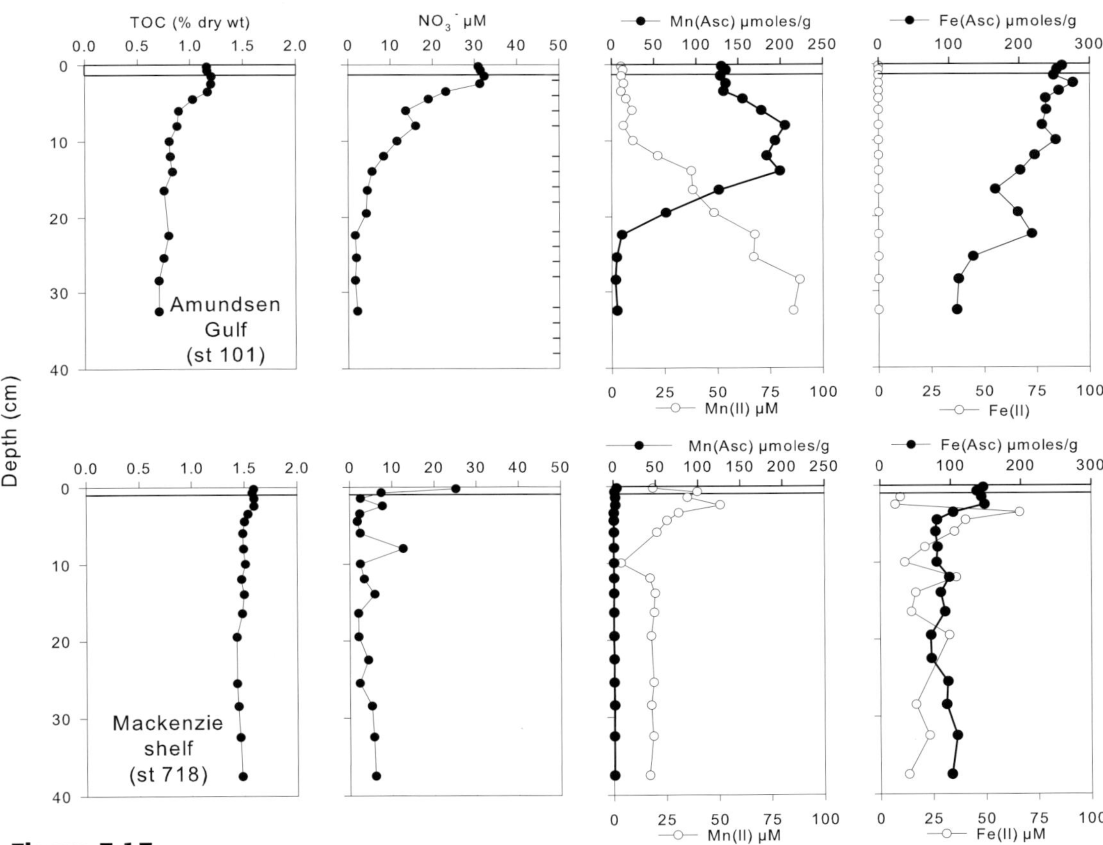

Figure 7.17

Profiles for typical sediment cores recovered on the Mackenzie Shelf and in Amundsen Gulf. The horizontal lines at 1.3 cm (101) and 1.0 cm (718) represent the depth at which dissolved O_2 became undetectable. Note the difference in Mn(Asc) contents.

7.3 Implication of this Work

The southern Beaufort Sea and Mackenzie Shelf region appears to have been a net sink for atmospheric CO_2 during the open water periods of the CASES expedition. Preliminary data also indicated a net CO_2 drawdown from the atmosphere during the ice-covered season, implying that this area could be very important in regulating rising atmospheric CO_2 levels (Papakyriakou et al., 2006). Unfortunately, we don't know if these conclusions are representative of long-term climatological conditions in the Beaufort Sea. In fact, historical data collected by the Institute of Ocean Sciences have indicated that the inorganic carbon system in the southern Beaufort Sea can vary quite dramatically from year to year (Fransson et al., 2006).

From our single year of sampling (2003-2004), it is difficult to determine how robust the observed CO_2 sink is, particularly since patches of high CO_2 waters have often been observed near the surface. Although this complicates predictions regarding climate change feedbacks, we can speculate about possible future scenarios. A cold, well-mixed ocean which is not covered in ice could be an even greater CO_2 sink than what has been observed in the southern Beaufort Sea (Miller et al., 1999); though, as noted above, increased stratification due to increased river outflow could severely inhibit that sink. In contrast to other Arctic coastal areas, such as the Northeast Water and North Water polynyas (Yager et al., 1995; Miller et al., 2002), the surface waters observed under the ice during CASES remained undersaturated throughout winter and into spring. This implies limited winter respiratory activity, and that a loss in ice cover would not result in dramatic wintertime outgassing. Therefore, we might predict that the observed sink could be quite robust but, at this time, it is still impossible to predict how the total effect of climate change would impact the balance between winter respiration and summer primary production.

Determining relationships between particle characteristics, salinity, wave/current action and settling rates is an essential step in formulating predictive models for suspended sediment dynamics and budgets on the Mackenzie Shelf. The interaction between marine and estuarine conditions can be a useful analog for potential climate change scenarios of intensified sediment transport, coastal erosion and reduced snowmelt (Leont'yev, 2003). Therefore, obtaining measurements of suspended sediment flux on the Mackenzie Shelf is critical to understand the interactions between physical/biogenic carbon cycling across the shelf and variations in river discharge, storms and ice cover (Macdonald, 2000).

We report subtle changes in biological and oceanographic variables beneath the land fast ice in Kugmallit Bay, which suggests the onset of a spring melt occurring hundreds of km further south in the Mackenzie Valley. Estimates of sediment erosion rates in the Canadian Beaufort Sea (using the BEAST) helped us further understand the resuspension and transport of benthic organic material associated with sediments and aggregated particles. The rates of sediment erosion and resuspension, and the timing of these biological and oceanographic events are important if we are to discern how sediment and carbon transport to the Arctic Ocean impact primary production. Accordingly, the large amount of allochthonous and autochthonous material supplied by the Mackenzie River under seasonal ice cover has significant consequences for carbon and sediment sequestration/export (Macdonald et al., 1998). Moreover, climate change in the Arctic may translate into the earlier onset of spring blooms (and subsequently, increased primary production), which would critically impact the total carbon budget and trophic dynamics of the Mackenzie Shelf (O'Brien et al., 2006).

The re-design and calibration of BEAST by Walker et al. (submitted) has proven to be useful to quantify sediment erosion on the Beaufort Sea Shelf. Sediment integrity was maintained when using push cores from box cores, which allowed erosion assessments to be made between stations. The diversity of information obtained using BEAST (critical threshold, erosion rate, particle size distribution) provided much of the information required for assessing the role of resuspension in particle cycling and benthic-pelagic coupling in Arctic shelf environments.

Modelling of the Mackenzie River indicated that both water and sediment discharges are likely to increase under future climate scenarios. The peak flood season will likely occur earlier in the year. The implications of this are too many to list. It is unknown to what extent increased river discharge would affect salinity and water temperature in the Mackenzie Delta. Both parameters would impact the formation of sea ice; in fact, decreased salinity might promote the formation of sea ice while increased temperature might demote it.

Hydrodynamic modelling of the Beaufort Sea indicated that strong wave energy dissipation occurred along the Tuktoyaktuk Peninula and Mackenzie Bay, suggesting that these areas are subject to strong coastal retreat. Under a scenario of reduced ice cover, the model predicted wave heights up to 20 cm higher than those observed throughout a typical storm. Such climate-induced changes (combined with storm surge and higher sea levels) would alter geomorphologic processes taking place at the coast (such as coastal retreat) and have dramatic impact on coastal communities.

Vertical fluxes of particulate matter in the southeastern Beaufort Sea during Fall/Winter 2003-2004 originated from bottom resuspension on the Mackenzie Shelf as well as lateral transport within BNL/INLs across the shelf edge. This created a displacement of buried POC and subsequently contributed to the transfer of carbon from the continental shelf to the deep sea. Nevertheless, our knowledge on the residence time of the refractory POC transported into the Arctic basins remains uncertain and prevents any conclusion on its ultimate fate (e.g. trophic transfer, remineralization, burial).

The observed vertical particle mass flux generated by the BNL/INL events during Fall/Winter 2003-2004 was highest at the outer Franklin Bay site (CA-20; ~0.1 kg m^{-2}). This indicated that Franklin Bay might be a depot-center of particulate matter in the southeastern Beaufort Sea due to an interplay between its coastal topography and local environmental conditions (such as winds and current dynamics; Forest et al., 2008).

Vertical particle fluxes induced by BNL/INLs processes helped feed the pelago-benthic food web throughout 2003-2004 despite limited primary production during the darkness season. This suggested that the classical view by which photosynthesis forms the base of the food web might not apply to the CASES study area; in fact, resuspension and advection processes might play key roles in providing organic matter to all types of living organisms. For example, the enhanced particle flux measured in fall/winter at the outer Franklin Bay site may perhaps explain persistent benthic abundance and diversity along the eastern slope of the Mackenzie Shelf.

The mixing and downwelling sustained by thermohaline convection during ice formation was identified as

The bow of the CCGS Amundsen frozen in the ice. Photo: Bjorn Sundby.

Deploying a sediment trap.
Photo: Alexandre Forest.

a major driving force for shelf sediment resuspension and shelf-basin material flux. Most climate models anticipate that Arctic shelves will become free of ice during the summer months before the end of the century. This could mean that a larger volume of ice would need to be formed in the fall and winter in order to isolate the surface ocean from the atmosphere. Accordingly, increased ice formation would intensify the resuspension of shelf sediment and its transport to deeper areas in fall/winter.

Trap collected zooplankton (TCZ) swimmers provide valuable information on the composition and seasonal behaviour of zooplankton communities, particularly when no direct sampling via plankton nets is possible. The ability to discriminate between passively sinking (dead) and active TCZ swimmers, however, remains a challenge. This shortcoming may greatly impede the accurate assessment of particulate carbon export to the deep through sediment traps.

Diagenesis in Amundsen Gulf sediment is not driven by sedimentation of terrestrial organic matter but by the settling of marine OM produced locally by primary producers. Presently, production is strongly seasonal, and organic matter is delivered to the sediment as sharp pulses during the short ice-free season. If global warming leads to extended ice-free periods, primary production may potentially deliver organic matter to the sediment over much of the summer (when light is not limiting). This is likely to change the diagenetic regime and affect the exchange of solutes across the sediment-water interface.

7.4 Recommendations

At the very least, a comprehensive survey of the mixed layer pCO_2 should be carried out over the area covered by the CASES 2003-2004 study to ascertain the annual variability of CO_2 fluxes across the air-sea interface. Preferably, a similar study should be undertaken over several years so fluxes can be properly correlated to surface productivity, climatology, river discharge, and the extent of ice cover.

Measurements of sediment transport, sedimentation, resuspension and erosion rates provide us with a useful baseline with which we can monitor future shifts in the balance of sediment and carbon transport in the Arctic. With predictions of Arctic climate change becoming more pronounced in the coming decades, we forecast that there will be greater sediment transport to Arctic shelf sediments. However, without continuous monitoring of the impacts of increased sediment load on the benthic pelagic coupling, it will be difficult to predict the socio-economic costs for local communities or the global population.

With increased sediment transport, permafrost collapse, and storm events in the Mackenzie Basin, there will be an increase in the transport of nutrients and the release of chemically bound contaminants. All of these will have serious negative impacts on the marine Arctic ecosystem.

Coastal erosion and the erosion surrounding the Mackenzie River itself will produce problems for infrastructure and the future economic development of the area—particularly if development of the oil and gas sector expands in the region.

Computer models are cost-effective tools to understand river and coastal processes. Haaf and Lintern (submitted) described the deficiencies of the current Mackenzie River model. They recommend that work continue to improve the accuracy of predictions for the Mackenzie sub-basins. Models focused on coastal storms (Jakes and Lintern, submitted) have clearly indicated that wave heights will likely increase with decreasing sea ice cover due to increased fetch. The next objective for coastal modeling will be to understand how a lengthier ice-free season will impact the annual cycle of sediment resuspension and transport, and coastal retreat.

Our assessment of the role of BNL/INLs transport on vertical particle fluxes calls for inter-annual sediment trap studies to confirm the persistence of these mechanisms in the Beaufort Sea. A pan-Arctic shelf comparison would be necessary to take into account different input sources of particulate matter to BNL/INLs processes. Future research programs should include long-term samplings (for particle flux determination) and hydrographic cross-shelf surveys along the edge of the Arctic margins.

Linking the transport of resuspended matter within BNL/INLs to the pelago-benthic food web is essential to understand the sources and pathways of organic carbon in the Arctic Ocean. Modelling of the structure and function of the Arctic marine food web will only be possible if such links are elucidated.

The connection between thermohaline convection and shelf-basin transport requires more in-depth study. This could be achieved through the deployment of automatic CTD profilers equipped with turbidity probes on strategically located moorings, and/or by direct ice-breaker-based measurements over the wintertime.

The systematic sampling of TCZ swimmers should be standardized in sediment trap studies in order to allow adequate and comparable taxonomic investigations. A simple methodology should also be elaborated to discriminate between passive and active TCZs.

The nearly-complete absence of terrestrial organic matter in Amundsen Gulf sediments suggests that sediment-laden water from the Mackenzie River does not reach into the Gulf (in spite its proximity). It is unclear at present why this is so. It would be useful to focus future studies on factors that control water circulation in Amundsen Gulf, particularly forces which prevent the intrusion of suspended sediment from the Mackenzie River to the Amundsen Gulf.

Storm at the ice camp in Franklin Bay. Photo: Alexandre Forest.

References

ACIA (Arctic Impact Assessment), 2004. Impacts of a warming Arctic. Cambridge University Press, New York.

Arrigo, K.R. and G.L van Dijken, 2004. Annual cycles of sea ice and phytoplankton in Cape Bathurst polynya, southeastern Beaufort Sea, Canadian Arctic. Geophys. Res. Lett. 31: L08304, doi:10.1029/2003GL018978.

Barber, D.G. and J. Hanesiak, 2004. Meteorological forcing of sea ice concentrations in the southern Beaufort Sea over the period 1979-2000. J. Geophys. Res. 109: C06014, doi:10.1029/2003JC002027.

Bickert, T., 2000. Influence of geochemical processes on stable isotope distribution in marine sediments, p. 309-333, In H. D. Schulz and M. Zabel (eds.), Marine Geochemistry. Springer.

Bonnin, J., W. Van Raaphorst, G. J. Brummer, H.van Haren, and H. Malschaert, 2002. Intense mid-slope resuspension of particulate matter in the Faeroe-Shetland Channel: short-term deployment of near-bottom sediment traps. Deep-sea Res. I49:1485-1505.

Carmack, E.C. and R.W. Macdonald, 2002. Oceanography of the Canadian Shelf of the Beaufort Sea: A setting for marine life. Arctic 55: 29-45.

Carmack, E.C., R.W. Macdonald and S.E. Jasper, 2004. Phytoplankton productivity on Canadian shelf of the Beaufort Sea. Mar. Ecol. Prog. Ser. 277:37-50.

Carson, M.A., J.N. Jasper and F.M. Conly, 1998. Magnitude and sources of sediment input to the Mackenzie Delta, Northwest Territories, 1974-94. Arctic 51: 116-124.

Comiso, J.C., 2003. Warming trends in the Arctic from clear sky satellite observations. J. Clim. 16: 3498-3510.

Puig, P. and A. Palanques, 1998. Nepheloid structure and hydrographic control on the Barcelona continental margin, northwestern Mediterranean. Mar. Geol. 149:39-54.

Droppo, I.G., D. Jeffries, C. Jaskot, and S. Backus, 1998. The prevalence of freshwater flocculation in cold regions: a case study from the Mackenzie River Delta, Northwest Territories, Canada. Arctic 51: 155-164.

Edenborn, H. M., A. Mucci, N. Belzile, J. Lebel, N. Silverberg, and B. Sundby, 1986. A glove box for the fine-scale subsampling of sediment box cores. Sedimentology 33: 147-150.

Forbes, J.R., R.W. Macdonald, E.C. Carmack, K. Iseki, and M.O. O'Brien, 1992. Zooplankton retained in sequential sediment traps along the Beaufort Sea shelf break during winter. Can. J. Fish. Aquat. Sci. 49:663-670.

Forest, A., M. Sampei, H. Hattori, R. Makabe, H. Sasaki, M. Fukuchi, P.Wassmann, and L. Fortier, 2007. Particulate organic carbon fluxes on the slope of the Mackenzie Shelf (Beaufort Sea): Physical and biological forcing of shelf-basin exchanges. J. Mar. Syst. doi:10.1016/j.jmarsys.2006.10.008.

Forest, A., M. Sampei, R.Makabe, H.Sasaki, D. Barber, Y. Gratton, P. Wassmann, and L. Fortier 2008. The annual cycle of particulate organic carbon export in Franklin Bay (Canadian Arctic): Environmental control and food web implications J. Geophys. Res. C. Oceans. 113:C03S05. DOI: 10.1029/2007JC004262.

Fransson, A., M. Chierici, L.A. Miller, and R. W. Macdonald, 2006. The Carbon Dioxide System in the Surface Water of Arctic Shelf Seas: A Comparison between the Beaufort Sea and the Barents Sea. 2006 Ocean Sciences Meeting, Honolulu, 20-24 February, 2006.

Froelich, P., G.P. Klinkhammer, M.L. Bender, N.A. Leudtke, G.R. Heath, D. Cullen, P. Dauphin, D. Hammond, B. Hartman, and V. Maynard, 1979. Early oxidation of organic matter in pelagic sediments of the eastern equatorial Atlantic: suboxic diagenesis. Geochim. Cosmochim. Acta 43: 1075-1090.

Goñi, M.A., M.B. Yunker, R.W. Macdonald, and T.I. Eglinton, 2000. Distribution and sources of organic biomarkers in arctic sediments from the Mackenzie River and Beaufort Shelf. Mar. Chem. 71: 23-51.

Gosselin, M., M. Levasseur, P.A. Wheeler, R.A Horner, and B.C. Booth, 1997. New measurements of phytoplankton and ice algal production in the Arctic Ocean. Deep-Sea Res. Part II, 44: 1623-1644.

Grasshoff, K., M. Ehrhardt, K. Kremling, and L.G. Anderson, 1999. Methods of Seawater Analysis, 3rd, completely rev. and extended ed. Wiley-VCH.

Guo, L., I. Semiletov, Ö Gustafsson., J. Ingri, P. Andersson, O. Dudarev, and D. White, 2004. Characterization of Siberian Arctic coastal sediments: Implications for terrestrial organic carbon export. Global Biogeochem. Cycle 18: GB1036, doi:10.1029/2003GB002087.

Haaf, J. and D.G. Lintern. Modeling the Mackenzie River Basin under current conditions and climate change scenarios. GSC Open File 5531, 101 p. Submitted.

Hayes, J. M., 1993. Factors controlling C-[13] contents of sedimentary organic-compounds - principles and evidence. Mar, Geol. 113: 111-125.

Hill, P.R., S.M. Blasco, J.R. Harper, and D.B. Fissel, 1991. Sedimentation on the Canadian Beaufort Shelf. Cont. Shelf Res. 11: 821-842.

Hill, P.R., C.P. Lewis, S. Desmarais, V. Kauppaymuthoo, and H. Rais, 2001 . The Mackenzie Delta: sedimentary processes and facies of a high-latitude, fine-grained delta. Sedimentology 48(5):1047-1078.

Holland, M.M., C.M. Bitz, and B. Tremblay, 2006 . Future abrupt reductions in the summer Arctic sea ice. Geophys. Res. Lett. 33: L23503, doi:10.1029/2006GL028024.

Inthorn, M., V. Mohrholz, and M. Zabel, 2006 . Nepheloid layer distribution in the Benguela upwelling area offshore Namibia. Deep-Sea Res. 53:1423-1438.

IPCC (Intergovernmental Panel on Climate Change), 2001. Climate Change: The IPCC Assessment, eds: J.T. Houghton, G.J. Jenkens, and J.J. Ephraums, Cambridge University Press, New York.

Jakes, H. and D.G. Lintern. Evaluation of models to predict storm resuspension in the coastal Beaufort Sea, Canada. Geol. Survey Can., Open File 5509, 102 p. Submitted .

Johannessen, O.M., E. Bjørgo, and M.W. Miles, 1996. Global warming and the Arctic. Science. 271:129-129.

Johnson, K.M., K.D. Wills, D.B. Butler, W.K. Johnson, and C.S. Wong, 1993. Coulometric total carbon dioxide analysis for marine studies: maximizing the performance of an automated gas extraction system and coulometric detector. Mar. Chem. 44: 167-187.

Knap, A., A. Michaels, A. Close, H. Ducklow, and A. Dickson, 1996. Protocols for the Joint Global Ocean Flux Study JGOFS Core Measurements. JGOFS Report Nr 19:155–162.

Kostka, J. and G.Luther, 1994. Partitioning and speciation of solid phase iron in saltmarsh sediments. Geochim. Cosmochim. Acta 58: 1701-1710.

Laxon, S., N. Peacock, and D. Smith, 2003. High interannual variability of sea ice thickness in the Arctic region. Nature 425: 947-949.

Leont'yev, I.O., 2003. Modeling erosion of sedimentary coasts in the western Russian Arctic. Coastal Engin. 47(4): 413-429.

Linder, C. A., G.G. Gawarkiewicz, and R.S. Pickart, 2004. Seasonal characteristics of bottom boundary layer detachment at the shelfbreak front in the Middle Atlantic Bight. J. Geophys. Res. (C Oceans) 109, C03049, doi:10.1029/2003JC002032.

Lintern, D.G., E. Carmack, T.R. Walker, and P.R. Hill, 2006a. Storm resuspension in the shallow Beaufort Sea, Canada. Proceedings of the Canadian Arctic Shelf Exchange Study, Science Meeting, Winnipeg, February 2006.

Lintern, D.G., T. Turkington, and P.R. Hill, 2006b. Exploring the Mackenzie Delta plume using satellite ocean data. Proceedings of the Canadian Arctic Shelf Exchange Study, Science Meeting, Winnipeg, February 2006.

Lintern, D.G., H. Jakes, and J. Haaf, 2006c. Modeling the Mackenzie River System and Coastal Beaufort Sea. Presented at the US NSF MARGINS Source to Sink workshop, California, September 2006.

Lintern, D.G., H. Jakes, J. Haaf, J. Grant, and P.R. Hill, 2006d . Modeling sediment transport through the Mackenzie River System and Coastal Beaufort Sea, EOS Trans. AGU, 87 52 , Fall Meet. Suppl., Abstract OS14A-02.

Lintern, D.G., P.R. Hill, S. Solomon, T.R. Walker, and J. Grant, 2005. Erodibility, sediment strength and storm resuspension in Kugmallit Bay, Beaufort Sea. In: Canadian Coastal Conference, Dartmouth, Nova Scotia. 13 pp.

Macdonald, R.W. and D.J. Thomas, 1991. Chemical interactions and sediments of the western Arctic Shelf. Cont. Shelf Res. 11:843-863.

Macdonald, R.W., 2000. Arctic estuaries and ice: a positive-negative estuarine couple. In: E.L Lewis, E. P. Jones, P. Lemke, T. D. Prowse and P. Wadhams (ed.). The Freshwater Budget of the Arctic Ocean. Kluwer Academic Publishers, Netherlands. pp. 383-40.

Macdonald, R.W., S.M. Solomon, R.E. Cranston, H.E. Welch, M.B. Yunker, and C. Gobeil, 1998. A sediment and organic carbon budget for the Canadian Beaufort Shelf. Mar. Geol. 144: 255-273.

Macdonald, R.W., A.S. Naidu, M.B. Yunker, and C. Gobeil, 2004. The Beaufort Sea: Distribution, sources, fluxes and burial of organic carbon.

In: Stein, R., and RW Macdonald RW (eds.) The Organic Carbon Cycle in the Arctic Ocean. Springer Publishing Company, Berlin – New York, pp 177-193.

McPhee-Shaw, E., 2006. Boundary-interior exchange: Reviewing the idea that internal-wave mixing enhances lateral dispersal near continental margins. Deep-Sea Res. II 53: 42-59.

McPhee-Shaw, E. E., and E. Kunze, 2002. Boundary layer intrusions from a sloping bottom: A mechanism for generating intermediate nepheloid layers. J. Geophys. Res. (C Oceans), 107, 3050, doi:10.1029/2001JC000801.

McPhee-Shaw, E.E., R.W. Sternberg, B. Mullenbach, and A.S. Ogston, 2004. Observations of intermediate nepheloid layers on the northern California continental margin. Cont. Shelf Res. 24:693-720.

Meyers, P.A., 1994. Preservation of elemental and isotopic source identification of sedimentary organic matter. Chem. Geol. 114: 289-302.

Meyers, .A., 1997. Organic geochemical proxies of paleoceanographic, paleolimnologic, and paleoclimatic processes. Org. Geochem. 27: 213-250.

Miller, L.A., M. Chierici, T. Johannessen, T.T. Noji, F. Rey, and I. Skjelvan, 1999. Seasonal dissolved inorganic carbon variations in the Greenland Sea and implications for atmospheric CO_2 exchange. Deep-Sea Res. II 46: 1473-96.

Miller, L.A., P.L. Yager, K.A. Erickson, D. Amiel, J. Bâcle, J.K. Cochran, M.-È. Garneau, M. Gosselin, D.J. Hirschberg, B. Klein, B. LeBlanc, and W.L. Miller, 2002. Carbon distributions and fluxes in the North Water, 1998 and 1999. Deep-Sea Res. II 49: 5151-70.

Müller, A. and M. Voss, 1999. The palaeoenvironments of coastal lagoons in the southern Baltic Sea, - II. delta C-13 and delta N-15 ratios of organic matter - sources and sediments. Palaeogeogr. Palaeoecol. 145: 17-32.

Mueller, D.R., W.F. Vincent, and M.O. Jeffries, 2003. Break-up of the largest Arctic ice shelf and associated loss of an epishelf lake. Geophys. Res. Lett. 30: 2031, doi:10.1029/2003 GL017931.

Naidu, A.S., R.S. Scalan, H.M. Feder, J.J. Goering, M.J. Hameedi, P.L. Parker, E.W. Behrens, M.E. Caughey, and S.C. Jewett, 1993. Stable organic carbon isotopes in sediments of the North Bering South Chukchi Seas, Alaskan Soviet Arctic Shelf. Cont. Shelf Res. 13: 669-691.

Naidu, A.S., L.W. Cooper, B.P. Finney, R.W. Macdonald, C. Alexander, and I. P. Semiletov 2000 . Organic carbon isotope ratios (delta C-13) of Arctic Amerasian continental shelf sediments. Int. J. Earth Sci. 89: 522-532.

O'Brien, M.C., R.W. Macdonald, H. Melling, and K. Iseki, 2006. Particle fluxes and geochemistry on the Canadian Beaufort Shelf: Implications for sediment transport and deposition Cont. Shelf Res. 26: 41-81.

Oliveira, A., J.Vitorino, A. Rodrigues, J. M. Jouanneau, J. A. Dias, and O. Weber, 2002. Nepheloid layer dynamics in the northern Portuguese shelf. Progr. Oceanogr. 52:195-213.

Ota, Y, H. Hattori, R. Makabe, M. Sampei, A. Tanimura, H. Sasaki, (2008) Seasonal changes in nauplii and adults of Calanus hyperboreus (Copepoda) captured in sediment traps, Amundsen Gulf, Canadian Arctic, Polar Science. In Press, doi:10.1016/j.polar.2008.08.002

Papakyriakou, T., L. Miller, and O. Owens, 2006. Spring CO_2 fluxes over Arctic first-year sea ice: Results from CICE02, Paper presented at the Annual Canadian Geophysical Union Meeting, Banff, AB., 14-17 May 2006.

Payette, S., A. Delwaide, M. Caccianiga, and M. Beauchemin, 2004. Accelerated thawing of subarctic peatland permafrost over the last 50 years. Geophys. Res. Lett., 31: L18208, doi:10.1029/2004GL020358.

Peterson, B.J. and R.W. Howarth, 1987. Sulfur, carbon, and nitrogen isotopes used to trace organic matter flow in the salt-marsh estuaries of Sapelo Island, Georgia. Limnol. Oceanogr. 32: 1195-1213.

Peterson, B.J., R.M. Holmes, J.W. McClelland, C.J. Vörösmarty, R.B. Lammers, A.I. Shiklomanov, and S. Rahmstorf, 2002. Increasing river discharge to the Arctic Ocean. Science 298: 2171-2173.

Pickart, R. S., 2000. Bottom boundary layer structure and detachment in the shelfbreak jet of the Middle Atlantic Bight. J. Phys. Oceanogr. 30:2668-2686.

Puig, P., J.B. Company, F. Sardà, and A. Palanques, 2001. Responses of deep-water shrimp populations to intermediate nepheloid layer detachments on the Northwestern Mediterranean continental margin. Deep-Sea Res.I 48:2195-2207.

Puig, P., A.S. Ogston, B.L. Mullenbach, C.A. Nittrouer, J.D. Parsons, and R.W.Sternberg, 2004. Storm-induced sediment gravity flows at the head of the Eel submarine canyon, northern California margin. J. Geophys. Res. 109, C03019, doi:10.1029/2003JC001918.

Rachold, V., Eicken, H., Gordeev, V.V., Grigoriev, M. N., Hubberten, H. W., Lisitzin, A. P., Shevchenko, V. P., and Schirrmeister, L. 2004. Modern terrigenous organic carbon input to the Arctic Ocean. Pages 33–55 in The Organic Carbon Cycle in the Arctic Ocean (R. Stein, and R. W. MacDonald, eds. Springer-Verlag, New-York.

Reeburgh, W.S., 1967. An improved interstitial water sampler. Limnol. Oceanogr. 12: 163-165.

Rothrock, D.A., Y. Yu, and G.A. Maykut, 1999. Thinning of the Arctic sea-ice cover. Geophys. Res. Lett. 26: 3469-3472.

Ruttenberg, K.C. and M. A. Goñi, 1997. Phosphorus distribution, C:N:P ratios, and delta C-13 (oc) in arctic, temperate, and tropical coastal sediments: Tools for characterizing bulk sedimentary organic matter. Mar. Geol. 139: 123-145.

Sampei, M., H. Sasaki, H. Hattori, A. Forest, M. Fukuchi, S. Kudoh, and L. Fortier. Morphological discrimination of passively sinking metazoans (PSM) from swimmers in sediment trap samples in the Arctic waters. In preparation.

Schell, D.M., 1983. Carbon-13 and Carbon-14 abundances in Alaskan aquatic organisms: delayed production from peat in Arctic food webs. Science 219: 1068-1071.

Serreze, M.C., J.A. Maslanik, T.A. Scambos, F. Fetterer, J. Stroeve, K. Knowles, C. Fowler, S. Drobot, R.G. Barry, and T.M. Haran, 2003. A record minimum arctic sea ice extent and area in 2002. Geophys. Res. Lett. 30: 1110 doi:10.1029/2002GL016406.

Shindell, D.T., R.L. Miller, G.A. Schmidt, and L. Pandolfo, 1999. Simulation of recent northern winter climate trends by greenhouse-gas forcing. Nature 399: 452-455.

Stein, R., 2000. Circum-Arctic river discharge and its geological record: an introduction. Int. J. Earth Sci. 89: 447-449.

Stein, R. and R. W. Macdonald, 2004. Organic carbon budget: Arctic Ocean vs. Global Ocean. Pages 315–322 in The Organic Carbon Cycle in the Arctic Ocean (R. Stein, and R. W. Macdonald, (eds.) Springer-Verlag, New-York.

Stroeve, J.C., M.C. Serreze, F. Fetterer, T. Arbetter, W. Meier, J. Maslanik, and K. Knowles, 2005. Tracking the Arctic's shrinking ice cover: another extreme September minimum in 2004. Geophys. Res. Lett., 32: L04501, doi:10.1029/2004GL021810.

Telang, S.A., R. Pocklington, A.S. Naidu, E.A. Romankevich, I.I. Gitelson, and M.I. Gladyshev, 1991. Carbon and mineral transport in major North American, Russin Arctic, and Siberian rivers: the St. Lawrence, the Mackenzie, the Yukon, the Arctic Alaskan rivers, the Arctic Basin rivers in the Soviet Union, and the Yenisei. In: Degens E.T., Kempe S. and Richey J.E. (eds.) SCOPE 42: Biogeochemistry of major world rivers. John Wiley & Sons, London, p. 75-104.

US Corps of Engineers, 1984. Shore Protection Manual. Department of the Army, Waterways Experiment Station. Corps of Engineers. Coastal Engineering Research Center. Fourth Edition.

van Weering, T. C. E., H. C. De Stigter, W. Balzer, E. H. G. Epping, G. Graf, , I. R. Hall , W. Helder, A. Khripounoff, L.Lohse, I. N. McCave, L.Thomsen, and A.Vangriesheim, 2001. Benthic dynamics and carbon fluxes on the NW European continental margin. Deep-Sea Res. II 48:3191-3221.

Venegas, S.A and L.A. Mysak, 2000. Is there a dominant timescale of natural climate variability in the Arctic? J. Climate 13: 3412-3434.

Wadhams, P. and N.R. Davis, 2000. Further evidence of ice thinning in the Arctic Ocean. Geophys. Res. Lett. 27: 3973-3975.

Walker, T.R., J. Grant, and P. Jarvis, 2008b. Approaching freshet beneath landfast ice in Kugmallit Bay on the Canadian Arctic Shelf: Evidence from sensor and ground truth data. Arctic 61(1): 76-86.

Walker T.R., J. Grant, P. Cranford, D.G. Lintern, P.S. Hill, P. Jarvis J. Barrell, and C. Nozais, (2008a) Sediment erosion rates and suspended particulate matter properties in Kugmallit Bay and Beaufort Sea in the Canadian Arctic during ice-free conditions. J. Mar, Syst. (CASES Special Issue). J. Mar. Sys. 74: 794-809.

Walker, T.R., J. Grant, P.S. Hill, D.G. Lintern and B. Schofield. BEAST-A portable device used for quantification of erosion in undisturbed sediment cores. Limnol. Oceanogr. Meth. Submitted.

Wegner, C., J. A. Hölemann, I. Dmitrenko, S. Kirillov, K.Tuschling, E.Abramova, and H. Kassens, 2003. Suspended particulate matter on the Laptev Sea shelf (Siberian Arctic) during ice-free conditions. Est. Coast. Shelf Sci. 57:55-64.

Weiss, R.F., 1974. Carbon dioxide in water and seawater: the solubility of a non-ideal gas. Mar. Chem. 2: 203-215.

Witbaard, R., R. Daan, M. Mulder, and M. Lavaleye, 2005. The mollusc fauna along a depth transect in the Faroe Shetland Channel: Is there a relationship with internal waves? Mar. Biol. Res. 1:186-201.

Yager, P.L., D.W.R. Wallace, K.M. Johnson, W.O. Smith Jr., P.J. Minnett, and J.W. Deming, 1995. The Northeast Water Polynya as an atmospheric CO_2 sink: a seasonal rectification hypothesis. J. Geophys. Res. 100 (C3): 4389–4398.

Yu, Y., G.A. Maykut, and D.A. Rothrock, 2004. Changes in the thickness distribution of Arctic sea ice between 1958-1970 and 1993-1997. J. Geophys. Res. 109: C08004, doi:10.1029/2003JC001982.

Yunker, M.B., R.W. Macdonald, D.J. Veltkamp, and W.J. Cretney, 1995. Terrestrial and marine biomarkers in a seasonally ice-covered Arctic estuary – integration of multivariate and biomarker approaches. Mar. Chem. 49: 1-50.

Yunker, M.B., L.L. Belicka, H.R. Harvey, and R.W. Macdonald, 2005. Tracing the inputs and fate of marine and terrigenous organic matter in Arctic Ocean sediments: A multivariate analysis of lipid biomarkers. Deep-Sea Res. II 52: 3478-3508.

CHAPTER 8

Historical Variability—Paleoclimates

David B. Scott[1]*, André Rochon[2], Trecia M. Schell[1],
Guillaume St. Onge[2], Steve Blasco[3], Robbie Bennett[3]

8.1 Introduction and Rationale

This part of the CASES project focused on the calibration and use of micropaleontological and sedimentological proxies for paleoceanographic studies which could be related to paleoclimate and millennial/decadal changes in sea ice cover. Our study area was on the Canadian Beaufort Shelf (Fig. 8.1). Although extensive data were collected in this region in the early 1970's, paleoceanographic sampling and processing techniques have since changed to such an extent as to warrant new sample collections (see Scott et al., 2008, Richerol et al., 2008). To that end, a large suite of samples were collected between 2003 and 2005 using boxcores (surface and short cores) for the last 100 years and piston cores (representing longer records which take us back to the last glacial period). The proxies which were of most use to this study were microfossils:

a) Dinoflagellates and Pollen

Dinoflagellates are protists that are common in all marine environments. About half are phototrophic and the other half are heterotrophic, feeding on ciliates, diatoms and other dinoflagellates. During their life cycle, they produce highly resistant cysts that are fossilizable and representative of the

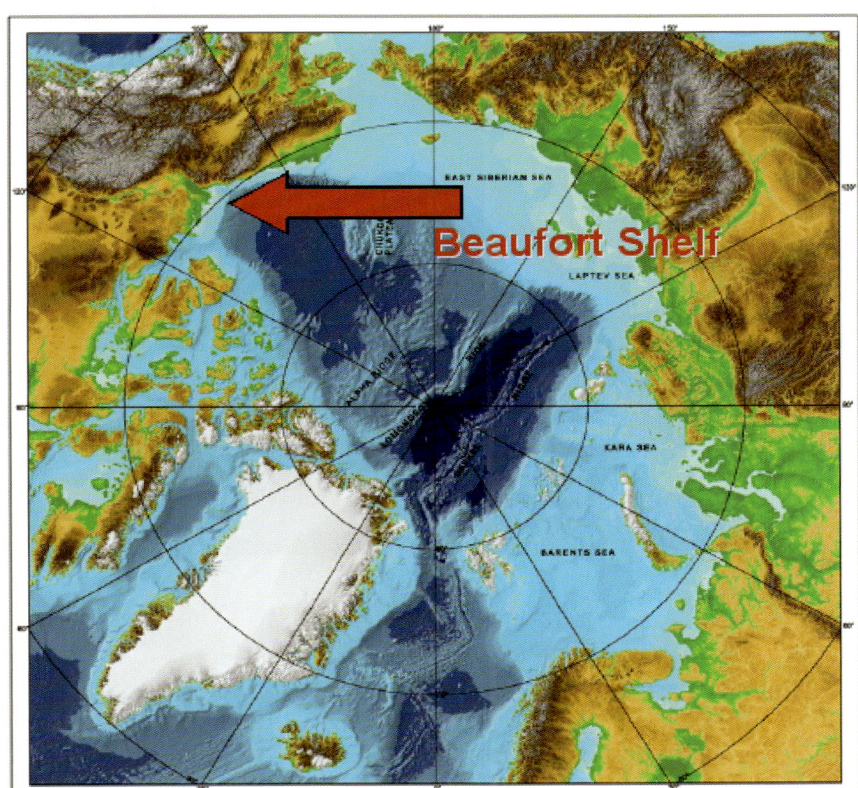

Figure 8.1

Bathymetric map of the Arctic Ocean including the Beaufort Shelf (arrow). LEFT: Inuit hunter. Photo: James Ford/ArcticNet.

[1]Centre for Environmental and Marine Geology, Dalhousie University, Halifax, NS, B3H 3J5, Canada

[2] Institut des sciences de la mer de Rimouski (ISMER), Université du Québec à Rimouski, 310 allée des Ursulines, Rimouski, QC, G5L 3A1, Canada

[3] Natural Resources, Canada, 1 Challenger Drive, Dartmouth, NS, B2Y 4A2, Canada

*Corresponding author: dbscott@dal.ca

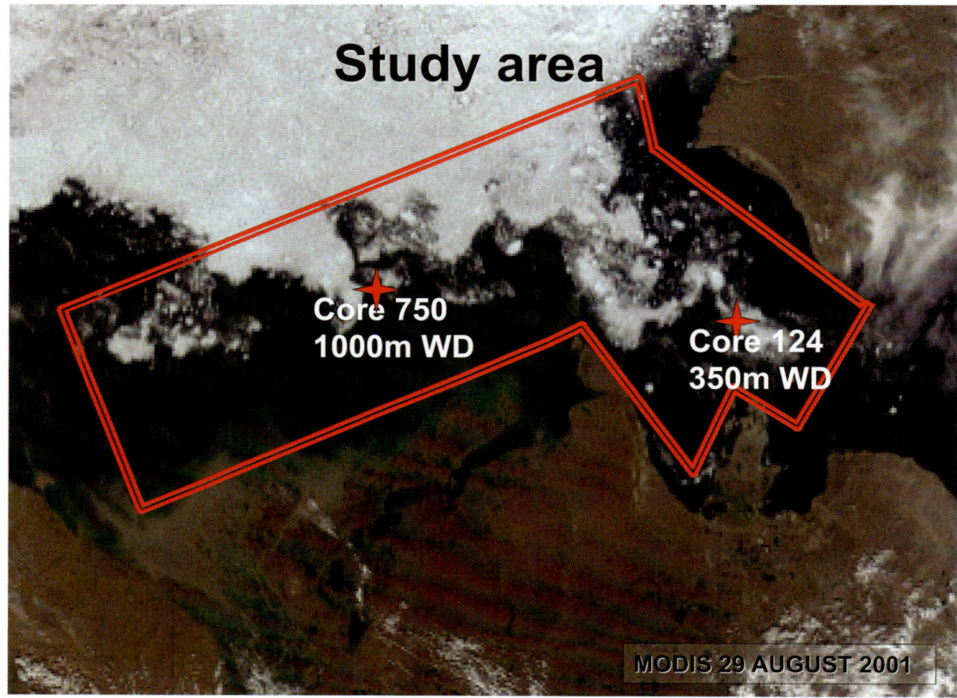

Figure 8.2

Modis satelite photograph of the Beaufort Shelf/Amundsen Gulf in August showing core positions and typical ice pattern. This photo was taken in 2001 but the ice coverage was very similar to that in 2004 when the samples for this study were collected.

seawater conditions in which they were formed. Some dinoflagellate life cycles are closely dependant upon ice cover, and as such, their cysts constitute excellent indicators of paleo-sea ice conditions. Dinoflagellate cysts, with the use of transfer functions, provide quantitative estimates of sea surface parameters such as temperature, salinity and sea ice cover. Pollen grains are usually examined in the same samples as dinoflagellate cysts because they are both organic-walled and concentrated during palynological processing. They provide a proxy for terrestrial vegetation, and therefore the continental climate adjacent to the water body being sampled;

b) Foraminifera-planktonic and Benthic, Stable Isotopes (oxygen-18 and carbon-13)

Foraminifera are singled-cell organisms that are found throughout the water column and at the water-sediment interface (pelagic and benthic). Their shells (or tests) can be either made of secreted calcium carbonate or small pieces of sediment cemented (agglutinated) together. Their potential as a proxy emerges when the carbonate of their tests dissolves (i.e. in cold, relatively fresh waters, like those of Arctic shelf ecosystems) and they leave behind a distinct fauna of agglutinated material. Benthic forms of foraminifera are found in all marine environments and provide information on deep conditions, while planktic forms reside largely in open waters and provide information on surface conditions. Stable isotope measurements derived from the calcareous foraminifera tests (both planktic and benthic) provide more specific information on changes in salinity and water temperature for both the surface and bottom environments;

c) Tintinnids

These are ciliates that live in the surface waters. The ones presented in this section have an agglutinated shell and leave a fossil record. Their presence indicates brackish environment and/or an abundance of suspended particulates. They were particularly useful in this study for examining the paleo-extent of the Mackenzie River plume;

d) Thecamoebians

These are freshwater rhizopods (the freshwater correspondent to foraminifera) and provide an indicator of particles transported by freshwater into a marine system;

e) Diatoms (limited extent)

We tried to use diatoms, but because they are composed of silica it was almost impossible to separate them out from the surrounding silt sediment (they are roughly the same size, i.e. <63 microns). It became too expensive and time consuming to deal with diatoms.

Illustrated examples of each of these organisms are shown on Plate 8.1.

8.2 Overview of the Results

Limited chronology was achieved through the radiocarbon (^{14}C) dating of wood from the Mackenzie Trough. Instead, the majority of our chronological results were derived from the carbonate shells of molluscs and foraminifera via accelerator mass spectrometry (AMS) and ^{14}C analysis. These provided information on a time period spanning up to 30 000 yr BP. For information on the last 1000 years (including detail on recent versus pre-industrial climate conditions), we used a combination of AMS-^{14}C and ^{210}Pb dating (some of which was provided by D. Amiel and K. Cochran, SUNY). Other aspects of our study included paleomagnetic analysis [i.e. low field volumetric magnetic susceptibility (k)] and anhysteretic remanent magnetization (ARM), as well as sediment size sorting and mean grain size analysis (in piston core 750). In particular, Ice Rafted Debris (IRD) were clearly identified and allowed direct comparisons between the Beaufort Slope and Amundsen Gulf. Together, these proxies provided sufficient information to fulfill our commitment toward determining long-term (10,000 years) changes within the water masses and sea ice of the Beaufort Shelf-Slope/Mackenzie Trough/Amundsen Gulf system. These data also provided a pre-industrial context from which the last 50 years of direct observations could be compared. This will allow us to answer some important questions, such as whether or not climate-related changes in the Arctic Ocean are unique to the last 10,000-15,000 years.

We began our research efforts by collecting modern proxy data (Fig. 8.3 and 8.4). This allowed us to "groundtruth" our flora and fauna samples and

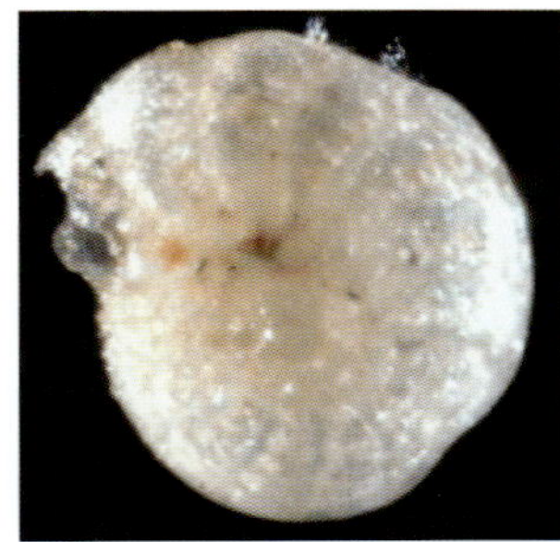

Elphidium sp.-foraminiferid, benthic, calcareous

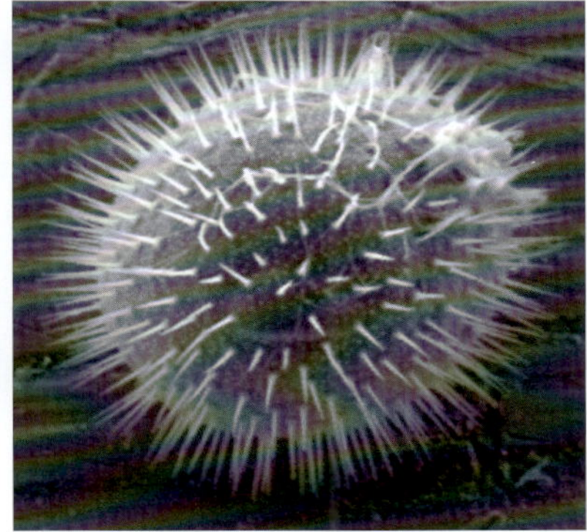

Islandinium minutum- dinoflagellate cyst organic-walled, pelagic

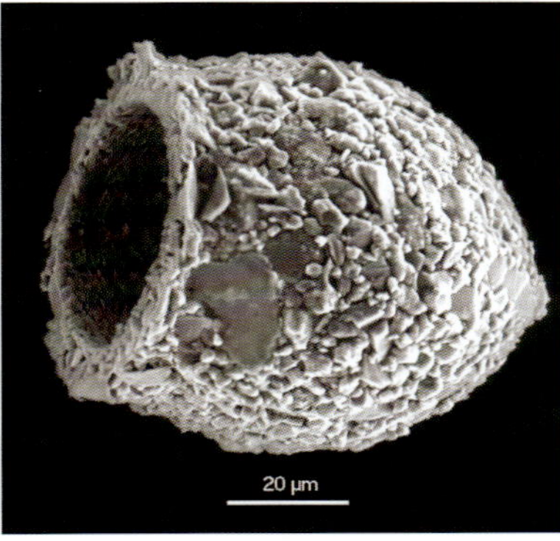

Tintinnid-agglutinated, pelagic

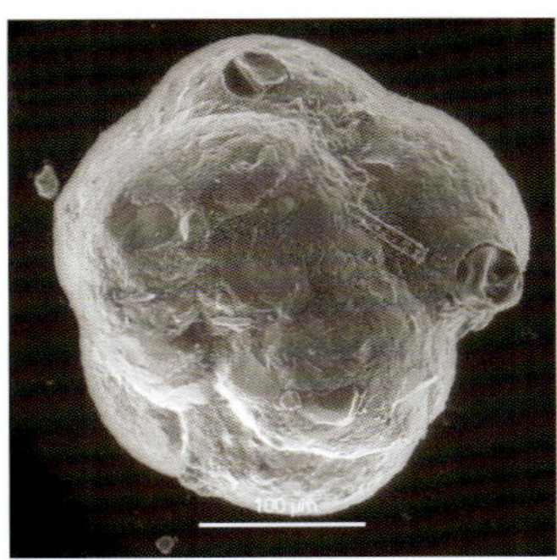

Recurvoides sp.-foraminiferid, benthic, agglutinated

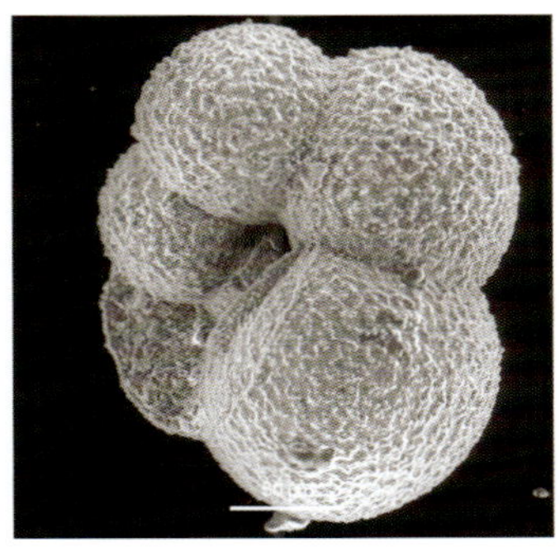

Neogloboquadrine pachyderma-foraminiferid, calcareous, planktonic (only polar plantic species)

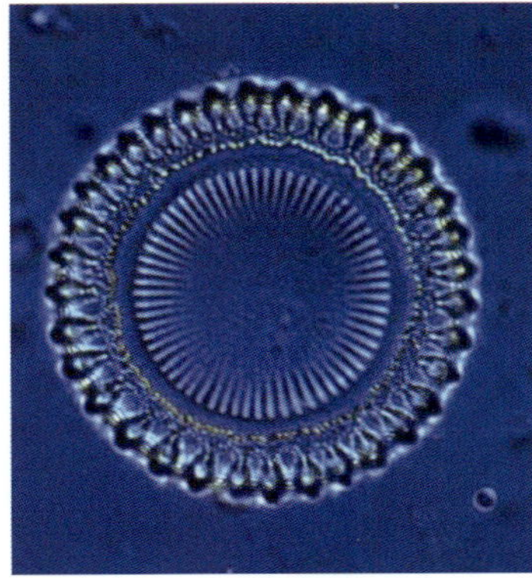

Diatom-siliceous, pelagic

Plate 8.1

Photographs of the different types of microfossils used to detect oceanographic changes in the cores from the Beaufort Sea. Top row (left to right)-benthic foraminiferid (calcareous), dinoflagellate, tintinnid; Lower row (left to right)-benthic foraminiferid (agglutinated), planktic foraminiferid (calcareous), and diatom (siliceous).

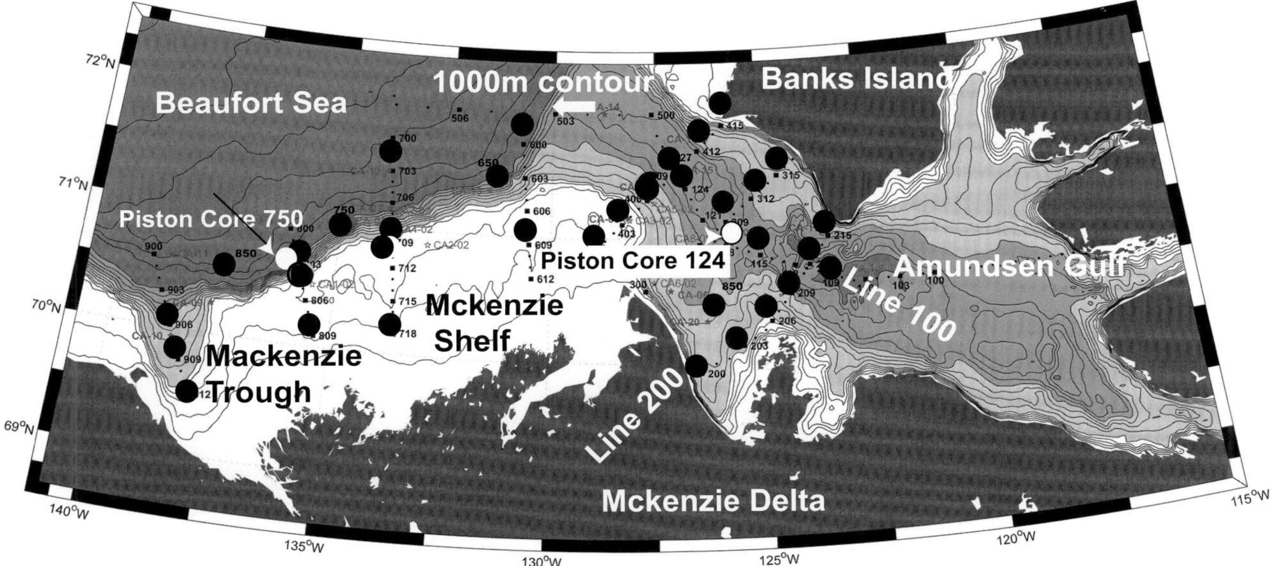

Figures 8.3

Areal map of the CASES field area with bathymetry and station locations.

LEFT: Icebergs. Photo: Melanie Simard/ ArcticNet.

provided a framework for accurately interpreting the subsequent paleo-record (Scott et al., 2008; Richerol et al., 2008). Dinoflagellate concentrations were high throughout the study region, and exhibited maximum abundance (> 6000 cysts/g of dry sediment) within the Cap Bathurst Polynya, south of Banks Island. The maximum relative abundance of cysts from heterotrophic species was located within the Mackenzie River plume area, presumably because of high particulate matter concentrations. However, cysts from autotrophic species were dominant outside of the plume area. One species, *Islandinium minutum,* is usually associated with sea ice and its minimum relative abundance was located within the polynya.

A number of papers have been published regarding the paleoceanography of the deep waters of the Arctic Ocean (e.g. Scott et al., 1989; Scott and Vilks, 1991; Poore et al., 1994). However, relatively few have examined data within the Canadian margins. There has been extensive work done on the Siberian Shelf (e.g. Polyak et al., 2002), where fauna is influenced by the North Atlantic and a larger river system than the Canadian Arctic. Data has also been presented from an ice island located near Ellesmere Island (Schröder et al., 1990). This proved to be a sediment-starved area with little freshwater influx and little comparative value to the Beaufort Sea Shelf environment. Vilks (1989) was the first to look at foraminiferal fauna on the Mackenzie Shelf and his findings were among the most useful found for comparing our work. Finally, recent studies (Rochon et al., 1999; Mudie and Rochon, 2001; Voronina et al., 2001; Kunz-Pirrrung, 2001; Grøsfjeld

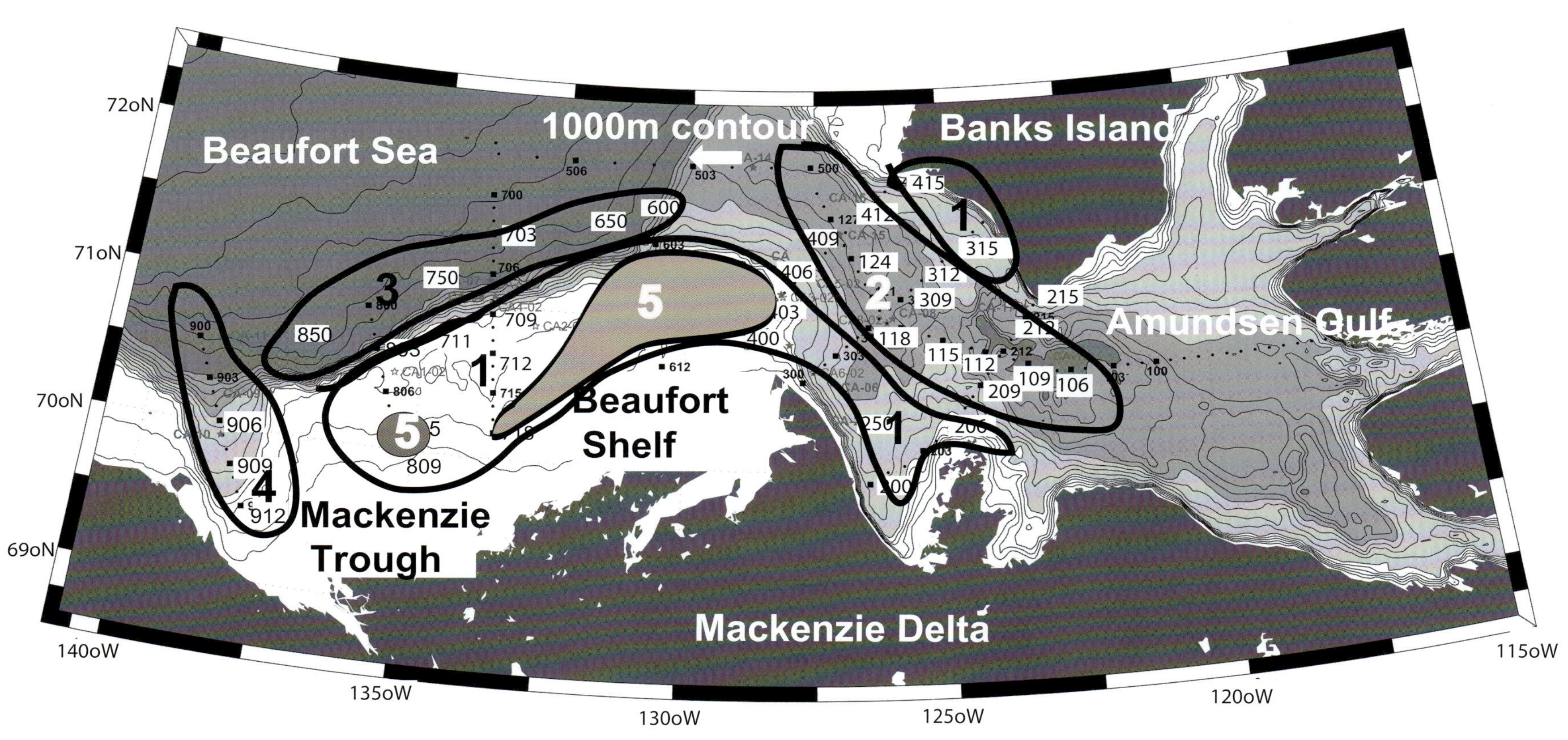

Figure 8.4

Areal position of the 5 foraminiferal zones on the Beaufort Shelf/Amundsen Gulf: 1) Beaufort Shelf (35-200 m), 2) central Amundsen Gulf, 3) deep sea stations (>1000 m), 4) Mackenzie Trough, and 5) Pingo-like-areas.

and Harland, 2001; Boessenkool et al., 2001; Radi et al., 2001) have illustrated that dinoflagellates are common to most surface sediments in the Arctic and that their distribution can be useful in reconstructing recent paleoceanographic conditions (specifically, temperature, salinity and sea ice cover conditions).

Several papers originating from the CASES Research Network were recently published in the Journal of Geophysical Research and the Journal of Marine Systems. Two papers (Richerol et al., 2008; Scott et al., 2008) focused on the modern distribution of the proxies we sampled and are pivotal to our understanding of the pre-industrial record. Other papers focused on the Holocene record, specifically the paleoceanography and sea ice cover for the last 10,000 cal BP in the Mackenzie Trough (Schell et al., in press), the paleoceanography and sea ice cover for the last 14,000 cal BP on the Beaufort Slope and Amundsen Gulf Gulf (Scott et al., in press; Schell et al., in press 2008), and sea ice history (1 000 cal BP) in a transect from the Beaufort Shelf to Banks Island.

Richerol et al. (2008) reported the surface distribution of dinoflagellate cysts from surface sediments sampled in the Mackenzie Trough and the Beaufort Shelf. Cysts are a resting stage for dinoflagellates, and while not all dinoflagellates produce cysts that preserve in sediments, several species leave a cyst record which can be related back to the original environmental conditions at their deposition. The dinoflagellates respond to high nutrients and irradiance, although some are heterotrophic. Richerol et al. (2008) reported that the heterotrophic forms were most abundant in the Mackenzie Trough where there was abundant particulate matter to feed on. The autotrophic forms were most common away from the plume on the Beaufort Shelf, and especially in the Amundsen Gulf where there was little river

influence. Clearly, the abundance of photosynthetic dinoflagellates relates to sea ice cover since more sea ice implies lower irradiance to the surface ocean. Dinoflagellates, on the other hand, are composed of very resistant organic cysts and can be concentrated using hydrofluoric acid to dissolve the mineral fraction. Although little core work has been done in the Arctic using dinocysts up to this point, they have proven useful in other parts of the Canadian margins for paleo-sea ice determinations (e.g. Levac et al., 2001).

As with dinocysts, plankic foraminifera are also most abundant where there is less sea ice since there is more productivity in the water column. Paradoxically however, with more productivity, there is more carbon flux to the seafloor, which creates unfavorable conditions for carbonate preservation (Wollenburg et al., 2000, 2001, 2004). Hence, in the Arctic, more carbonate in the deep sea suggests more sea ice (as observed in the central Arctic; Scott et al., 1989) and this is one of our major proxies for paleo-sea ice cover. Coincidently, however, dinoflagellates and calcareous foraminifera proxies can be complementary since many of one may mean less of the other.

The tintinnids and thecamoebians provide some information on freshwater sources and amounts. Tintinnids are common across the Beaufort Shelf since there is a lot of organic-rich freshwater coming across the shelf; they decrease in abundance once we are in the Amundsen Gulf. Thecamoebians, if found in the marine environment, suggest transport of fine material from a freshwater source—in our case, this source would be one of the rivers feeding the shelf.

In core 750 (Fig. 8.5) the multi-beam data showed that there may have been slumping on the upper slope; however, seismic data indicated continuous deposition. No Atlantic deep water species were found in

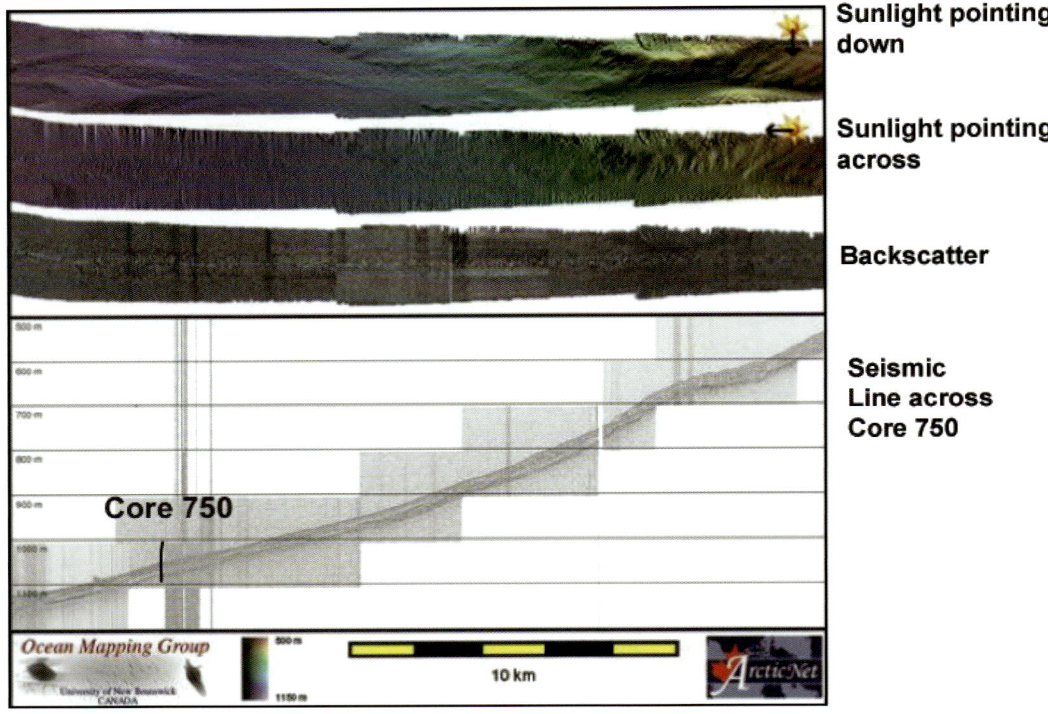

Figure 8.5

Seismic profile and multibeam images over core site 750. A 3.5 KHz subbottom profiler was used for seismics. Image provided and enhanced during Leg-8 of CASES (June 2004) by Jason Bartlett, Canadian Hydrographic Service.

core 124, possibly due to shallower water or lower sedimentation rates there. In the boxcore sample from this site there was an organic unit at about 15 cm (Fig. 8.6), which was barely noticeable at the top of the piston core. X-rays of the lower part of the core indicated IRD in the 300-400 cm zone and also in the 170-230 cm zone (Fig. 8.7). Thinly laminated sediments between the IRD units suggested seasonal influxes of sediment.

Deposition characteristics at site 124 were quite different from those at site 750. Multibeam images showed striking seafloor morphology changes near the core site. Initially, the site was chosen because seismic profiles suggested that it was seaward of the glacial margin (Fig. 8.8), as opposed to the drumlin areas which

are believed to be sub-glacial features. The scale of the Holocene/late glacial transition in core 124 was compressed such that the interval over which the IRD was observed in core 750 (beginning at 170 cm) began at 100 cm. IRD dates for both cores suggested that they occurred at approximately the same time period and would have started in the Amundsen Gulf as the glaciers started to shed icebergs. They were also found to be of similar origin (based on double-peak magnetic susceptibility patterns), and were largely composed of crystalline igneous or metamorphic gravel and sand size particles (Figs. 8.10 to 8.12).

Our results (i.e. Scott et al., 2008, Schell et al., 2008) suggest that in the last few years (based on ^{210}Pb measurements) that sea ice may have been more extensive because of a wel-developed calcareous fauna not present in the previous few thousand years. In core 750 (Fig.8.5) the multibeam data show that there may be slumping on the upper slope but the seismic and sedimentological data indicate continuous deposition for the section of core presented here. The first ancient carbonates appear at ~11,500 cal BP in core 750 (radiocarbon dated foraminifera Fig. 8.6). The IRD in core 750 coincided foraminiferal units and suggested lower sedimentation rates during this period and/or better carbonate preservation with extensive sea ice cover. Also of interest in core 750 is the limited occurrence of the Atlantic deep-water species *Oridorsalis umbonatus*. This species occurs in the central Arctic CESAR site as well as in surface and core samples from the Eurasian Basin (Jackson et al., 1985, Scott et al., 1989, Scott and Vilks, 1991). In the CESAR site, because the sedimentation rate is low (1 cm/10 Kyr), it is impossible to know if the *O.umbonatus* occurred in the entire interval or just for a short time. Given the higher sedimentation rates in core 750 it appears that at least at this site, the Atlantic influence

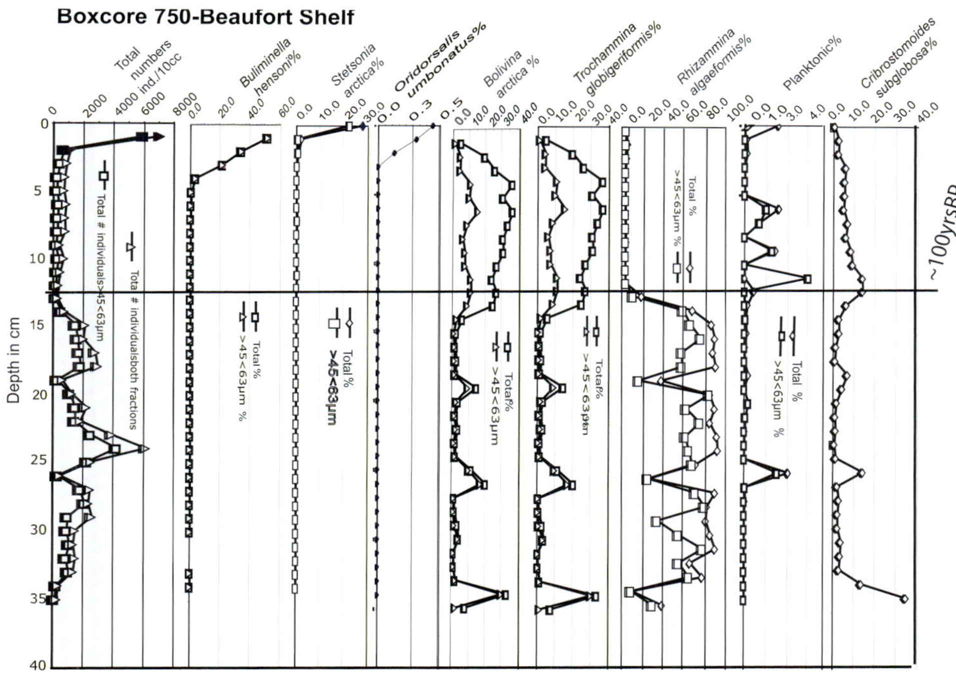

Figure 8.6

Distribution of selected foraminiferal species for the >63 μm fraction of Piston Core 750 versus depth in cm with indications for radiocarbon dates and IRD occurrences as well as the narrow "Atlantic species" interval.

Figure 8.7

X-Ray photographs from Piston Core 750. Ice rafted sediments occurred within 170 & 210 cm and 320 & 360 cm.

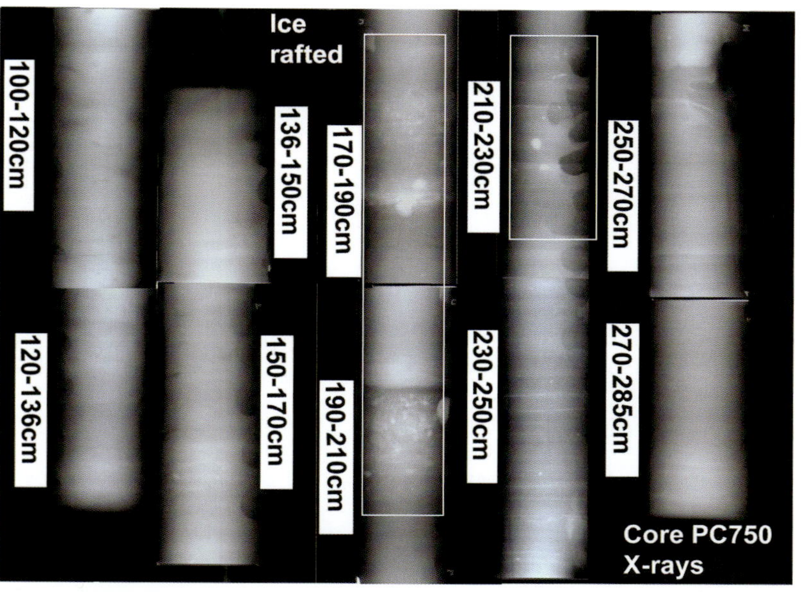

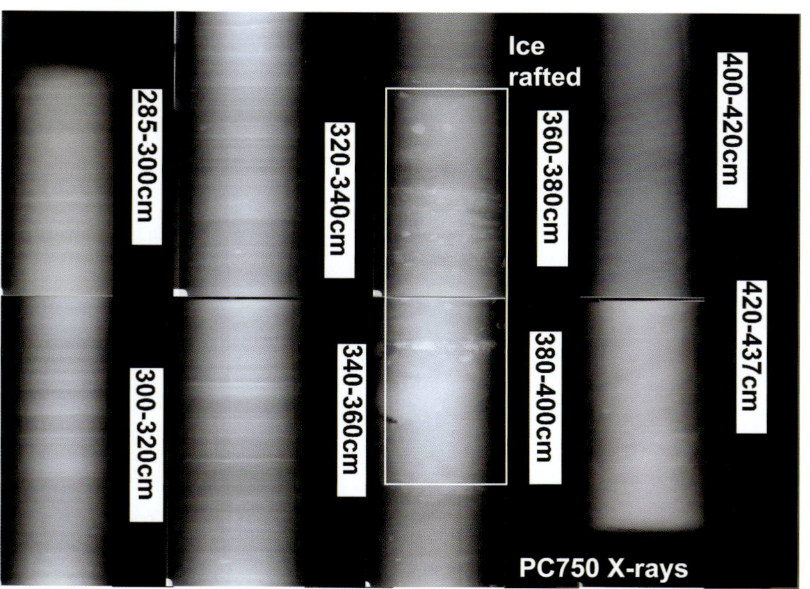

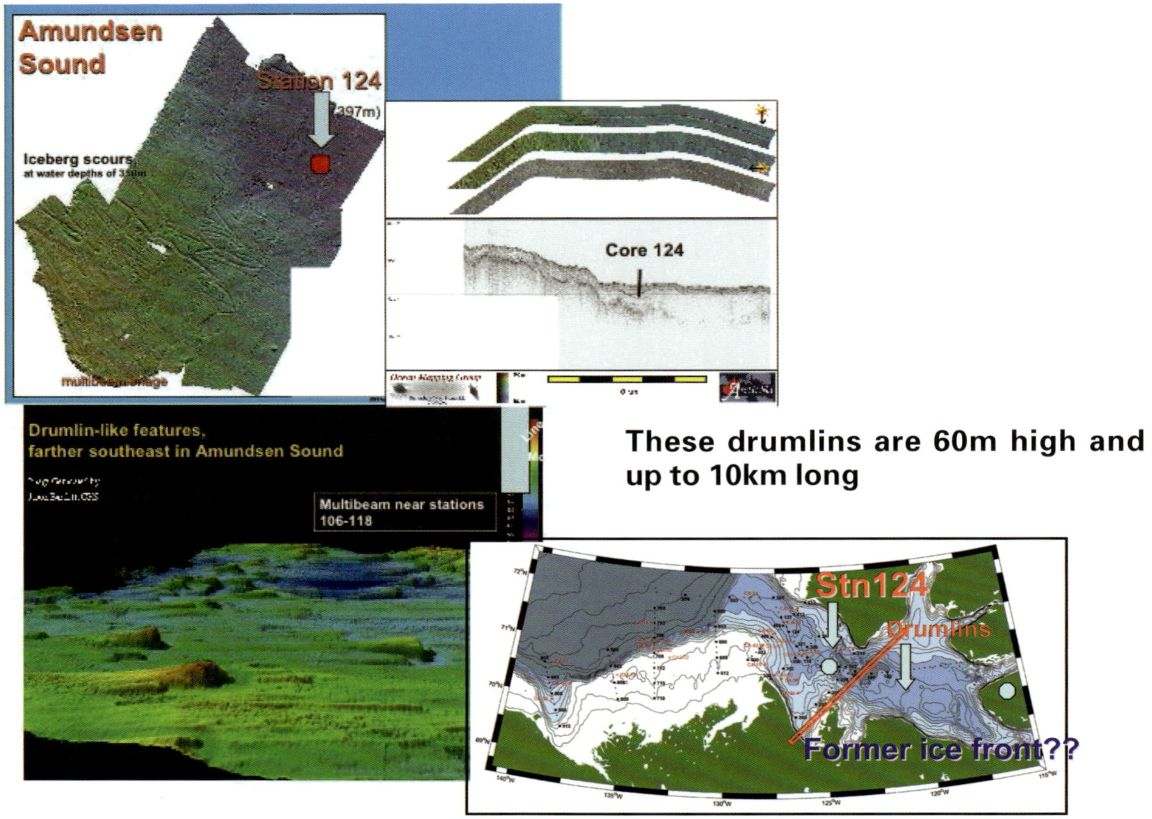

Figure 8.8

Multibeam and subbottom seismic images from core site 124 and southeast of the site. The core site shows the relatively undisturbed sediments while the other shows the glacial features characteristic of most of the inner part of the Amundsen Gulf. The core location is indicated on the seismic section, and it is possible to see the transition between sub-glacial and glacial marine sediments. The areal map shows the approximate area where drumlins are located. Multibeam images collected during Leg-8 of CASES (June 2004) by Jason Bartlett, Canadian Hydrographic Service.

was limited to the early postglacial period although a few *O. umbonatus* specimens do occur at the very surface of the boxcore in site 750. *Oridorsalis umbonatus* was not observed in core 124 even with a strong influence of Arctic Deep Water as evidenced by the presence of Arctic deep water foraminiferal species (*S. arctica, B. hensoni*).

The ^{14}C chronology suggested that Glacial ice began retreating much earlier than previously thought. Stokes et al. (2005, 2006) discussed ice streams coming into the upper Amundsen Gulf around 9000-11,000 cal BP. Stokes et al. (2005) also suggested ice streams coming out of M'Clure Strait at older ages, but this was based on correlations with IRD from Fram Strait in the eastern Arctic. The date of 13,286 cal BP from core 750 and the date of 12,376 cal BP from the younger IRD unit of core 124 pre-date these values and suggest glacial ice movement in the Gulf occurred much earlier than on land.

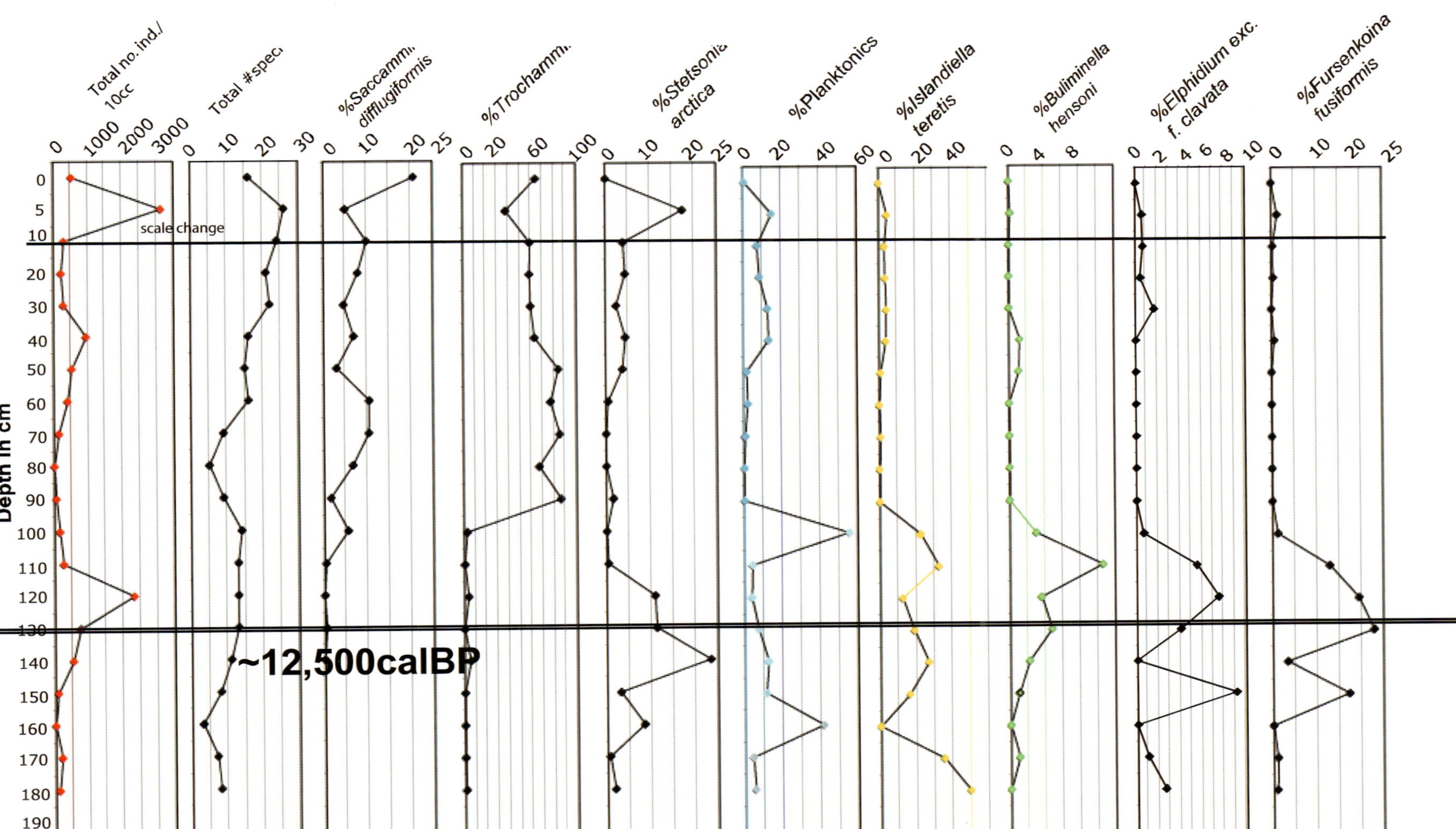

Figure 8.9

Distribution of selected foraminiferal species for the >63 µm fraction of Piston Core 124 versus depth with indications for approximate radiocarbon age.

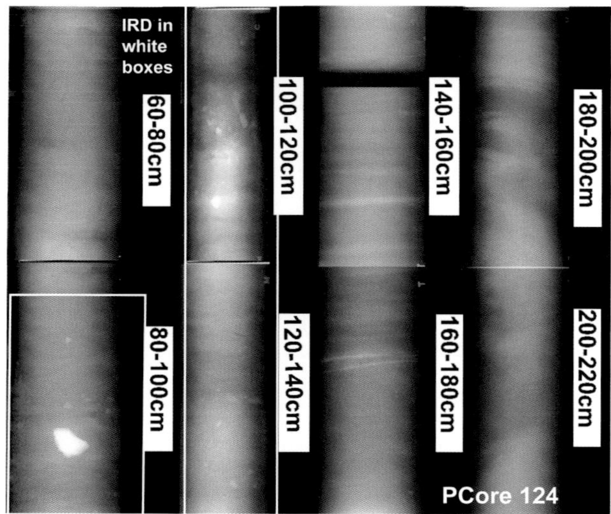

Figure 8.10

X-ray Photographs from Piston Core 124; 60-220 cm. Ice-rafted sediments occur between 80 and 180 cm.

Figure 8.11

Low field volumetric magnetic susceptibility (k), mean grain size and sorting of Piston Core 750. The two grey areas highlight zones of poorly sorted coarser sediments at the base and upper part of the IRD zone discussed in this paper (see text for details). Stars indicate discrete intervals of IRDs as observed on the X-rays.

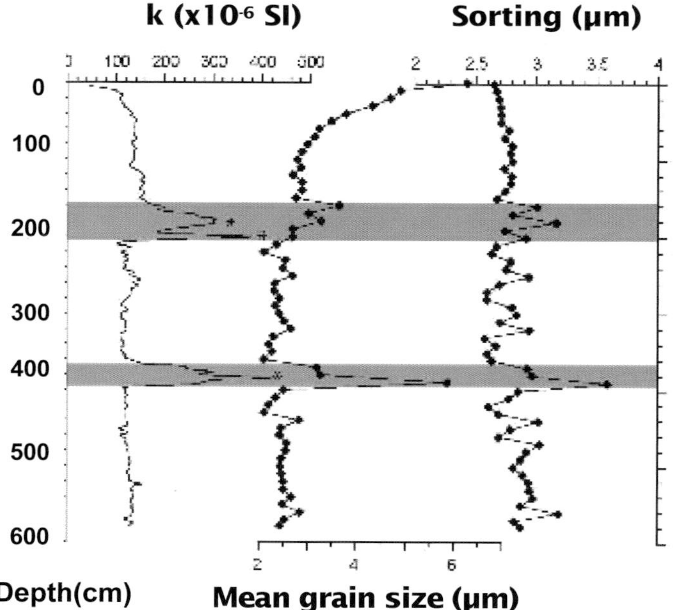

8.3 Implications of this work

This work has several implications:

a) We can, using micropaleontological and sedimentological proxies, delineate several distinct faunal/floral/sedimentological zones within the study area. This provides information on the distribution of micropaleontological proxies, which in turn allows us to determine paleoceanographical conditions, sea ice cover, freshwater input, and glacial activity for the Holocene period.

b) Atlantic deep water seems to be restricted to the deep water core, as well as to a very narrow interval in early postglacial times when there were calving icebergs.

c) We can identify periods of higher and lower sea-ice cover, higher freshwater input, early glacial movement, as well as recent changes.

d) We can now begin to determine a glacial history for the Beaufort Shelf and its relation to glacial activity further South, with Glacial ice retreating much earlier in marine areas than on land (at least near Banks Island).

e) With the more recent data on sea-ice cover, we can also begin to compare present conditions with pre-industrial and Holocene history (e.g. Dyke et al., 1996; Dyke and Savelle, 2002; Fisher et al., 2006).

8.4 Recommendations

It appears that the Holocene was a far more active period climate-wise than previously assumed, and most of the paleo-climate work done in the Arctic confirms this. However, paleoclimatic records are very limited even with the advent of projects like CASES and ArcticNet. There are several key areas which require more investigation: the NW Passage, particularly M'Clure Strait and the adjacent channels, in order to determine when they were open in the past; land-sea connections, which could provide answers to discrepancies such as the time difference between glacial movements on Banks Island and in Amundsen Gulf; the outer Beaufort Sea (from which we presently do not possess cores deeper than 1000 m), in order to determine the context of former multiyear ice margins (i.e. whether or not those margins were further past the present ice margin in the early Holocene); and deep water records, in order to determine how much and how often there is Atlantic deep water entering the Arctic basins. Currently, only the CESAR work can determine something on this latter issue. However, the record is of such low resolution that it provides no real timeline for the duration of events or how long the Atlantic water resides in the basins. These are just a few things that could be addressed using the paleo-climate records.

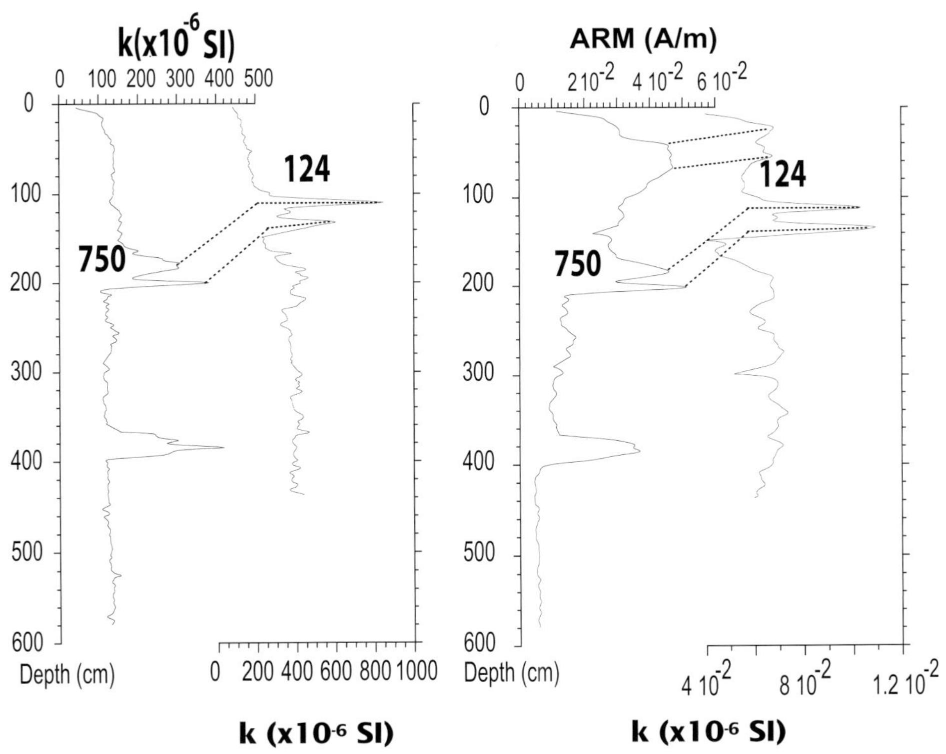

Figure 8.12

Comparison of low field volumetric magnetic susceptibility (k) and anhysteretic remanent magnetization (ARM) of Piston Cores 750 and 124. Dashed lines indicate correlative features.

References

Boessenkool, K.P., M.J. van Gelder, and S.R. Troelstra, 2001. Distribution of organic- walled dinoflagellate cysts in surface sediments from transects across the Polar Front offshore southeast Greenland. J. Quater. Sci., 16: no. 7, p. 661-666.

Dyke, A.S., J. Hooper, and J.M. Savelle, J.M., 1996. A history of sea ice in the Canadian Arctic Archipelago based on postglacial remnants of the bowhead whale *(Balaena mysticetus)*. Arctic 49: 235-255.

Dyke, A.S. and J.M. Savelle, 2001. Holocene history of the Bering sea bowhead whale *(Balaena mysticetus)* in its Beaufort Sea summer grounds off southwestern Victoria Island, western Canadian Arctic. Quater. Res. 55: 371-379.

Fisher, D., A. Dyke, R. Koerner, J. Bourgeois, C. Kinnard, C. Zdanowicz, A. de Vernal, C. Hillaire-Marcel, C., J. Savelle, and A. Rochon, 2006. Natural Variability of Arctic Sea Ice over the Holocene. EOS 87: 273- 275.

Grøsfjeld, K., and R. Harland, 2001. Distribution of modern dinoflagellate cysts from inshore areas among the coast of southern Norway. J. Quater. Sci.16: 651-659.

Jackson, H.R., P.J. Mudie, and S.M. Blasco, 1985. Initial Geological Report on CESAR-the Canadian expedition to study the Alpha Ridge, Arctic Ocean. Geological Survey of Canada, Paper 84-22, 177p.

Kunz-Pirrrung, M., 2001. Dinoflagellate cyst assemblages in surface sediments of the Laptev Sea rgion (Arctic Ocean) and their relationship to hydrographic conditions. J. Quater. Sci. 16: 637-649.

Levac, E., A. deVernal, and W. jr. Blake, 2001. Holocene palynology of cores from the North Water Polyna, Baffin Bay. J. Quater. Sci. 16: 353-363.

Mudie, P.J., and A. Rochon, 2001. Distribution of dinoflagellate cysts in the Canadian Arctic marine region. J. Quater. Sci. 16: 603-620.

Polyak, L., S. Korsun, L. Febo, V. Stanovoy, T. Khusind, M. Hald, B.E. Paulsen, and Lubinski, 2002. Benthic foraminiferal assemblages from the southern Kara Sea, a river- influenced Arctic marine environment. J. Foram. Res. 32: 252-273.

Poore, R.Z., S.E. Ishman, R.L. Phillips, and D.H. McNeil, 1994. Quaternary stratigraphy and paleoceanography of the Canada Basin, Western Arctic Ocean. United States Geological Survey Bulletin 2080, 32p.

Radi, T., A. deVernal, and O. Peyron, 2001. Relationships between dinoflagellate cyst assemblages in surface sediment and hydrographic conditions in the Bering and Chukchi Seas. J. Quater. Sci. 16: 667-680.

Richerol, T., A. Rochon, S. Blasco, D.B. Scott, T. Schell, and R. Bennett, R., 2008. Distribution of dinoflagellate cysts in surface sediments of the Mackenzie Shelf and Amundsen Gulf, Beaufort Sea (Canada). J. Mar. Sci. 74: 825-839.

Rochon, A., A. de Vernal, J-L Turon, J. Matthiessen, and M. Head, 1999. Distribution of recent dinoflagellate cysts in surface sediments from the North Atlantic Ocean and adjacent seas in relation to sea surface parameters. Contr. Series number 35, American Association of Stratigraphic Palynologists Foundation: Dallas, Texas., 152p.

Schell, T.M., D.B. Scott, A. Rochon, S. Blasco. Holocene paleoceanography of the mid Western Canadian Arctic Ocean, and paleo-sea ice conditions in the Mackenzie Trough and Canyon, Beaufort Sea. Can. J. Earth. Sci. In press.

Schell, T. M., T.J. Moss, D.B. Scott, and A. Rochon, 2008. Paleo-sea ice conditions of the Amundsen Gulf, Canadian Arctic Archipelago: indications from the foraminiferal record of the last 200 years. J. Geophys. Res. – Ocean, 113, C03S02, doi:10.1029/2007JC004202.

Schröder-Adams, C.J., F.E. Cole, F.S. Medioli, P.J. Mudie, D.B. Scott, and L. Dobbin, 1990. Recent Arctic shelf foraminifera: seasonally ice covered vs. perennial ice covered areas. J. Foram. Res. 20:8-36.

Scott, D.B., P.J. Mudie, V. Baki, K.D. MacKinnon, and F.E. Cole, 1989. Biostratigraphy and late Cenozoic paleoceanography of the Arctic Ocean: foraminiferal, lithostratigraphic, and isotopic evidence. Geol. Soc. Am. Bull. 101: 260-277.

Scott, D.B., and G. Vilks, G., 1991. Benthonic foraminifera in the surface sediments of the deep-sea Arctic Ocean. J. Foram. Res. 21: 20-38.

Scott, D.B., T.M. Schell, A. Rochon, and S. Blasco, 2008. Benthic foraminifera in the surface sediments of the Mackenzie Shelf and Slope, Beaufort Sea, Canada: applications and implications for past sea-ice conditions. J. Mar. Sys. J. Mar. Sys 74: 840-863.

Scott, D.B., T. Schell, G. St-Onge, A. Rochon, and S. Blasco. Foraminiferal assemblage changes over the last 15,000 years on the Mackenzie/ Beaufort sea slope and Amundsen Gulf, Canada: implications for past sea-ice conditions. Paleoceanography. In press.

Stokes, C.R., C.D. Clark, and D.A. Darby, 2005. Late Pleistocene ice export events into the Arctic Ocean from the M'Clure Strait Ice Stream, Canadian Arctic Archipelago. Global Plan. Change 49:139-162.

Stokes, C.R., C.D. Clark, and C.M. Winsborrow, 2006. Subglacial bedform evidence for a major palaeo-ice stream and its retreat phases in Amundsen Gulf, Canadian Arctic Archipelago. J. Quater. Sci. 21:399-412.

Vilks, G., 1989. Ecology of Recent Foraminifera on the Canadian Continental Shelf of the Arctic Ocean *in* Herman, Y., ed., The Arctic Seas: Climatology, Oceanography, Geology, and Biology. p. 497-569.

Vilks, G., F.G.E. Wagner, and B.R. Pelletier, 1979. The Holocene marine environment of the Beaufort shelf: Geol. Surv. of Can. Bul. 303, 43p.

Voronina, E., L. Polyak, A. deVernal, and O. Peyron, 2001. Holocene variations of sea-surface conditions in the southeastern Barents Sea, reconstructed from dinoflagellate cyst assemblages. J. Quarter. Sci. 16: 717-726.

Wollenburg, J.E., and W. Kuhnt, 2000. The response of benthic foraminifers to carbon flux and primary production in the Arctic Ocean. Mar. Micropaleo. 40: 189-231.

Wollenburg, J.E., W. Kuhnt, and A. Mackensen, 2001. Changes in Arctic Ocean paleoproductivity and hydrography during the last 145kyr: the benthic foraminiferal record. Paleoceanogra. 16: 65-77.

Wollenburg, J.E., J. Knies, and A. Mackensen, A., 2004. High-resolution paleoproductivity fluctuations during the past 24kyr as indicated by benthic foraminifera in the marginal Arctic Ocean. Palaeogeogr., Palaeoclim., Palaeoecol. 204: 209-238.

Glacier. Photo: Ramon Terrado/ArcticNet.

CHAPTER 9

The Benthic Environment

Alec E. Aitken[1], Kathleen Conlan[2], Paul E. Renaud[3, 4], Ed Hendrycks[2], Christine McClelland[2], Philippe Archambault [5,6], Mathieu Cusson[1,7], Nathalie Morata[3]

9.1 Introduction and Rationale

Arctic benthic communities are essential food resources for diving sea birds (Dickson and Gilchrist, 2002) and large mammals such as gray whales, walruses and bearded seals (Frost and Lowry, 1984). Strong pelagic-benthic coupling on Arctic continental shelves suggests that benthic communities may play an important role in carbon cycling and the regeneration of nutrients. Soft sediment benthic communities affect the micro-biology of sediments (e.g. Papaspyrou et al., 2006) and the compostion of meiofaunal and macrofaunal communities (e.g. Peachey and Bell, 1997; Callaway, 2006). They can also modify sediment chemistry (e.g. Norling et al., 2007), alter sediment stability and near-bed hydrodynamics (e.g. Norkko et al., 2001), release and draw-down oxygen, nutrients and particulates from the water column (e.g. Kamp and Witte, 2005), supply and consume pelagic organisms (e.g. Snelgrove et al., 2001), degrade and recycle detritus and primary production (e.g. Duchêne and Rosenberg, 2001; Renaud et al., 2007), and rework sediment to >20 cm depth (Dauwe et al., 1998).

Although Arctic benthic communities are abundant and diverse, they can be sensitive to disturbances like ice scour (Conlan and Kvitek, 2005). The loss of large epifauna—particularly ophiuroid echinoderms—to sea ice scouring may significantly impact the nature of carbon cycling in Arctic continental shelf sediments (Piepenberg et al., 1995; Ambrose et al., 2001; Renaud et al., 2007).

In the Canadian Beaufort region, benthic studies were spurred by hydrocarbon exploration during the 1970's

LEFT: The CCGS Amundsen in an Arctic Basin. Photo: Ramon Terrado/ ArcticNet.

[1] Department of Geography & Planning, 117 Science Place, University of Saskatchewan, Saskatoon, Saskatchewan, Canada S7N 5C8, Canada

[2] Canadian Museum of Nature, Canadian Museum of Nature, P.O. Box 3443, Station D, Ottawa, Ontario, K1P 6P4, Canada

[3] Department of Marine Sciences, University of Connecticut, Groton, Connecticut, United States of America 06340

[4] Present address: Akvaplan-niva, Polarmiljøsenteret, N-9296 Tromsø, Norway

[5] Maurice Lamontagne Institute, Fisheries and Oceans Canada, C.P. 1000 Mont-Joli, Québec, G5H 3Z4, Canada

[6] Present address: Institut des sciences de la mer (ISMER), Université du Québec à Rimouski, 310 Allée des Ursulines, Rimouski, QC, G5L 3A1, Canada

[7] Present Address: Département des sciences fondamentales, Université du Québec à Chicoutimi, 555 boulevard de l' Université, Chicoutimi, QC, G7B 2B1, Canada

* Corresponding author: alec.aitken@arts.usask.ca

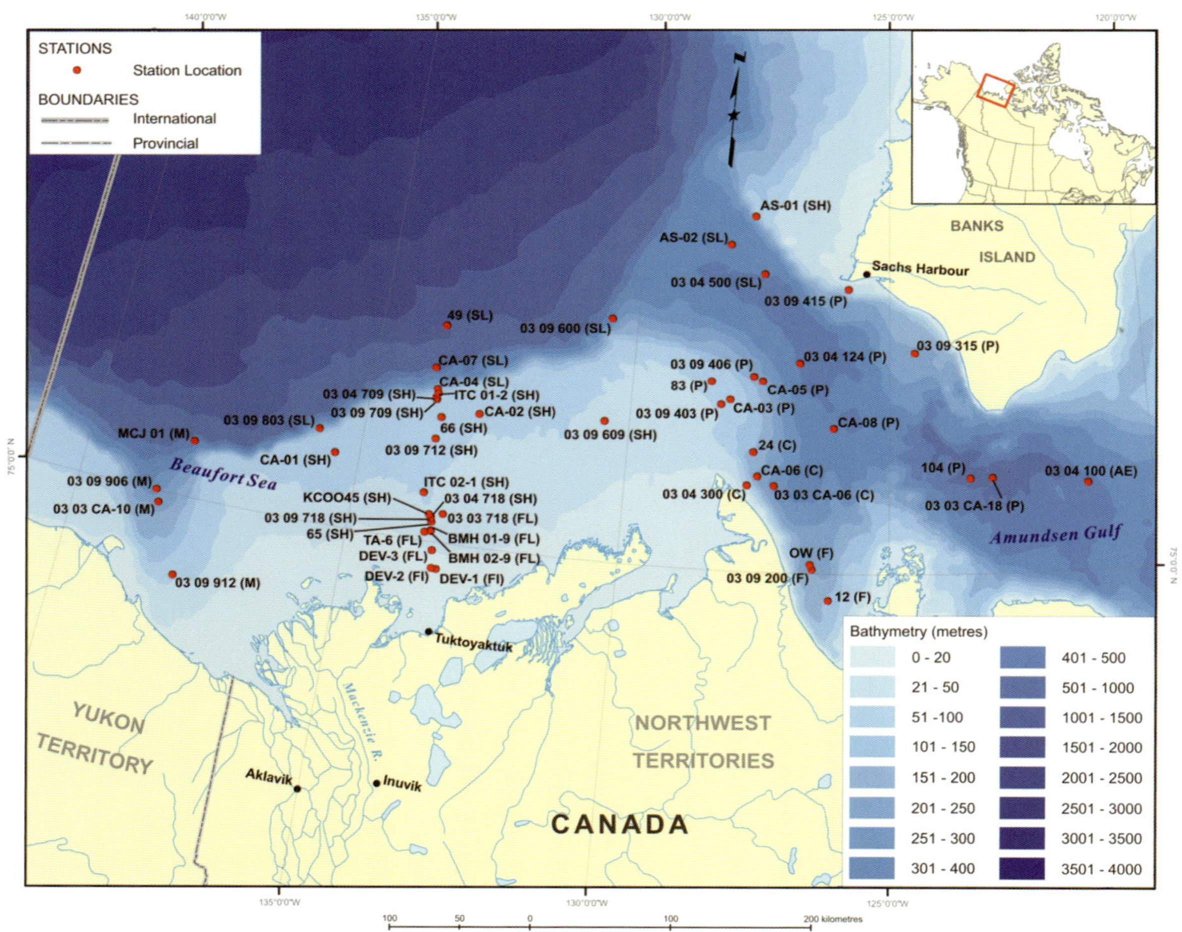

Figure 9.1

Macrofaunal sampling sites throughout the Beaufort Sea Shelf and Amundsen Gulf. M: Mackenzie Canyon; FI: fast ice; FL: flaw lead; SH: main shelf; SL: slope; C: Cape Bathurst; P: Cape Bathurst polynya; F: Franklin Bay; AE: Amundsen Gulf East. Base map produced by Elise Pietroniro (GIServices, University of Saskatchewan). Map data source: National Atlas of Canada 1:7.5 million scale, Natural Resources Canada, Government of Canada.

and 1980's (Wacasey, 1975; Wacasey et al., 1977; Atkinson and Wacasey, 1989). However, syntheses of these studies in the primary literature are sparse (Carey, 1991; Cusson et al., 2007). After the collapse of the Mackenzie gas pipeline development (following the Berger Enquiry of 1977), benthic studies fell into a lapse until the current decade (with the advent of Inuvialuit land claims and the resumption of hydrocarbon exploration). Now, large scientific endeavors (e.g. CASES; the Joint Western Arctic Climate Studies, JWACS; the *CCGS Nahidik* sampling programs) along with improved multivariate analytical methods (Clarke and Warwick, 2001) make it possible to statistically test hypotheses pertaining to local benthic community patterns and their relation to environmental variation.

The Canadian Beaufort Shelf is a broad platform extending over 64,000 km^2 to the 200 m isobath (O'Brien et al., 2006). The shelf is bordered by the Mackenzie Canyon to the west, the Amundsen Gulf to the east, the Mackenzie River Delta to the south, and the Beaufort Sea to the north (Fig. 9.1). Immense inputs of freshwater from the Mackenzie River (333 km^3/y; Macdonald et al., 1998) make it the most estuarine of all circumpolar shelves. The Mackenzie River's sediment load (127 million Mt/y) exceeds the combined annual sediment load of all other rivers entering the Arctic Ocean (O'Brien et al., 2006). Most of this load is delivered between late May and the end of August. Riverine discharge contributes to both particulate (POC) and dissolved (DOC) organic carbon (Dunton et al., 2006); however, the fate of this terrigenous carbon in the shelf ecosystem remains poorly understood. Beaufort Shelf sediments consist essentially of silts and clays discharged by the Mackenzie River or released by coastal erosion (Macdonald et al., 1998; Forest et al., 2007). Sand and gravel are largely confined to depths <10 m. Coarse-grained sediments are found at greater depths and are derived from drowned beaches or by ice-rafting (Héquette et al., 1995; Carmack and Macdonald, 2002). Waters between the surface and 220 m depth are mainly supplied by the relatively nutrient-rich Pacific Ocean, while those below 220 m are of Atlantic origin (Carmack et al., 2004).

Ice cover on the Beaufort Shelf is markedly variable on an inter-annual scale (Carmack and Macdonald, 2002). Generally, freeze-up begins in mid-October, break-up begins in late May, and the shelf can be clear of ice by mid-July (depending on winds). Winter landfast ice extends approximately to the 20 m isobath and is bordered by a stamukhi zone of grounded ice and pressure ridges. Intermittent open flaw leads can be found offshore the stamukhi zone, while pack ice (which tends to drift westward with the Beaufort Gyre) can be found on the outer shelf. The flaw lead to the east of the region widens into the Cape Bathurst polynya system, which is centered at the mouth of Amundsen Gulf (Fig. 9.1). The flaw lead and polynya system are vital to marine mammals and migratory birds (Harwood and Stirling, 1992; Dickson and Gilchrist, 2002). The polynya, however, exhibits marked variability in the timing, extent and persistence of its open water (Arrigo and van Dijken, 2004). This results in associated variability in the intensity, timing and duration of ice algae and phytoplankton blooms, and the subsequent flux of carbon to the benthos. Increased vertical mixing due to wind and brine production can also impact the benthos within the polynya. Specifically, these result in a greater export of carbon to the seafloor than during stratified summer conditions (when carbon production is intercepted by zooplankton; Arrigo and van Dijken, 2004). Grebmeier and Cooper (1995) documented the flux of organic carbon to the benthos associated with brine release and convective mixing within a polynya

in the central Bering Sea. Renaud et al. (2006) reported that the benthos along the Beaufort Shelf was responsive to surface processes by measuring a >10-fold increase in the oxygen demand of benthic sediment in the spring of 2004 at the time of the ice algae bloom. It was later concluded that Beaufort Shelf benthic communities mineralized approximately 60% of annual new production (Renaud et al., 2007), an amount indicative of their local importance in the carbon cycle.

The circumpolar region is experiencing rapid environmental changes. Recent studies have documented significant air and ocean temperature increases as well as reductions in sea ice cover—and it is predicted that these changes will accelerate over the next 50-100 years (Vinnikov et al., 1999; IPCC, 2001; Moritz et al., 2002; ACIA, 2004; Johannessen et al., 2004). It is unclear how these changes will impact ecosystems on circumpolar continental shelves. However, qualitative and quantitative shifts in the trophic structure and pathways of the carbon cycle are likely to occur (Carmack and Macdonald, 2002; Walsh et al., 2004; Wassmann et al., 2004; Tremblay et al., 2006). Climate warming may profoundly alter biogeochemical fluxes on circumpolar continental shelves by increasing the photosynthetic fixation of atmospheric carbon (through a reduction of ice cover and by increasing the riverine discharge of carbon; Carmack and McLaughlin, 2001). This would affect the export of carbon to the pelagic/benthic food webs, as well as to the deep basins (where it can ultimately be sequestered).

Given regional differences in sea ice cover, carbon supply, riverine influence, sedimentation, and upwelling throughout the Beaufort Sea Shelf region, we can hypothesize the following regarding the structure and respiration of its benthic community:

1. The composition of the benthic community within the Cape Bathurst polynya (Amundsen Gulf) is distinctive from its immediate surroundings at similar depth;

2. The composition of the benthic community is likely distinctive in upwelling regions like Cape Bathurst and the Mackenzie Canyon;

3. The Beaufort Shelf possesses a cross-shelf gradient in benthic community composition associated with depth-related changes in ice cover, ice scour, water masses (including input from the Mackenzie River), and upwelling at the shelf edge;

4. The respiration of the benthic community in the region reflects the seasonality of local sea ice, water mass input, and circulation/mixing cycles.

9.2 Overview of Results

9.2.1 Benthic community structure

9.2.1.1 Sample collection and species identification

Nine regions within the Beaufort Shelf and Amundsen Gulf study area (Table 9.1) were identified based on geography and ice regime (after Carmack and Macdonald, 2002, and O'Brien et al., 2006): The Mackenzie Canyon offshore of the Mackenzie delta; the Beaufort Shelf encompassing the inshore fast ice area (< 20 m depth); the flaw lead (20-35 m); the main part of the shelf (> 35 m to 200 m); the slope (> 200 m); and four regions in Amundsen Gulf (Cape Bathurst, the Cape Bathurst polynya, Franklin Bay and the part of the gulf sampled east of the polynya). Sampling occurred over six ship cruises in 2002-2004. *CCGS Sir Wilfrid Laurier* samples were collected in Sept 2002; *CCGS Radisson* samples in Oct 2002; and *CCGS Amundsen* (Legs 1, 2, 6 and 8) samples in Sept-Oct 2003, Oct-Nov 2003, April 2004 and July 2004 (respectively).

A 0.25 m^2 box corer was used for sampling aboard *CCGS Radisson* and *CCGS Amundsen* and a 0.1 m^2 van Veen grab was used aboard *CCGS Laurier*. The constraints of shared ship use minimized replication to three replicates per station, however, only one sample per station was recovered on *CCGS Radisson*. Shared sediment requirements also limited collection of biota to 0.063 - 0.13 m^2 of the box corer. The depth sampled was mostly 15 cm. Despite these constraints, 134 samples at 52 stations were collected, which comprised 30,802 individuals > 0.4 mm at a mean abundance of 230.1 ± 18.7 individuals per sample.

Macrofauna were elutriated from the sediment, fixed in buffered 5% formalin and seawater, and preserved in 70% ethanol. Organisms traditionally considered meiofauna (i.e. nematodes and harpacticoid copepods) were excluded from the data. The small mesh size used in this study (0.4 mm) did not capture additional species but augmented the abundance of smaller bodied species, particularly tanaids (Conlan et al., 2008). A total of 497 taxa were identified in the samples (227 crustaceans, identified by E. Hendrycks and C. McClelland; 111 molluscs, identified by A. Aitken; 122 polychaetes, identified by P. Pocklington and K. Conlan; 26 echino-

TABLE 9.1 AT RIGHT

Sampling characteristics of each region. Ship cruises: *Amundsen* (A) legs 1, 2, 6 and 8, *Laurier* (L), *Radisson* (R). Collectors: Aitken (A), Archambault (Ar), Conlan (C), McClelland (M), Morata (Mo), Simard (S). Sampler: Box core (1) or Van Veen grab (2). Since sample size varied, abundances were converted to no./m² for diversity calculations. ND = no data. (after Conlan et al., 2008).

REGION	MACKENZIE CANYON		BEAUFORT SHELF							
			INSHORE FAST ICE		FLAW LEAD		MAIN SHELF		SLOPE	
No. sites	4		2		5		12		7	
No. sites re-sampled seasonally	0		0		0		2		0	
No. samples	11		6		15		37		18	
Ship cruise	A1&8, L		L		A1, L		A2&8, L, R		A2&8, L, R	
Collector	A, Ar, C		C		Ar, C		A, C, M, S		A, C, M, S	
	MEAN	S.E.	MEAN	S.E.	MEAN	S.E.	MEAN	S.E.	MEAN	S.E.
Day	6.82	0.85	13.00	0.00	14.47	1.82	13.62	1.13	18.00	1.51
Month	8.36	0.41	9.00	0.00	9.07	0.07	8.54	0.21	8.50	0.31
Year (2: 2002; 3:2003; 4: 2004)	3.18	0.26	2.00	0.00	2.20	0.11	2.73	0.15	2.72	0.21
Latitude (North)	6981.76	11.15	6950.41	0.01	6996.92	5.32	7051.59	9.08	7125.70	12.92
Longitude (West)	13812.10	9.02	13320.92	1.20	13328.41	1.48	13275.59	31.36	13157.31	73.25
Water depth (m)	291.09	49.55	11.00	0.00	32.79	1.16	71.37	4.96	384.56	44.96
Sampler (1: box core; 2: van Veen grab)	1.27	0.14	2.00	0.00	1.80	0.11	1.49	0.08	1.50	0.12
Sample area (m^2)	0.08	0.01	0.10	0.00	0.11	0.01	0.09	0.00	0.10	0.01
Sample depth (cm)	13.91	0.58	15.00	0.00	13.53	0.79	15.00	0.00	14.44	0.38
Sieve mesh size (mm)	0.40	0.00	0.40	0.00	0.40	0.00	0.40	0.00	0.40	0.00
Bottom water fluorescence (µg/l)	0.08	0.01	0.50	0.04	0.49	0.05	0.40	0.06	0.07	0.01
Bottom water temp (°C)	-0.19	0.22	0.19	0.00	-0.80	0.04	-1.32	0.03	-0.01	0.13
Bottom water salinity (PSU)	34.26	0.26	22.65	0.07	27.07	1.48	32.24	0.07	34.57	0.09
Bottom water oxygen (ml/l)	6.06	0.11	7.60	0.00	7.97	0.04	7.06	0.12	6.05	0.10
Sediment % nitrogen	0.23	0.03	0.06	0.00	0.13	0.01	0.15	0.01	0.15	0.01
Sediment % organic carbon	2.09	0.60	0.58	0.02	0.66	0.02	1.10	0.12	1.43	0.24
Sediment C/N ratio	8.39	1.71	10.06	1.05	5.32	0.37	6.83	0.58	6.91	0.91
Sediment δ^{13}C (ppt)	-25.20	0.00	-26.25	0.02	-26.18	0.03	-25.03	0.19	-24.40	0.35
Sediment % clay	48.64	2.78	13.63	1.10	26.05	2.75	32.26	2.68	37.79	1.61
Sediment % silt	50.24	2.50	82.30	1.15	71.59	2.46	46.36	2.67	59.24	1.44
Sediment % vf sand	1.28	0.00	1.56	0.26	0.56	0.11	2.52	0.38	1.32	0.21
Sediment % f sand	0.43	0.00	0.48	0.02	0.29	0.02	10.38	2.58	0.64	0.16
Sediment % m sand	0.53	0.00	0.17	0.00	0.53	0.07	5.90	1.96	0.93	0.18
Sediment % c sand	0.00	0.00	1.02	0.14	0.58	0.17	1.77	0.61	0.48	0.15
Sediment % vc sand	0.00	0.00	0.84	0.18	0.40	0.12	1.27	0.51	0.19	0.14
Sediment F&W phi mean	8.39	0.17	7.06	0.09	8.20	0.23	6.83	0.51	8.43	0.13
Sediment F&W phi sorting	2.37	0.13	1.95	0.03	2.33	0.07	2.52	0.09	2.43	0.11
Sediment F&W phi skewness	0.22	0.02	0.25	0.00	0.36	0.01	0.26	0.04	0.23	0.05
Sediment F&W phi kurtosis	1.49	0.24	1.50	0.03	1.18	0.10	1.10	0.11	1.00	0.08
No./m^2	2562.82	362.62	1708.33	115.05	3261.00	1029.76	1945.92	160.06	882.56	150.71
Taxonomic diversity	72.75	3.48	59.52	6.90	44.47	4.12	71.81	1.92	77.89	1.60
Taxonomic distinctness	87.97	1.75	92.10	0.99	79.75	1.45	83.95	0.86	89.61	0.71

TABLE 9.1 (CONTINUED)

REGION	AMUNDSEN GULF							
	CAPE BATHURST POLYNYA		CAPE BATHURST		FRANKLIN BAY		AMUNDSEN GULF EAST	
No. sites	3		11		3		1	
No. sites re-sampled seasonally	1		0		0		0	
No. samples	10		28		6		3	
Ship cruise	A1&2, L, R		A1,2&8, L, R		A6&8, R		A2	
	MEAN	S.E.	MEAN	S.E.	MEAN	S.E.	MEAN	S.E.
Day	15.30	1.33	19.71	0.96	20.17	2.90	31.00	0.00
Month	9.90	0.28	8.39	0.24	6.50	0.92	10.00	0.00
Year	2.60	0.16	3.00	0.18	3.67	0.33	3.00	0.00
Latitude (North)	7037.78	1.30	7109.46	5.88	6994.56	8.86	7035.91	0.01
Longitude (West)	12730.83	4.03	12653.99	29.72	12625.47	3.27	12117.06	0.06
Water depth (m)	165.80	29.61	219.18	32.47	219.73	14.97	484.00	2.00
Sampler	1.30	0.15	1.32	0.09	1.00	0.00	1.00	0.00
Sample area (m²)	0.09	0.01	0.09	0.00	0.09	0.01	0.13	0.00
Sample depth (cm)	13.65	0.69	14.66	0.24	13.33	1.05	15.00	0.00
Sieve mesh size (mm)	0.40	0.00	0.40	0.00	0.40	0.00	0.40	0.00
Bottom water fluorescence (µg/l)	0.10	0.02	1.28	0.63	0.05	0.01	0.02	0.00
Bottom water temp (°C)	-0.70	0.21	-0.57	0.14	-0.49	0.23	0.24	0.00
Bottom water salinity (PSU)	33.54	0.38	33.53	0.23	34.07	0.28	34.72	0.00
Bottom water oxygen (ml/l)	6.04	0.37	6.43	0.23	5.80	0.26	5.18	0.00
Sediment % nitrogen	0.13	0.00	0.18	0.02	0.03	0.02	ND	ND
Sediment % organic carbon	0.70	0.00	1.72	0.18	2.13	0.09	2.77	0.32
Sediment C/N ratio	5.60	0.00	11.97	4.70	46.90	7.92	ND	ND
Sediment δ^{13}C (ppt)	-25.70	0.00	-24.30	0.14				
Sediment % clay	18.15	5.60	31.04	2.67	27.64	9.35	37.63	5.51
Sediment % silt	57.78	5.97	57.32	2.43	45.85	15.31	45.91	6.34
Sediment % vf sand	10.66	2.98	1.61	0.36	0.89	0.45	1.79	0.16
Sediment % f sand	12.36	6.20	5.92	2.58	0.16	0.08	1.48	0.08
Sediment % m sand	3.44	2.45	5.47	2.26	0.70	0.24	0.90	0.45
Sediment % c sand	0.73	0.18	0.61	0.20	12.79	12.40	4.66	4.65
Sediment % vc sand	0.85	0.24	0.17	0.09	20.73	20.73	7.63	7.63
Sediment F&W phi mean	6.26	0.61	7.33	0.35	6.17	2.16	8.80	0.06
Sediment F&W phi sorting	2.29	0.09	2.28	0.10	1.86	0.60	2.68	0.03
Sediment F&W phi skewness	0.24	0.05	0.26	0.06	0.20	0.10	0.41	0.02
Sediment F&W phi kurtosis	1.17	0.13	0.84	0.05	0.84	0.07	0.71	0.03
No./m²	7513.60	2140.74	2519.75	432.49	4747.33	740.66	490.67	62.37
Taxonomic diversity	71.72	3.04	74.60	1.36	80.00	1.65	77.00	3.87
Taxonomic distinctness	83.85	1.94	86.13	0.66	88.36	1.26	85.86	0.59

derms, identified by P. Lambert, C. McClelland and K. McKendry; and 11 others, identified by P. Lambert, K. Conlan and C. McClelland). Of the 497 taxa, 3.8% were identified to phylum or order, 6.2% to family, 12.5% to genus, and 77.5% to species. All samples are housed at the Canadian Museum of Nature, Ottawa.

Bottom water characteristics were measured at the time of faunal and sediment collection. Seasonal variation in temperature, salinity and current speed were recorded from fixed moorings at stations BMH02-9, BMH01-9, ITC02-1, 03 09 709, ITC01-2 and AS-01 (see figure 9.1). Sediment samples taken from a separate section of the macrofaunal sample were frozen and analyzed for grain size, % N, % organic C and $\delta^{13}C$. Analytical procedures are presented in Conlan et al. (2008).

9.2.1.2 Multivariate analyses

The common practice of converting small sample abundances to organism abundance over 1 m^2 can often lead to over-estimation if the sample area is small (De Grave et al., 2001). Our sample area always exceeded the recommended 0.0137 m^2 required for the accurate estimation of organism abundance in muddy habitats. Therefore, we only converted our values to number m^{-2} for the purpose of illustrating abundance distributions; otherwise, values were standardized by totals to adjust for variable sample size, and later transformed (fourth root, Cusson et al., 2007; square root, Conlan et al., 2008) to prevent over-domination by any abundant species. Abundances varied from 0 to 508 individuals per sample and so required only light transformation.

Because of the variability in sample sizes, only the 'taxonomic diversity index' of Warwick and Clarke (1995) was used for diversity comparisons. This measure is not influenced by sample size while commonly reported indices such as the 'Shannon-Wiener index'

and 'species richness index' are (Clarke and Warwick, 1998). Taxonomic diversity is also a more sensitive univariate measure of community structure because it considers taxonomic relationships. Furthermore, it includes an evenness component (a form of 'Simpson's index') (Clarke and Warwick, 1998).

Resemblances were calculated using the Bray-Curtis measure for similarity (based on species and abundances in common among the sites). Resemblances were mapped using multidimensional scaling and differences among regions were tested for significance using the one-way ANOSIM procedure in Primer 6.1.5 (Clarke and Green, 1988; Clarke, 1993; Clarke and Gorley, 2006). The taxa which characterized or distinguished each region were then identified using the SIMPER procedure in Primer. This routine determines each species' contribution to similarity within a group and compares it to its contribution to distinguishing groups.

The samples differed in their method of collection, date, location, and bottom water and sediment characteristics. The BEST procedure was applied to examine which single or multiple variables most highly correlated with biotic abundance patterns. Percent sand fractions were summed. The abiotic data were individually transformed ($\log x + 1$) or square root transformed if skewed (for percentage data); then all variables were normalized (x-mean sd^{-1}) and site similarities were calculated. The biotic data were standardized by total and square root transformed. The Spearman rank correlation was calculated between the species resemblance matrix and the normalized abiotic resemblance matrix (using different combinations of up to five environmental variables). Readers are referred to Cusson et al. (2007) and Conlan et al. (in press) for further information about the statistical analyses applied to these data.

TABLE 9.2

List of dominant taxa that contributed a total of 75% of the numerical abundance in each of the three zones on the Beaufort Sea Shelf (Cusson, unpublished).

Zone 1 (4-10 m)	%	Zone 2 (10-30 m)	%	Zone 3 (30-200 m)	%
Minuspio cirrifera	16.3	*Micronephthys minuta*	18.9	*Diastylis rathkei*	26.4
Ampharete vega	11.1	*Minuspio cirrifera*	14.4	*Micronephthys minuta*	13.4
Nemata 2008	8.4	*Tharyx acutus*	7.5	*Tharyx acutus*	8.6
Chaetozone sp.	7.2	Nemata	6.3	*Levinsenia gracilis*	6.0
Halicryptus spinulosus	7.1	*Montacuta maltzani*	5.8	Nemata	4.2
Yoldiella intermedia	6.6	*Portlandia arctica*	5.2	*Maldane sarsi*	4.1
Oligochaeta	5.7	*Cythereis* sp. A	2.7	*Aricidea suecica*	3.1
Macoma balthica	4.4	*Aricidea* sp.	2.4	Oligochaeta	2.4
Micronephthys minuta	4.2	*Cylichna occulta*	2.3	*Ampelisca eschrichti*	2.0
Peloscolex sp.	3.9	*Ampharete vega*	2.3	*Philomedes brenda*	1.9
		Artacama proboscidea	2.1		
		Levinsenia gracilis	2.0		
		Cythereis dunelmensis	1.9		

9.2.1.3 Regional variation in benthic community structure

Wacasey et al. (1977) described the benthic community of the Beaufort Sea Shelf in terms of three depth zones (Zone 1: 4-10 m depth; Zone 2: 10-30 m depth; Zone 3: 30-200 m depth). These zones corresponded closely to the inshore fast ice region (<20 m), the flaw lead region (20-35 m), and the main Shelf region (>35 m to 200 m) of the shelf. Figure 9.2 shows a Multidimensional Scaling (MDS) plot of all the benthic sites reported by Wacasey et al. (1977) based on the Bray-Curtis index of similarity (Cusson, unpublished). Results from similarity analyses showed that these assemblages were dif-ferent (global R statistic: 0.232; p = 0.0002; see Clarke and Warwick, 2001): *Minuspio cirrifera*, *Malacoceros fuliginosus*, and *Ampharete vega* contributed to most of the similarity within zone 1 (explaining 12.4, 11.5 and 8.4% of similarities among stations, respectively); *Micronephthys minuta*, *Tharyx acutus*, and *Portlandia arctica* contributed to most of the similarity within zone 2 (explaining 22, 14.3 and 7.5% of similarities among stations, respectively); *Tharyx acutus*, *Micronephthys minuta*, and *Aricidea suecica* contributed to most of the similarity within zone 3 (explaining 15.8, 12.6 and 8.5% of similarities among stations, respectively); and *Micronephthys minuta*, *Tharyx acutus*, and *Minuspio cirrifera* contributed to most of the general dissimilarities among the 3 zones. Dominant species in each zone are identified in Table 9.2. Data from 76 stations were used to link environmental variables and biotic assemblages. Mean Chlorophyll *a* concentration and bottom temperature best explained the pattern of macrobenthic assemblages (Table 9.3).

Using the method of Spearman rank correlation to analyse the environmental and biological characteristics of the 9 regions sampled during the CASES program, we were able to assess that water depth was the abiotic environmental factor which shared the highest relationship (correlation=0.631) with macrofaunal abundance site similarities. In contrast, species distributions correlated weakly with fluorescence, bottom salinity, oxygen concentration, and various measures of grain size distributions (Conlan et al., in press). Figure 9.3 and Figure 9.4 illustrate the spatial variation of macrobenthos abundance and taxonomic diversity recorded at the time of faunal sampling. Regional macrofaunal abundance varied between 490.7 m^{-2} (in eastern Amundsen Gulf) and 6884.7 ± 3459.4 m^{-2} (off Cape Bathurst; Table 9.1). The highest value for abundance (17,950 m^{-2}) was recorded at station 03 04 300-3 (Figure 9.3), the station closest inshore at Cape Bathurst. Taxonomic diversity (which incorporates an evenness component) ranged from 47.2 ± 5.8 m^{-2} (in the flaw lead region of the Beaufort Shelf) to 78.3 ± 1.9 m^{-2} (on the shelf slope).

Maximum fluorescence values for bottom water were observed in the Cape Bathurst polynya region (1.05 ± 0.93 µg l^{-1}; Table 9.1). Bottom water temperature was lowest on the Beaufort Shelf (-1.31 ± 0.05 °C, averaged over 15 stations) while salinity was lowest close to the Mackenzie River Delta (22.65 ± 0.15 ‰ averaged over 2 stations in the fast ice region). Bottom water oxygen levels were high throughout the area (5.18 – 7.97 ± 0.07 ml l^{-1}).

TABLE 9.3

Combinations of environmental variables (taken *k* at a time) which gave the largest rank correlation ρ_s between biotic and abiotic similarity matrices (Based on data from 76 stations). Bold type indicates the best combination overall: Chla corresponds to mean Chlorophyll *a* concentration (measured by SeaWIFS); T corresponds to bottom temperature (°C); Sal corresponds to salinity (‰); % Silt, % Clay, and % Sand correspond to the percentage of silt, clay or sand in sediment, respectively (Cusson, unpublished).

κ	**Best variable combinations (ρ_s)**			
1	T (.54)	Sal (.51)	Chla (.38)	...
2	Chla, T (.56)	T, Sal (.53)	Chla, Sal (0.51)	Chla, T, Sal (.38) ...
3	**Chla, T, Sal (.53)**	T, Sal, %Silt (.40)	...	

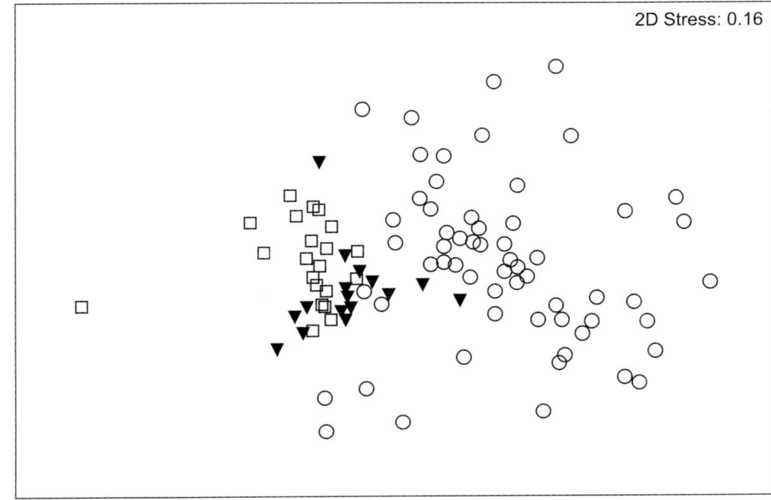

Figure 9.2

Non-metric multidimensional scaling plot of bottom community similarities from three depth zones on the Beaufort Sea Shelf. Data are from Wacasey et al. (1977). Open circles: From 4 to 10 meters; black triangles: from 10 to 30 m; open squares: from 30 to 200 m (Cusson, unpublished).

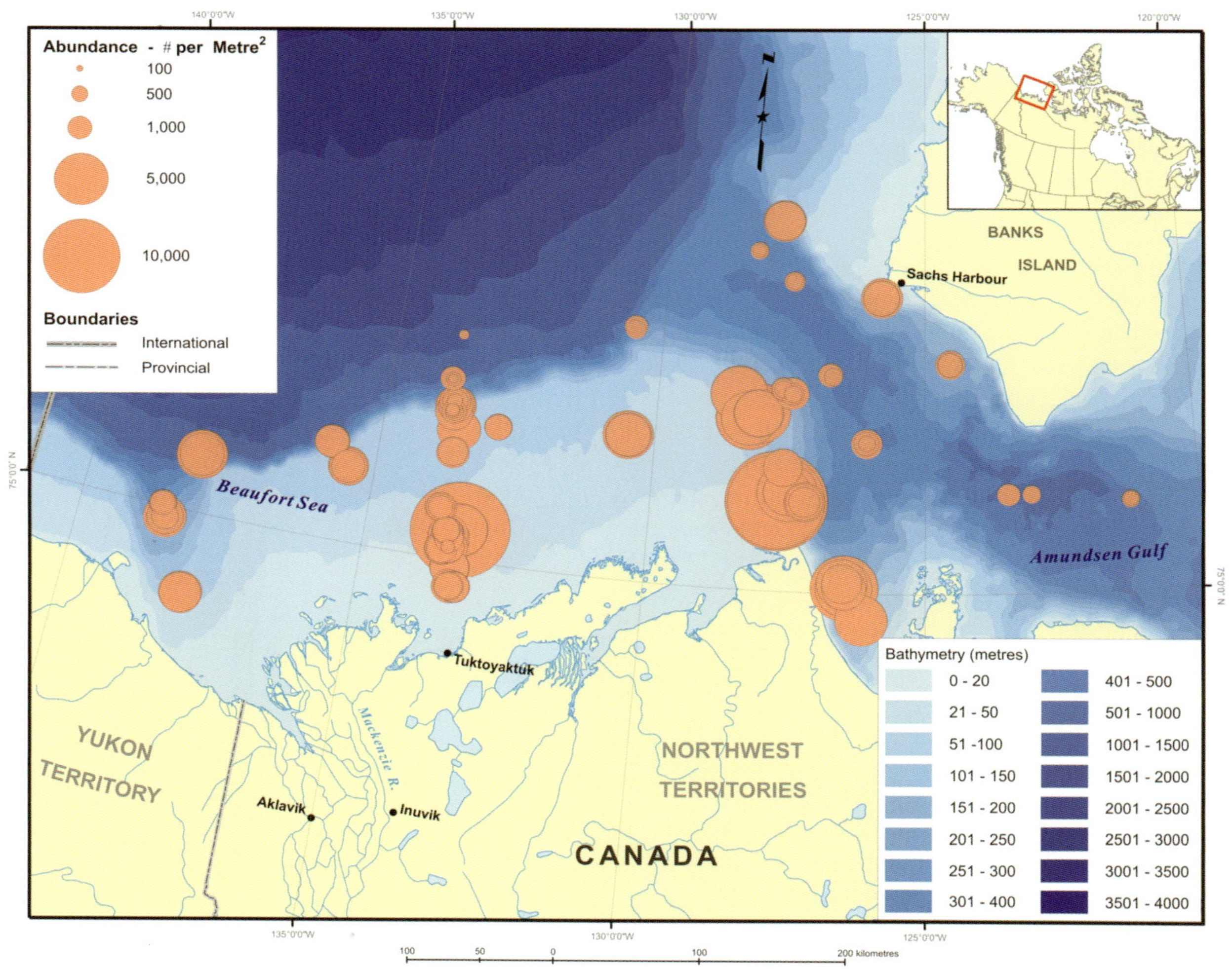

Figure 9.3

Macrofaunal community abundance (individuals/m²) on the Beaufort Sea Shelf and in Amundsen Gulf. Base map produced by Elise Pietroniro (GIServices, University of Saskatchewan). Map data source: National Atlas of Canada 1:7.5 million scale, Natural Resources Canada, Government of Canada.

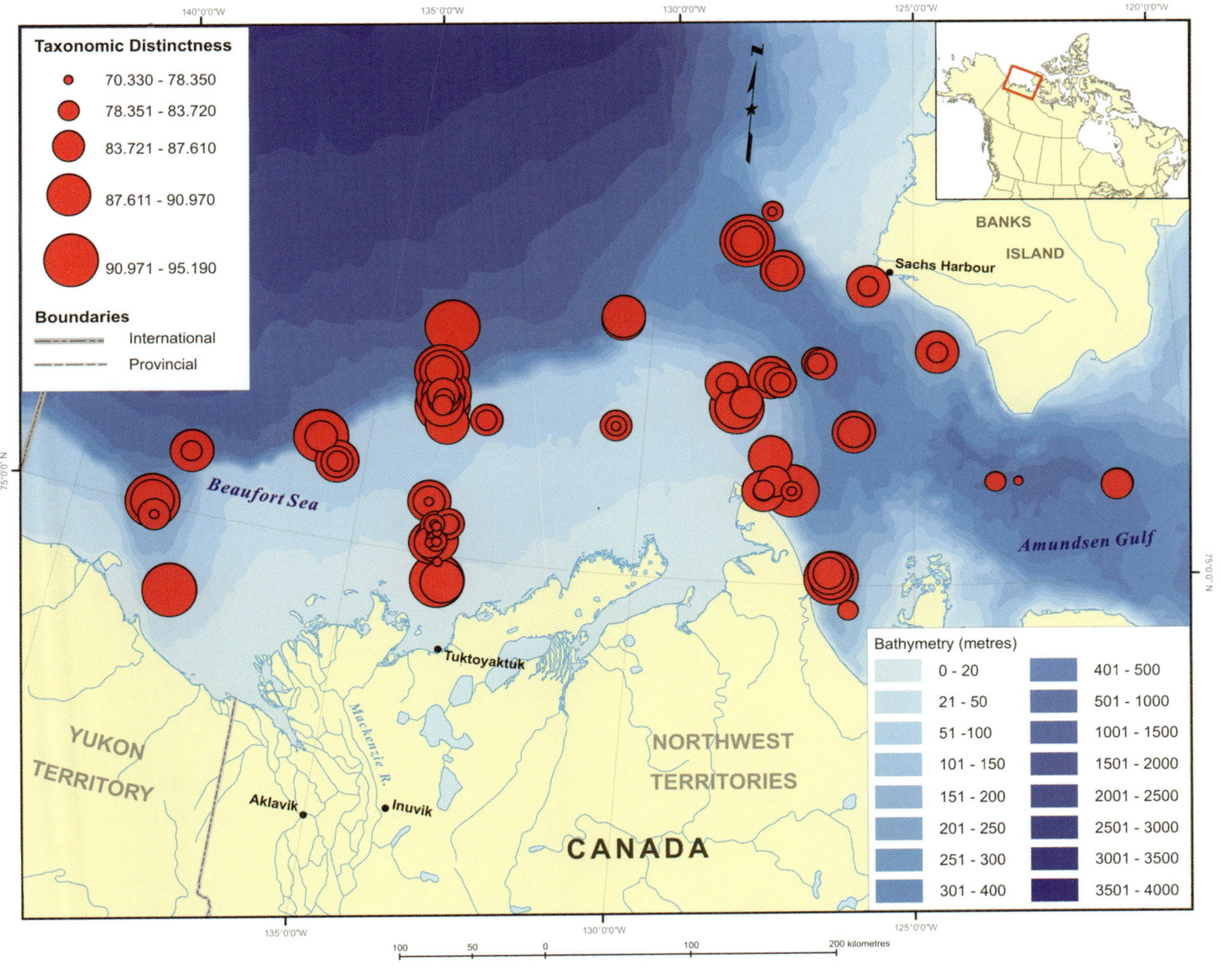

Figure 9.4

Taxonomic distinctness of macrofaunal communities on the Beaufort Sea Shelf and in Amundsen Gulf. Base map produced by Elise Pietroniro (GIServices, University of Saskatchewan). Map data source: National Atlas of Canada 1:7.5 million scale, Natural Resources Canada, Government of Canada.

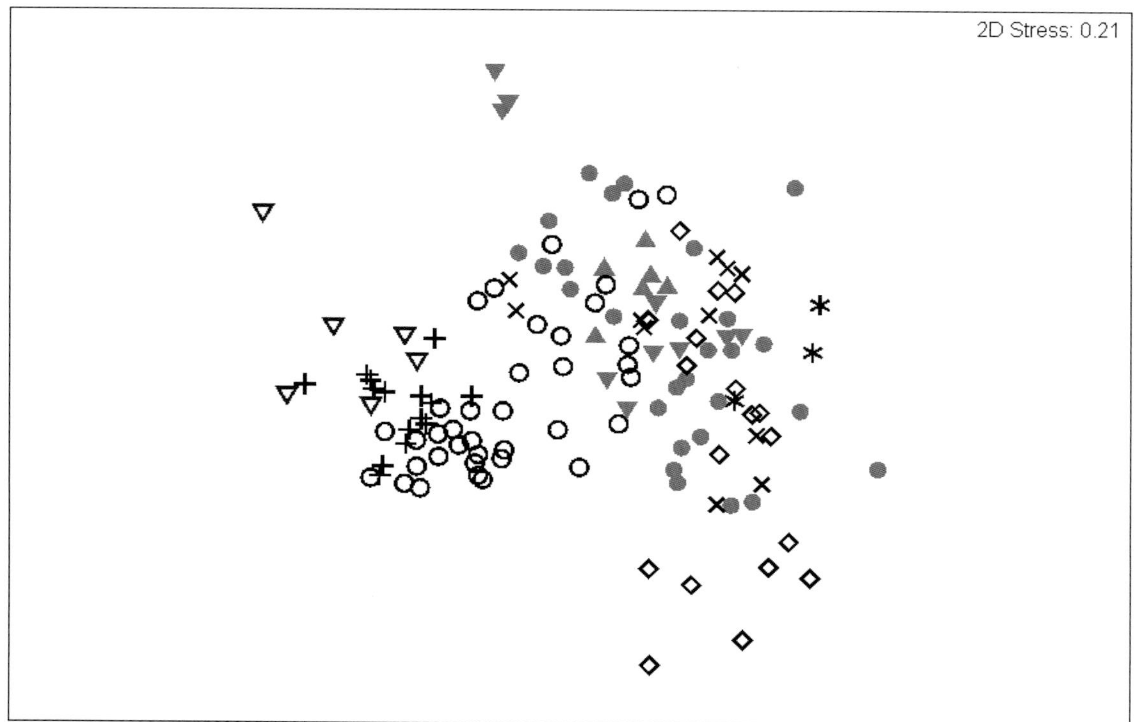

Figure 9.5

Non-metric multidimensional scaling plot of community similarities (replicates averaged) in each region of the Beaufort study area (after Conlan et al., 2008).

Sediment C/N ranged from 5.32 ± 0.71 (flaw lead) to 10.06 ± 2.34 (fast ice). Sediment $\delta^{13}C$ was more negative inshore on the Beaufort Shelf (-26.25 ± 0.05 ‰) and less negative offshore (-24.40 ± 0.70 ‰). The sediments in each region were well sorted silt-clays (Folk and Ward mean range from 6.61 ± 1.00 phi to 8.34 ± 0.31 phi). However, higher sand fractions were found at some stations on the shelf (for example, station AS01 and station 03 09 709), off Cape Bathurst (station 83 and station 03 04 300) and Franklin Bay (station 03 09 200).

A Multidimensional Scaling (MDS) plot of all the macrobenthos samples collected was plotted (Fig. 9.5) based on the Bray-Curtis index of similarity. Here, polynya samples intermingled with samples from neighbouring Cape Bathurst, Franklin Bay and Amundsen Gulf East. The three replicates taken at the shallowest station at Cape Bathurst (station 03 04 300) clustered tightly and were dissimilar to other samples taken in this region. Other regional samples were broadly cohesive. Mackenzie Canyon samples mixed with all other samples but the fast ice and flaw lead samples.

TABLE 9.4

Taxa typifying (cumulatively accounting for 75% of the within-group similarity) each region.
The region Amundsen Gulf East is not listed because it comprises only one site (after Conlan et al., 2008).

Taxa	% Contribution to within-group similarity
Mackenzie Canyon. Average within-group similarity: 27.84	
Thyasira flexuosa (Montagu, 1803) (Bivalve)	10.86
Tharyx kirkegaarde + *T. marioni* St. Joseph, 1894 (Polychaete)	10.03
Lumbrineris impatiens (Claparède, 1868) (Polychaete)	6.86
Terebellides stroemi Sars, 1835 (Polychaete)	5.95
Prionospio cirrifera Wiren, 1883 + *P. steenstrupi* Malmgren, 1867 (Polychaete)	5.93
Paraleptognathia gracilis (Kroyer, 1842) (Tanaid)	5.41
Sipunculids	4.86
Nemerteans	3.84
Maldane sarsi Malmgren, 1865 (Polychaete)	3.72
Ektonodiastylis nimia (Hansen, 1920) (Cumacean)	3.27
Pholoe minuta (Fabricius, 1780) (Polychaete)	3.17
Aricidea albatrossae Pettibone, 1957 (Polychaete)	2.79
Ehlersia cornuta (Rathke, 1843) (Polychaete)	2.11
Micronephthys minuta (Theel, 1879) (Polychaete)	2.08
Philomedes brenda (Baird, 1850) (Ostracod)	1.84
Trochochaeta multisetosa (Oersted, 1844) (Polychaete)	1.66
Rabilimis mirabilis (Brady, 1868) (Ostracod)	1.63
Fast ice. Average within-group similarity: 52.58	
Micronephthys minuta (Polychaete)	32.21
Portlandia arctica (J. E. Gray, 1824) (Bivalve)	12.6
Podocopid 3b (Ostracod)	10.74
Xestoleberis depressa G.O. Sars, 1866 (Ostracod)	7.51
Heteromastus sp. (Polychaete)	6.79
Normanicythere leioderma (Norman, 1869) (Ostracod)	5.99
Flaw lead. Average within-group similarity: 34.19	
Micronephthys minuta (Polychaete)	45.05
Tharyx kirkegaarde + *T. marioni* (Polychaete)	7.81
Levinsinea gracilis (Tauber, 1879) (Polychaete)	6.21

TABLE 9.4 (CONTINUED) **Taxa**	**% Contribution to within-group similarity**
Prionospio cirrifera + steenstrupi (Polychaete)	5.12
Pontoporeia femorata Kroyer, 1842 (Amphipod)	3.13
Saduria sabini (Kroyer, 1849) (Isopod)	2.68
Aceroides latipes (G.O. Sars, 1883) (Amphipod)	2.56
Rabilimis mirabilis (Brady, 1868) (Ostracod)	2.52
Beaufort shelf. Average within-group similarity: 31.12	
Micronephthys minuta (Polychaete)	21.79
Tharyx kirkegaarde + marioni (Polychaete)	15.15
Maldane sarsi (Polychaete)	8.54
Levinsinea gracilis (Polychaete)	6.77
Paraleptognathia gracilis (Tanaid)	4.4
Ophiocten sericeum (Forbes, 1852) (Ophiuroid)	3.89
Lumbrineris impatiens (Polychaete)	2.67
Prionospio cirrifera + steenstrupi (Polychaete)	2.63
Leucon nasicus (Kroyer, 1841) (Cumacean)	2.35
Pectinaria hyperborea (Malmgren, 1866) (Polychaete)	1.55
Pontoporeia femorata (Amphipod)	1.42
Barantolla americana Hartman, 1963 (Polychaete)	1.3
Saduria sabini (Isopod)	1.26
Aricidea catherinae Laubier, 1967+ *A. nolani* Webster and Benedict, 1887 + *A. suecica* (Polychaete)	1.17
Crenella faba (Muller, 1776) (Bivalve)	1.04
Beaufort deep. Average within-group similarity: 29.30	
Maldane sarsi (Polychaete)	15.5
Tharyx kirkegaarde + marioni (Polychaete)	11.69
Thyasira flexuosa (Bivalve)	8.4
Sipunculids	7.25
Portlandia sp. (Bivalve)	6.37
Lumbrineris tenuis (Verrill, 1873) (Polychaete)	5.09
Lumbrineris impatiens (Polychaete)	3.67
Aricidea albatrossae (Polychaete)	2.32
Gnathia sp. (Isopod)	1.82
Paraleptognathia gracilis (Tanaid)	1.77
Indeterminate bivalves	1.6

Taxa	% Contribution to within-group similarity
Prionospio cirrifera + steenstrupi (Polychaete)	1.54
Thyasira sp. (Bivalve)	1.41
Nemerteans	1.37
Haploops sp. (nr. *tubicola*) (Amphipod)	1.36
Ehlersia cornuta (Polychaete)	1.33
Philomedes brenda (Ostracod)	1.33
Ektonodiastylis nimia (Hansen, 1920) (Cumacean)	1.3

Cape Bathurst. Average within-group similarity: 27.47

Taxa	% Contribution
Maldane sarsi (Polychaete)	7.5
Paraleptognathia gracilis (Tanaid)	7.39
Ektonodiastylis nimia (Cumacean)	7.06
Prionospio cirrifera + steenstrupi (Polychaete)	6.47
Thyasira flexuosa (Bivalve)	6.19
Philomedes brenda (Ostracod)	5.89
Cypridina megalops G.O. Sars, 1872 (Ostracod)	4.49
Micronephthys minuta (Polychaete)	4.48
Tharyx kirkegaarde + marioni (Polychaete)	4.1
Lumbrineris impatiens (Polychaete)	4.07
Terebellides stroemi M. Sars, 1835 (Polychaete)	3.39
Paraleptognathia manca (G.O. Sars, 1882) (Tanaid)	3.2
Heteromastus sp. (Polychaete)	2.42
Harpinia sp. (Amphipod)	2.2
Portlandia frigida (Torell, 1859) (Bivalve)	2.03
Nemerteans	1.88
Ophelina cylindricaudata (Hansen, 1878) (Polychaete)	1.72
Levinsinea gracilis (Polychaete)	1.44

Polynya. Average within-group similarity: 27.83

Taxa	% Contribution
Maldane sarsi (Polychaete)	14.11
Tharyx kirkegaarde + marioni (Polychaete)	9.75
Prionospio cirrifera + steenstrupi (Polychaete)	7.81
Thyasira flexuosa (Bivalve)	6.1
Philomedes brenda (Ostracod)	4.45
Pholoe minuta (Polychaete)	3.51

TABLE 9.4 (CONTINUED) **Taxa**	% Contribution to within-group similarity
Terebellides stroemi (Polychaete)	3.19
Paraleptognathia gracilis (Tanaid)	2.99
Ophelina cylindricaudata (Polychaete)	2.66
Ektonodiastylis nimia (Cumacean)	2.64
Lumbrineris impatiens (Polychaete)	2.31
Sipunculids	2.21
Aricidea albatrossae (Polychaete)	1.83
Portlandia sp. (Bivalve)	1.8
Micronephthys minuta (Polychaete)	1.62
Sabellidae (Polychaete)	1.38
Eudorella sp. (Cumacean)	1.34
Cypridina megalops (Ostracod)	1.33
Haploops sp. (Amphipod)	1.26
Gnathia sp. (Isopod)	1.12
Gnathia elongata (Kroyer, 1846) (Isopod)	1.03
Pseudosphyrapus anomalus (G.O. Sars, 1869) (Tanaid)	1.01
Franklin Bay. Average within-group similarity: 45.08	
Tharyx kirkegaarde + marioni (Polychaete)	7.89
Podocopid 1a (Ostracod)	7.11
Lumbrineris impatiens (Polychaete)	6.85
Prionospio cirrifera + steenstrupi (Polychaete)	6.28
Philomedes brenda (Ostracod)	4.94
Maldane sarsi (Polychaete)	4.65
Rabilimis mirabilis (Ostracod)	4.58
Thyasira flexuosa (Bivalve)	4.26
Barantolla americana (Polychaete)	4.01
Pholoe minuta (Polychaete)	3.94
Aricidea albatrossae (Polychaete)	3.74
Portlandia intermedia (Sars, 1865) (Bivalve)	3.73
Cypridina megalops (Ostracod)	3.45
Micronephthys minuta (Polychaete)	3.39
Ektonodiastylis nimia (Cumacean)	2.87
Cossura longocirrata Webster and Benedict, 1887 (Polychaete)	2.38
Normanicythere leioderma (Ostracod)	2.32

Macrofauna which characterized each of the 9 regions of the study area are listed in Table 9.4. Polychaetes dominated the communities, with *Maldane sarsi, Tharyx kirkegaarde/marioni, Lumbrineris impatiens* and *Prionospio cirrifera/steenstrupi* being common to most regions. The deep-burrowing bamboo worm, *Maldane sarsi,* was one of the most abundant species at depth. The small, shallow-burrowing polychaete *Micronephtys minuta* occurred there as well but was more abundant in the inshore fast ice and flaw lead regions. The tanaid *Paraleptognathia gracilis,* the cumacean *Ektonodiastylis nimia* and the ostracod *Philomedes brenda* were the most widespread and common crustaceans. The isopods *Saduria sabini, S. entomon* and *S. sibirica* were common in the flaw lead and on the Beaufort Shelf, but given their large size they were more often captured as wandering megafauna than macrofauna. Deposit feeders *Thyasira flexuosa, Portlandia* spp., and *Bathyarca* spp. were the dominant bivalves in the region.

Based on the relative abundance of taxa, the benthic community beneath the polynya did not differ significantly from adjacent communities in Amundsen Gulf East, Franklin Bay and Cape Bathurst (Conlan et al., 2008). The species composition of the polynya was significantly different from the distant Beaufort Shelf and inshore regions. However, it did not differ significantly from communities on the Beaufort slope and in the Mackenzie Canyon. Figure 9.6 shows an artist's reconstruction of the benthic community composition observed throughout the polynya zone in Amundsen Gulf.

The most significant difference between the polynya community and that of the main Beaufort Shelf lay in the relative abundance of several large fauna (Table 9.5). The polychaetes *Maldane sarsi, Pholoe minuta, Barantolla americana, Ophelina cylindricaudata* and *Terebellides*

TABLE 9.5

Mean abundances (s.e.) m^{-2} of the dominant macrofauna in the Cape Bathurst polynya compared with the Beaufort shelf, flaw lead, and fast ice regions (after Conlan et al.,2008).

	Polynya	Shelf	Flaw lead	Fast ice
Philomedes brenda	520.9 (329.4)	14.3 (10.6)	153.3 (153.3)	1.9 (1.9)
Maldane sarsi	212.4 (55.0)	128.3 (48.3)	1.5 (0.9)	0
Tharyx kirkegaarde + T. marioni	133.7 (41.3)	160.1 (23.2)	264.9 (108.8)	38.9 (38.9)
Cypridina megalops	112.9 (69.7)	14.8 (10.1)	0	0
Pholoe minuta	93.0 (61.2)	11.8 (7.8)	0.7 (0.7)	0
Thyasira flexuosa	73.3 (22.9)	0.6 (0.4)	0	0
Prionospio cirrifera + P. steenstrupi	69.1 (18.8)	49.8 (27.5)	39.7 (16.8)	31.5 (27.8)
Levinsinea gracilis	49.0 (29.4)	83.5 (20.3)	104.1 (35.0)	5.6 (5.6)
Micronephthys minuta	43.7 (19.9)	536.5 (133.8)	2341.0 (1158.4)	735.2 (227.8)
Barantolla americana	31.4 (17.6)	16.8 (4.5)	7.3 (3.1)	7.4 (3.7)
Lumbrineris impatiens	30.0 (14.9)	46.5 (15.6)	12.5 (5.3)	1.9 (1.9)
Normanicythere leioderma	29.9 (26.3)	0.8 (0.5)	0	31.5 (5.6)
Ektonodiastylis nimia	29.1 (11.7)	9.0 (6.2)	0	0
Paraleptognathia gracilis	28.3 (10.5)	75.8 (29.2)	47.5 (42.2)	13.0 (1.9)
Ophelina cylindricaudata	27.0 (9.3)	10.2 (5.3)	0	0
Portlandia sp.	24.7 (15.6)	5.6 (3.8)	2.6 (1.4)	0
Terebellides stroemi	22.2 (5.4)	5.5 (3.3)	5.7 (3.9)	1.9 (1.9)
Haploops sp.	18.4 (14.5)	0.3 (0.3)	8.9 (8.9)	33.3 (33.3)
Sipuncula	17.8 (6.1)	16.0 (9.3)	0.7 (0.7)	0
Xestoleberis depressa	15.9 (12.9)	14.0 (7.5)	14.1 (10.1)	101.9 (61.1)
Eudorella sp.	11.1 (5.5)	1.2 (0.7)	3.0 (1.8)	0
Cossura longocirrata	10.6 (6.7)	13.0 (7.3)	18.5 (6.7)	18.5 (3.7)
Ampharete balthica + A. vega	9.9 (4.0)	2.4 (1.6)	0	3.7 (0)
Nemertea	9.8 (2.7)	5.6 (1.3)	8.2 (3.0)	1.9 (1.9)
Heteromastus sp. + Mediomastus sp.	9.6 (4.5)	16.2 (8.0)	6.7 (3.0)	42.6 (31.5)
Sabellidae	9.5 (4.0)	2.6 (1.3)	0	0
Aricidea albatrossae	8.8 (3.0)	11.0 (5.2)	3.0 (3.0)	3.7 (3.7)
Leitoscoloplos sp. + Scoloplos sp.	8.3 (3.2)	15.9 (5.4)	10.4 (5.0)	11.1 (0)
Lumbrineris tenuis	8.3 (3.8)	2.6 (1.5)	2.1 (2.1)	27.8 (1.9)
Ophiocten sericeum	7.2 (3.1)	69.0 (29.1)	1.1 (0.7)	0
Gnathia sp.	6.1 (2.7)	0.5 (0.5)	0	0
Aricidea catherinae + A. nolani + A. suecica	4.7 (2.9)	20.4 (6.1)	70.1 (28.5)	7.4 (3.7)
Pontoporeia femorata	1.6 (1.1)	22.9 (13.8)	28.2 (14.5)	9.3 (9.3)
Saduria sabini	1.5 (1.5)	5.3 (1.8)	8.1 (2.9)	0
Haploops laevis	1.3 (1.3)	10.9 (4.8)	5.0 (3.6)	50.0 (24.1)
Leucon nasicus	1.0 (1.0)	20.1 (6.9)	14.1 (7.3)	0
Trochochaeta multisetosa	1.0 (1.0)	2.2 (1.5)	0.7 (0.7)	48.1 (11.1)
Portlandia arctica	0.9 (0.6)	0.6 (0.6)	23.7 (21.9)	133.3 (18.5)
Pectinaria hyperborea	0	37.0 (18.6)	109.0 (109.0)	0
Podocopid 3b	0	1.3 (1.3)	0	113.0 (24.1)

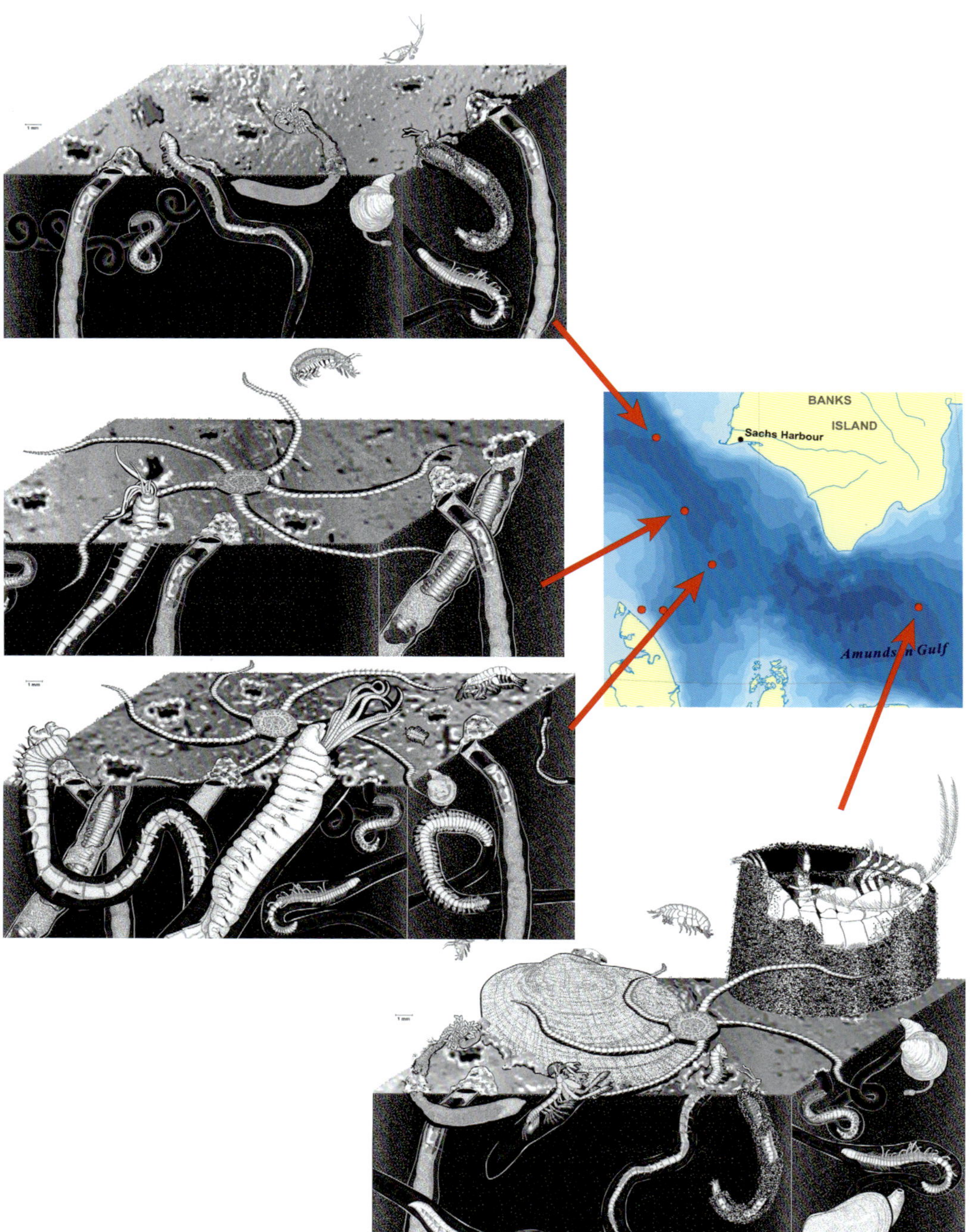

stroemi, the bivalves *Thyasira flexuosa* and *Portlandia sp.*, the ostracods *Cypridina megalops* and *Philomedes brenda* and the cumacean *Ektonodiastylis nimia*, and the isopod *Gnathia* sp. were relatively more abundant in the polynya than on the Beaufort Shelf. Conversely, the polychaetes *Levinsinea gracilis*, *Micronephthys minuta*, *Heteromastus* sp./*Mediomastus* sp., *Aricidea* spp. and *Pectinaria hyperborea*, the bivalve *Portlandia arctica*, the cumacean *Leucon nasicus*, the tanaid *Paraleptognathia gracilis*, the amphipod *Ponotporeia femorata*, and the ophiuroid echinoderm *Ophiocten sericeum* were more abundant on the shelf than in the polynya. This contrast in abundance was even more pronounced inshore, where the polynya species were rare or absent.

Figure 9.6

Observed community composition for the Cape Bathurst polynya. Stations from north to south: 0304 500-3 (396 m); 0304 124-1 (440 m); CA08-1 (390 m); 0304 100-1 (486 m). Species accounting for cumulatively 75% of the sample abundance are shown (scale bars = 1 mm). Deposit-feeding polychaetes were observed to dominate the community, including taxa which feed at the suface (Prionospio cirrifera, Terebellides stroemi and Melinna cristata) and taxa which feed at depth (Lumbrineris spp., Tharyx kirkegaarde and Maldane sarsi). Epifaunal dominants included the tubicolous amphipod Haploops sp., the bivalve Bathyarca sp. and the ophiuroid echinoderm Ophiocten sericeum. Bathyarca also provided an attachment and nestling substrate for organisms such as hydroids and foraminiferans. Base map produced by Elise Pietroniro, GIServices, University of Saskatchewan. Map data source: National Atlas of Canada 1:7.5 Million Scale, Natural Resources Canada, Government of Canada. Community illustrations by Susan Laurie-Bourque in consultation with Pat Pocklington (Arenicola Marine) and the authors.

The inshore seafloor at Cape Bathurst was characterized by extremely high faunal abundance (Table 9.1), reaching 17,127.7 ± 647.6 individuals m^{-2} at 38 m depth (station 03 04 300). The tube-dwelling amphipods *Ampelisca macrocephala* (8250.0 ± 113.4 individuals m^{-2}) and *Photis* spp. (927.8 ± 253.2 individuals m^{-2}), the ostracod *Philomedes brenda* (1016.7 ± 76.4 individuals m^{-2}, the cumacean *Brachydiastylis nimia* (644.4 ± 55.6 individuals m^{-2}) and the capitellid polychaete *Barantolla americana* (2005.6 ± 349.9 individuals m^{-2}) dominated this community. Faunal composition was distinct relative to shallow sites sampled on the Beaufort Shelf (Table 9.5). Offshore of Cape Bathurst, abundance declined and the faunal composition resembled that in nearby Franklin Bay and the polynya.

The abundance of fauna in the Mackenzie Canyon was similar to that observed offshore at Cape Bathurst, within the polynya, and on the Beaufort slope (Table 9.5). The dominant species (Table 9.4) comprised a range of small amphipod, ostracod and cumacean crustaceans, and surface feeding polychaetes (Figure 9.7). The dominant taxa were comparable to those of the deeper regions of the Beaufort Shelf and Amundsen Gulf, and included: the polychaetes *Maldane sarsi*, *Tharyx kirkegaarde/marioni*, *Lumbrineris impatiens*, *Terebellides stroemi* and *Prionospio cirrifera/steenstrupi*, the bivalve *Thyasira flexuosa*, the tanaid *Paraleptognathia gracilis*, cumacean *Ektonodiastylis nimia* and ostracod *Philomedes brenda*.

Figure 9.8 shows how faunal abundance and diversity are related to observed environmental variables measured in Sept–Oct (2002) across the Beaufort Shelf and slope (in the area of Kugmallit Valley, 133°18'N-133°53'N). Macrofaunal abundance in the fast ice zone, the flaw lead zone and on the shelf generally ranged from 1000 to 3000 individuals m^{-2} and only declined

to less than 200 individuals m^{-2} at the deepest station on the slope (station 49 at 1000 m depth). Taxonomic diversity was lowest in the flaw lead (32.7 ± 1.1) and highest at the shelf edge (82.8 ± 3.0). Water depth, bottom temperature, bottom oxygen concentration, sediment % sand and phi mean combined were most highly correlated (R = 0.828) with changes in faunal composition across the shelf. Changes in bottom water temperature, salinity, fluorescence and oxygen reflected the late summer influence of the Mackenzie River inshore (Polar-Mixed Layer <50 m), the Pacific Halocline (50-200 m) and Atlantic water (>200 m). Bottom water temperature decreased from 0.18 °C inshore to -1.45 °C on the shelf, and increased to 0.40 °C on the slope. Bottom water salinity increased from 16.0 ‰ inshore to 30.0 ‰ 75 km off shore, and thereafter increased to 34.9 ‰ at 1000 m depth. Fluorescence declined with increasing distance from shore (from a maximum of

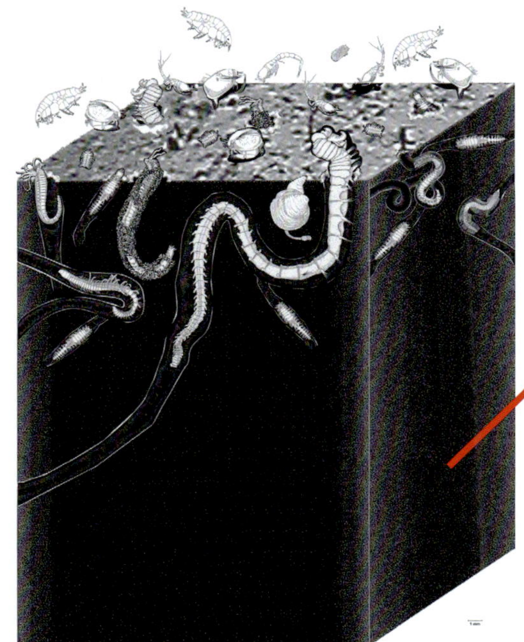

Figure 9.7

Community composition in the Mackenzie Canyon at station 0303 CA10-1 (240 m). Species accounting for cumulatively 75% of the sample abundance are shown; scale bar = 1 mm. Base map produced by Elise Pietroniro, GIServices, University of Saskatchewan. Map data source: National Atlas of Canada 1:7.5 Million Scale, Natural Resources Canada, Government of Canada. Community illustrations by Susan Laurie-Bourque in consultation with Pat Pocklington (Arenicola Marine) and the authors.

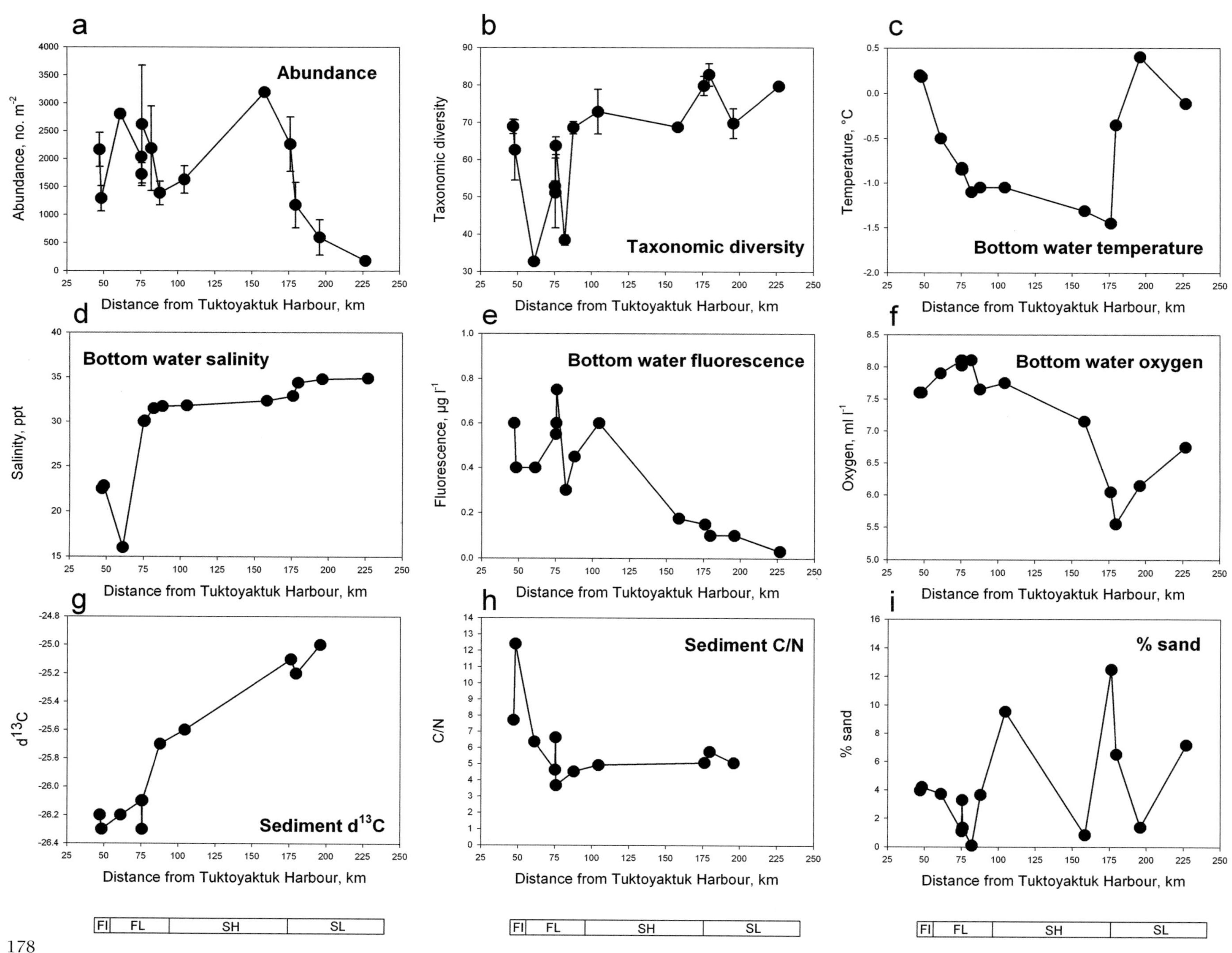
a
Abundance
Abundance, no. m⁻²
Distance from Tuktoyaktuk Harbour, km
b
Taxonomic diversity
Taxonomic diversity
Distance from Tuktoyaktuk Harbour, km
c
Bottom water temperature
Temperature, °C
Distance from Tuktoyaktuk Harbour, km
d
Bottom water salinity
Salinity, ppt
Distance from Tuktoyaktuk Harbour, km
e
Bottom water fluorescence
Fluorescence, µg l⁻¹
Distance from Tuktoyaktuk Harbour, km
f
Bottom water oxygen
Oxygen, ml l⁻¹
Distance from Tuktoyaktuk Harbour, km
g
Sediment d¹³C
d¹³C
Distance from Tuktoyaktuk Harbour, km
h
Sediment C/N
C/N
Distance from Tuktoyaktuk Harbour, km
i
% sand
% sand
Distance from Tuktoyaktuk Harbour, km
FI FL SH SL
FI FL SH SL
FI FL SH SL

0.75 µg l^{-1} in the flaw lead region to a minimum of 0.03 µg l^{-1} on the slope). Bottom water oxygen concentrations were high inshore (from 7.6 ml l^{-1} close to shore to 8.10 ml l^{-1} in the flaw lead region), were decreasing (to 5.55 ml l^{-1}) with off shore distance on the slope, and slightly higher (6.75 ml l^{-1}) at the sampling endpoint. Sediment $\delta^{13}C$ increased with distance from shore (from -26.3‰ in the fast ice region to -24.9‰ at the deepest offshore station on the slope) indicating a riverine supply of carbon inshore and an ocean-based supply offshore. Sediment C/N was highest inshore (with a maximum of 12.4 at 48.2 km from shore) and less off shore (4.5-5.0 at 75 km from shore). Mean sediment phi varied little throughout Kugmallit Valley: values range from 6.86 phi (fine silt according to the Wentworth scale) to 9.11 phi (clay according to the Wentworth scale). The proportion of sand was relatively low throughout Kugmallit Valley as well (1.1% to 12.6%). The fine-textured sediments of the Beaufort Shelf and slope reflect the influence of sedimentation from the Mackenzie River plume and deeper offshore water masses.

Figures 9.9 and 9.10 illustrate observed differences in the dominant bottom community fauna between the fast ice, flaw lead, and shelf/slope regions of Kugmallit Valley. The polychaete *Micronephthys minuta* was the most abundant species, reaching 2077 ± 116.0 individuals m^{-2} in the flaw lead region. Other species which were abundant inshore were the bivalve *Portlandia arctica*, the ostracods Podocopid 3b and *Xestoleberis depressa*, and the polychaetes *Cossura longocirrata* and *Tharyx kirkegaarde/T. marioni*. The polychaete *Levinsinea gracilis* and the brittle star *Ophiocten sericeum* were more abundant on the shelf than inshore or on the slope, while sipunculids and the polychaetes *Maldane sarsi* and *Lumbrineris impatiens* exhibited peak abundance on the slope. However, none of these species were found in abundance at the most offshore station.

9.2.1.4 Macrofaunal patterns on the Beaufort Shelf

Macrofaunal abundance (> 0.4 mm size) on the Beaufort Shelf and in Amundsen Gulf (Table 9.1) was comparable to that in the Bering and Chukchi Seas (> 0.1 mm size; Grebmeier and Cooper, 1995; Feder et al., 2007), in the eastern Canadian Arctic (> 0.5 mm; Cusson et al., 2007), on the northeast Greenland Shelf (> 0.5 mm; Ambrose and Renaud, 1995) and in the Barents Sea (> 0.5 mm; Denisenko et al., 2007), and an order of magnitude more abundant than that in the Barents and Kara Seas (> 0.1 mm; Grebmeier and Barry, 1991). The broad regional variations in species composition correlated highly with water depth (similar to the Laptev Sea; Steffens et al., 2006). However, the correlation was low (0.631) between the environmental and biotic similarity matrices. This could have been due to two reasons: (1) our observations were not able to capture the full range of variability in the marine environment experienced by the benthos (particularly that component of variability caused by the Mackenzie River); and (2) since many benthic species are long-lived, they may have been able to adapt to environmental variability.

The benthic macrofauna observed on the Beaufort Shelf and in Amundsen Gulf was essentially a mud-associated community dominated by the deposit-feeding polychaetes *Maldane sarsi* and *Tharyx kirkegaarde/maroni*, and the deposit-feeding bivalve *Thyasira flexuosa*. The inshore communities were dominated by the bivalve *Portlandia arctica*, the polychaete *Micronephthys minuta* (near the Mackenzie River), and by the amphipod crustacean *Ampelisca macrocephala* and the polychaete *Barantolla americana* (Cape Bathurst). These species occurred in small num-

Figure 9.8 AT LEFT

Changes in benthic macrofaunal abundance (a) and taxonomic diversity (b) with bottom water temperature (c), salinity (d), fluorescence (e) and oxygen (f) and sediment $\delta^{13}C$ (g), C/N (h) and phi mean (i) across the Beaufort Shelf in Sept.-Oct. 2002 in Kugmallit Valley. Replicate samples were averaged (mean and s.e.). FI: fast ice; FL: flaw lead; SH: shelf; SL: slope. (after Conlan et al., 2008).

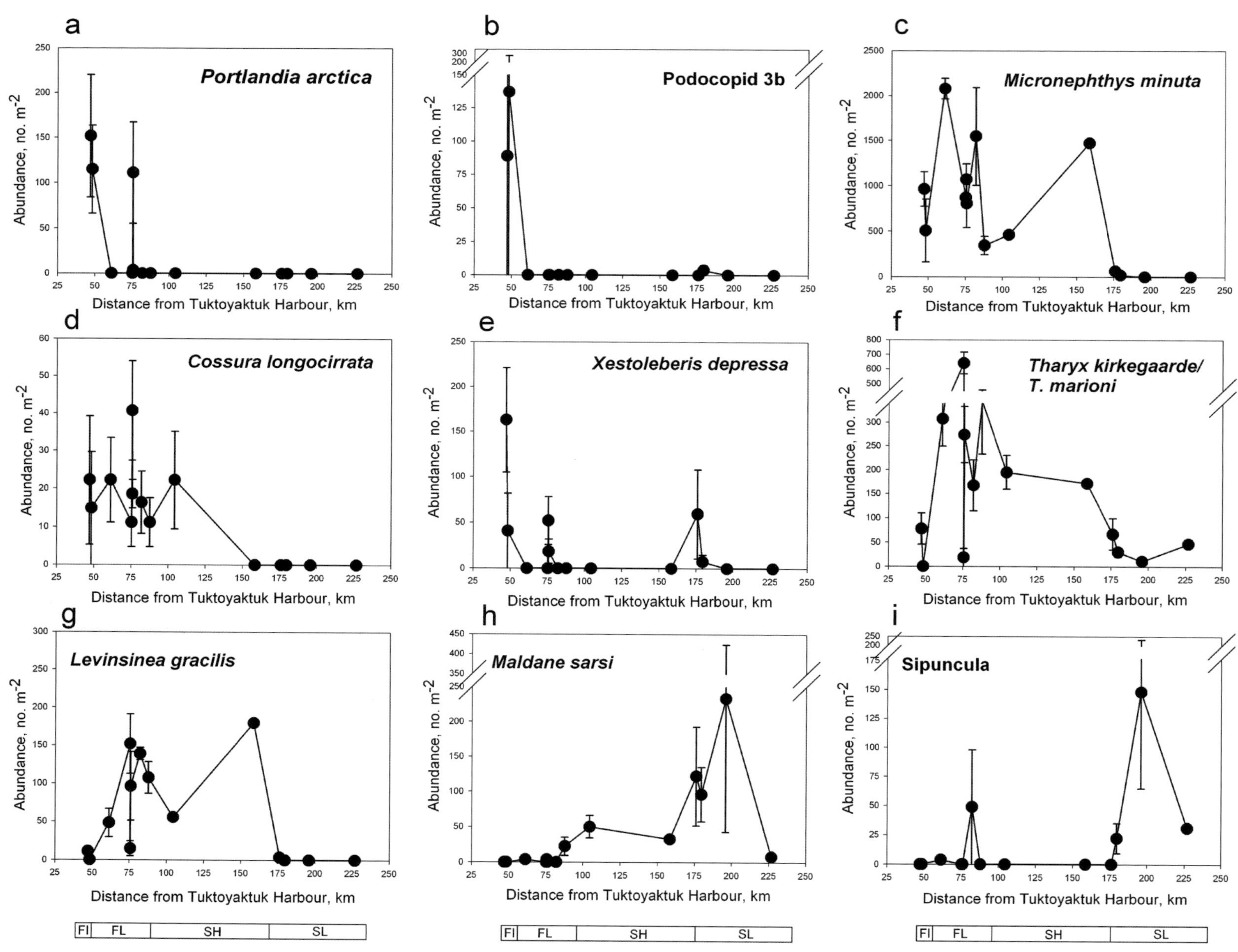

a
Portlandia arctica
Abundance, no. m⁻²
Distance from Tuktoyaktuk Harbour, km

b
Podocopid 3b
Abundance, no. m⁻²
Distance from Tuktoyaktuk Harbour, km

c
Micronephthys minuta
Abundance, no. m⁻²
Distance from Tuktoyaktuk Harbour, km

d
Cossura longocirrata
Abundance, no. m⁻²
Distance from Tuktoyaktuk Harbour, km

e
Xestoleberis depressa
Abundance, no. m⁻²
Distance from Tuktoyaktuk Harbour, km

f
Tharyx kirkegaarde/
T. marioni
Abundance, no. m⁻²
Distance from Tuktoyaktuk Harbour, km

g
Levinsinea gracilis
Abundance, no. m⁻²
Distance from Tuktoyaktuk Harbour, km
FI FL   SH   SL

h
Maldane sarsi
Abundance, no. m⁻²
Distance from Tuktoyaktuk Harbour, km
FI FL   SH   SL

i
Sipuncula
Abundance, no. m⁻²
Distance from Tuktoyaktuk Harbour, km
FI FL   SH   SL

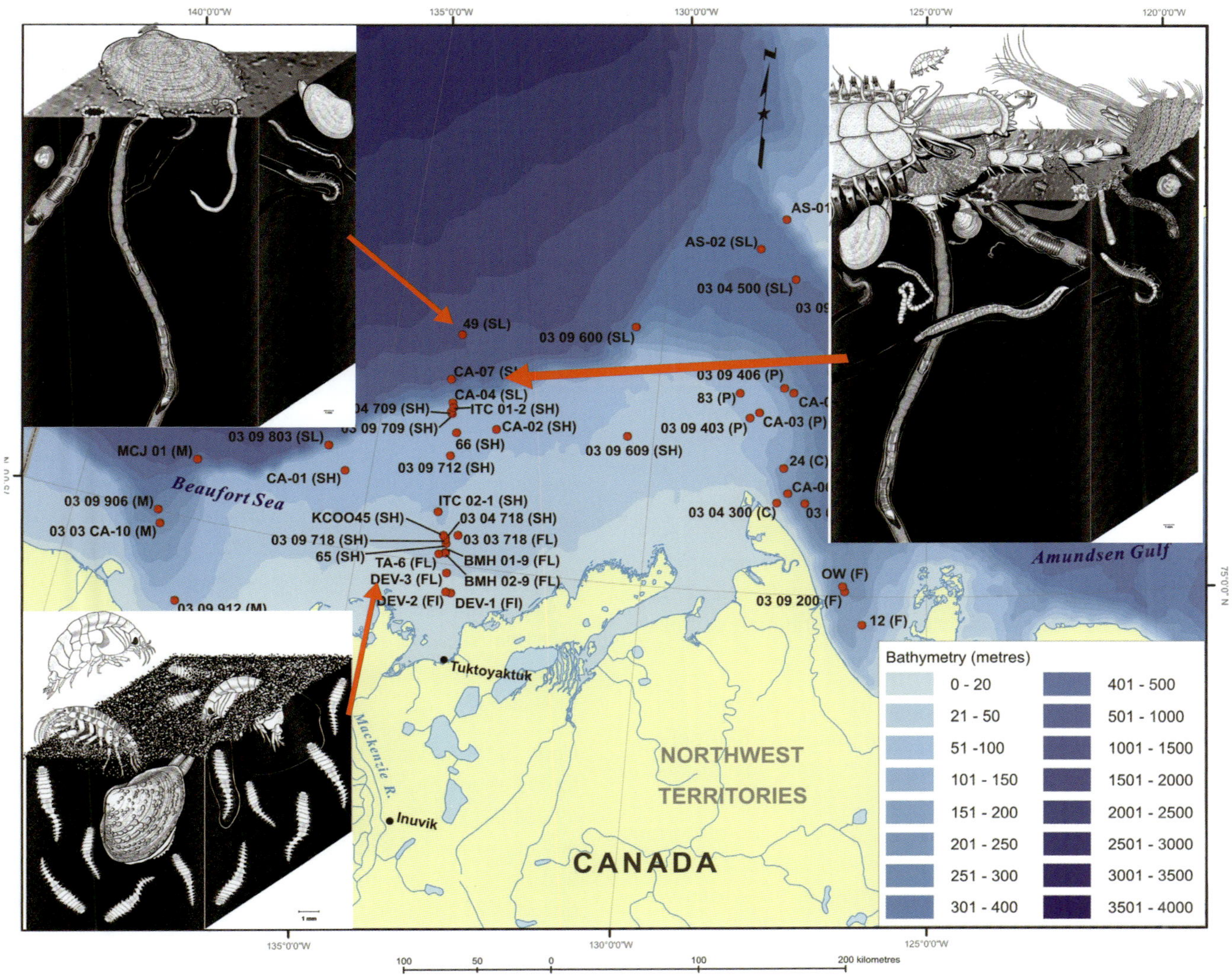

Figure 9.10

Community composition across the Beaufort Sea Shelf. Stations from north to south: Beaufort slope at 1000 m (station 49); Beaufort shelf edge at 214 m (station CA-04); inshore fast ice and flaw lead regions of the Beaufort Shelf at 20 m (near station DEV-3). Species accounting for cumulatively 75% of the sample abundance are shown; scale bars = 1 mm. Base map produced by Elise Pietroniro, GIServices, University of Saskatchewan. Map data source: National Atlas of Canada 1:7.5 Million Scale, Natural Resources Canada, Government of Canada. Community illustrations by Susan Laurie-Bourque in consultation with Pat Pocklington (Arenicola Marine) and the authors.

bers elsewhere, but respectively appeared to favour estuarine and upwelling conditions.

The Cape Bathurst polynya did not appear to influence the local benthic community structure in itself. In fact, community structure within the polynya region was was not very different from that on the Beaufort Shelf at similar depths. This suggested that processes within the polynya which influence carbon production and export to the benthos were not sufficient to produce a footprint on the benthos. Although benthic communities rely on organic carbon produced in surface waters, this supply could have been advected into the Cape Bathurst polynya region. Indeed, re-suspended shelf sediment and sediment from the Mackenzie River have been observed to reach Amundsen Gulf during onshore wind storms (O' Brien et al., 2006). The sills and basins within the polynya region might also act to isolate local bottom water masses and restrict their circulation; these conditions would not be conducive to supporting elevated benthic abundance within the polynya.

Barry et al., (2003) found that the spatial distribution of megabenthos under the Antarctic Ross Sea polynya correlated more strongly with benthic environmental variables than with the timing of local ice cover or primary productivity. However, tight coupling of benthic metabolism in the St. Lawrence Island, Northeast Water and North Water polynyas in the Arctic were observed by Grebmeier and Cooper (1995), Ambrose and Renaud (1995) and Grant et al. (2002), respectively. In addition, Piepenburg and Schmid (1996) reported higher benthic epifaunal abundance under the Northeast Water polynya compared to the benthos underlying the pack ice. A common conclusion to each of these studies was that benthic metabolism was not fuelled directly by the vertical sinking of ungrazed

organic matter. Instead, organic material produced at the margins of the polynyas began to sink and was subsequently advected into the regions. In the case of the Cape Bathurst polynya, there may be a tighter coupling in benthic activity than in species composition and abundance. Renaud et al. (2007), for instance, measured a rapid increase in the respiration of the macrofauna in Franklin Bay at the time of the spring ice algae bloom. This increase suggested similar activity in the nearby Cape Bathurst polynya, provided that organic production within the polynya was not intercepted pelagically (see Section 9.2.3 below). Seuthe et al. (2007) measured the initiation of feeding by major pelagic grazers within the polynya, *Calanus glacialis* and *C. hyperboreus*, and the associated increase in fecal pellet production. Lower δ^{13}C values in the sediment organic carbon on the eastern edge of the Beaufort Shelf (-23.62 ± 0.28‰, n = 4) compared to the northern and western edges (-25.12 ± 0.05‰, n = 4) suggested that the polynya's fauna was influenced by a relatively more ocean-based carbon supply. These results corresponded with those reported by Dunton et al. (2006).

Cape Bathurst was found to be a region of high macrofaunal abundance and diversity. Maximum abundance values (>17,000 m^{-2} at 38 m depth) were greater than any other recorded values in the Arctic (Grebmeier and Barry, 1991) with the exception of those found in portions of the Bering and Chukchi seas (Grebmeier and Cooper, 1995; Feder et al., 2007). The high abundance observed at Cape Bathurst was mainly due to the capitellid polychaete *Barantolla americana* and the tubicolous amphipod *Ampelisca macrocephala*. Capitellids are non-selective deposit feeders (Fauchald and Jumars, 1979), while ampeliscid amphipods can be suspension or surface deposit feeders (Mills, 1967; Highsmith and Coyle, 1990). *Ampelisca macrocephala* is a prime food resource for California gray whales (*Eschrichtius robustus*) in the Bering and Chukchi Seas (Oliver et al., 1983) and is highly productive (Highsmith and Coyle, 1990). Averaging 8250 ± 113.5 individuals m^{-2}, the ampeliscid abundance observed at Cape Bathurst was twice the value of the mean ampeliscid abundance in the Chirikov Basin (Bering Sea) which feeds the California gray whale (Highsmith and Coyle, 1990). Although gray whales do not occur in the Beaufort Sea, bowhead whales (*Balaena mysticetus*) forage as far east as Cape Bathurst and Amundsen Gulf in the summer (Frost and Lowry, 1984). In fact, this part of Cape Bathurst is locally known as 'Whale Bluff'. While bowhead whales are generally considered pelagic zooplankton feeders (Harwood and Smith, 2002; Lee et al., 2005), they are also known to feed benthically (Frost and Lowry, 1984). Hence, it's possible that the rich *Ampelisca macrocephala* and *Barantolla americana* communities at Cape Bathurst undergo summer foraging by bowhead whales. These two communities might also be significant resources for diving seabirds like the surf scoter (*Melanitta perspicillata*), the white wing scoter (*Melanitta fusca*), the common eider (*Somateria mollissima v nigra*), the king eider (*Somateria spectabilis*), and the long-tailed duck (*Clangula hyemalis*; Dickson and Gilchrist, 2002).

The transect we sampled across the Beaufort Shelf demonstrated variation in environmental conditions and bottom community composition with depth. Benthos in the inshore fast ice and flaw lead zones can be subject to frequent disturbances by ice scour (Gilbert and Pedersen, 1987; Lewis and Blasco, 1990; Myers et al., 1996). 97% of the area inside the 24 m isobath, for instance, is subject to disturbance in <100 years (S. Blasco, pers. comm., 2006). Observed variations in inshore benthic abundance and diversity (Figs. 9.3 & 9.4) appeared to reflect such disturbances. Additional

disturbances caused by storm effects and coastal erosion, and the various salinity, temperature and turbidity influences of the Mackenzie River would also likely shape community structure. The macrofauna observed here were taxonomically diverse, however, and some species (such as the small polychaete *Micronephthys minuta*) were highly abundant. Nematodes were very abundant on the Beaufort Shelf (referred to as Nemata in Table 9.2). Other dominant taxa noted on the inshore shelf were several ostracod species and the small deposit-feeding bivalve *Portlandia arctica*. The large scavenging and predaceous isopod *Saduria sabini* also proved to be a conspicuous component of the megabenthos on the inshore shelf. Overall, the inshore fauna composition bore strong resemblance to that of the shallow (<30 m depth) Laptev Sea Shelf; i.e. it was dominated by the bivalve *Portlandia arctica*, the mysid shrimp *Mysis oculata* and the isopod *Saduria entomon* (Schmid et al., 2006; Steffens et al., 2006).

A relatively large mix of species characterized the main portion of the Beaufort Sea Shelf including many species found in Amundsen Gulf and the Mackenzie Canyon. Some of the dominant offshore taxa observed in this region were deposit feeding polychaetes *Tharyx kirkegaarde/marioni*, *Levinsinea gracilis*, *Prionospio cirrifera* and *Maldane sarsi*, the tanaid *Paraleptognathia gracilis*, and the ophiuroid echinoderm *Ophiocten sericeum*. Since *Maldane sarsi* (a deep burrowing, head-down, non-selective deposit feeder) defecates at the surface, it is likely an important contributor to both sediment mixing and surface nutrient replenishment (Holte, 1998); it likely also benefits the surface microbial and meiofaunal community, which in turn benefits larger surface micropredators and deposit feeders, such as *Ophiocten sericeum*, *Paraleptognathia gracilis* and *Prionospio cirrifera* (Fauchald and Jumars, 1979; Jangoux and Lawrence, 1982; Blazewicz-Paszkowycz and Ligowski, 2002). *Maldane sarsi, Lumbrineris fragilis, Portlandia arctica, Levinsinea gracilis, Prionospio cirrifera, Paraleptognathia gracilis* and other species observed in our study were also found to dominate glacial fjords and estuarine areas in the eastern Canadian Arctic (Curtis, 1972; Aitken and Gilbert, 1996), Svalbard (Holte and Gulliksen, 1998) Spitsbergen (Blazewicz-Paszkowycz and Sekulska-Nalewajko, 2004; Wlodarska-Kowalczuk and Pearson, 2004) and Siberia (Golikov and Averintzev, 1977), indicating that these species were tolerant to high rates of clastic sedimentation.

Warm bottom water temperatures on the Beaufort Shelf's edge reflected the presence of Atlantic-origin (at depth) and Pacific-origin (above the Atlantic-origin) water masses. A relatively greater range in salinity was observed over the steeper slope of the outer shelf (>100 m) compared to the relatively flat middle shelf, consistent with the upwelling of deeper water masses across the shelf edge. Stations CA-04 and CA-07 exhibited sediments with high terrigenous carbon contents (25-60%) despite their distances from shore (177.6 and 195.9 km from Tuktoyaktuk Harbour, respectively). The winter supply of POC to the benthos likely originated from resuspension and advection of shelf bottom particles via the benthic nepheloid layer (ice algae may have also supplied a slight increase in POC during spring; Forest et al., 2007). Macrofaunal abundance on the Beaufort Shelf's edge and slope were quite variable (including, for example, low values at the deepest site despite high taxonomic diversity). Species typical of the main portion of the shelf were still present on the shelf's edge and slope, though in smaller numbers (Fig. 9.9).

The Mackenzie Canyon appeared to provide an excellent habitat for benthic communities. It's an area which has been reported to receive seven times more annual deposition than stations CA-04 and CA-07 on the Beaufort Shelf's edge (O'Brien et al., 2006). Moreover, the shape and location of the canyon tends to focus upwelling such that nutrient-rich water from more than 180 m depth frequently reaches shallow depths. During the short ice-free season, upwelling and downwelling of sub-surface waters on the Beaufort Shelf and its slope are forced directly by windstorms. For most of the year, however, they are driven by ice motion. The upwelling response to westward ice movement by easterly winds is actually stronger than that due to wind alone. Conversely, downwelling during westerly wind events is relatively weak since the strength of pack ice inhibits rapid and prolonged eastward ice drift (Williams et al., 2006). The effect is a net upwelling that is stronger than would result from the wind-stress without the intervention of pack ice. Moreover, the net upwelling in Mackenzie Canyon brings a flux of nitrate, phosphate, and silicate to the head of the trough and adjacent areas. The abundant and diverse fauna observed in the Mackenzie Canyon (Table 9.1 and Fig. 9.7) appeared to reflect such a rich food supply.

9.2.2 Macrobenthic secondary production

Energy flow studies are important to the understanding of benthic ecology (e.g. Petersen and Curtis, 1980; Welch et al., 1992; Frontier and Pichot-Viale, 1998). Secondary production (the amount of organic material produced through time) provides a broad picture of the success of a population in an ecosystem; i.e. it integrates multiple biotic variables (reproduction, biomass, recruitment, competition, predation, body mass, etc.) and environmental conditions (temperature, depth, food availability, presence of refuges, exploitation etc.) which influence individual growth and population mortality.

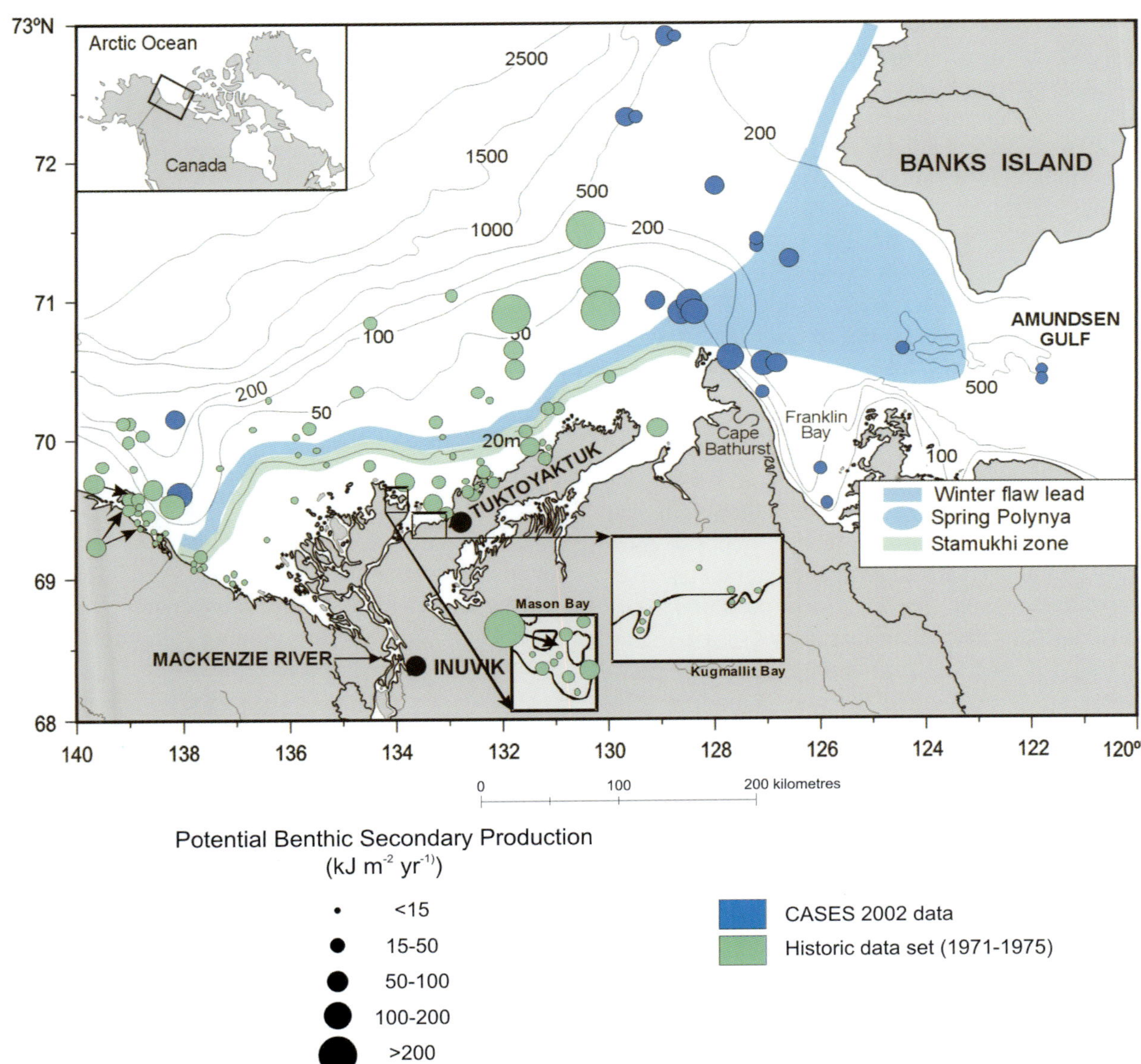

Figure 9.11

Map of Beaufort Sea and Mackenzie Shelf region showing categorical values of potential production from the benthic community for CASES 2002 sampling sites (n = 24) and from historical survey (n = 101; 1971-1975; Wacasey et al., 1975; Cusson, unpublished).

Stations sampled during the *CCGS Laurier* cruise in Sept-2002 and the *CCGS Radisson* cruise in Oct. 2002 were included in our study. The stations were grouped as follows: Mackenzie Canyon (stations: MCJ 01 3; 03 03 CA10-1); East Amundsen Gulf (stations: 03 04 100-1; 03 04 100-2); Cape Bathurst (station 24); Franklin Bay (stations: 12; 03 09 200-20; CA06 3); Polynya (stations: 83; 83Q; CA03-1; CA03-2; CA05-1; CA05-2; CA08-1; 03 04 124-1; 104) and West Admundsen Gulf (stations: 03 04 500-3; AS01-1; AS01-2; AS02-1; AS02-2). Historical data (1971-1975) from the Beaufort Sea and Mackenzie Shelf were derived from data reported by Wacasey et al. (1977).

Benthic production was assessed using indirect methods: the P/B ratio and an empirical equation. No biomass data were available for the 2002 samples; hence we made the following assumptions: 1) benthic communities were in steady state such that the mean biomass of specific taxa was considered constant; 2) the mean biomasses for specific taxa were similar to those from historical data sets; 3) for species with incomplete information relating to mean biomass: values from sister taxonomic groups were assumed to be similar; 4) specific values of production to biomass (P/B) ratios in polar environments from the literature were preferred to any empirical equation estimates.

We used an equation derived by Brey (2004) to assess potential production:

$$\mathrm{Log}P = 7.947 + \log B - 2.294 \log W_{mean} - 2409.856\,(1/T + 273) + 0.168\,(1/D) + 582.851 \log W_{mean}\,(1/T + 273)\;;$$

Where,

P: potential production (kJ m^{-2});
B: mean annual biomass (kJ m^{-2});
W_{mean}: mean body mass (kJ ind^{-1});
T: mean annual temperature (°C);
D: site depth (m)

The water depths of each sampling site were used in the equation but the annual bottom temperatures were assumed to be 1°C (i.e. a temperature slightly warmer than those measured on the Beaufort Shelf, which ranged from -1.4 °C to 0 °C; Table 9.1). All body mass units were converted to kilojoules (kJ) per m^2 using conversion factors provided by the original source or by other literature sources (Ricciardi and Bourget, 1998; Brey, 2004).

ANOVAs were performed to compare potential production among regions for 1971-1975 and 2002. Tukey-Kramer (SAS, 1999) multiple comparison tests were performed to determine mean differences among regions. Regression analysis was used to link depth to potential production estimates (both variables were $\log_{10}$ transformed to normalize the data). Normality was verified using the Shapiro-Wilk's test (Zar, 1999) and homoscedasticity was confirmed by graphical examination of the residuals (Scherrer, 1984; Montgomery, 1991). A significance threshold $\alpha = 0.05$ was adopted for all statistical tests.

A total of 2,944 specific potential secondary production values were estimated from 520 different taxa from 125 different sampling sites (~24 estimates per station). The potential production of benthic communities on the Beaufort Shelf and in Amundsen Gulf was found to be spatially variable (Fig. 9.11). Some areas showed high values (for example, sites close to Cape Bathurst and the Mackenzie Trough) while others were

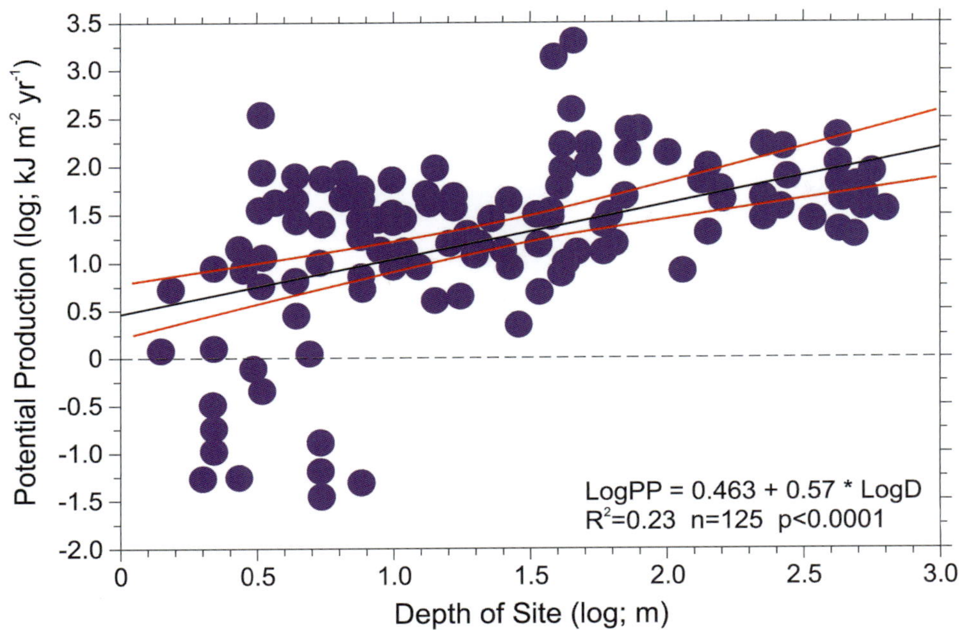

Figure 9.12

Regression between the potential production (KJ·m⁻²·yr⁻¹; log-scale) and depth of site (meters; log-scale; Cusson, unpublished).

low but variable (shallow areas < 20 m depth). Analyses of variances revealed that there were no significant differences (p = 0.467) in potential secondary production estimates among regions of the Beaufort Shelf and Amundsen Gulf. Benthic potential production (PP) was positively related to the depth at sampled sites, reflecting observed trends in benthic community structure (Fig. 9.12). Regression analysis showed that 23% of the potential secondary production variability was explained by site depth. This trend, however, was influenced by the low species abundance (Table 9.1) and benthic biomass recorded at depths < 10 m (1 in the log scale) within the inshore fast ice environment. High values of potential production were recorded between 50-200 m depth on the eastern Mackenzie Shelf and at the head of Mackenzie Trough (Fig. 9.11). This may have been due to wind-driven ice motion and westward surface currents which contribute to the upwelling of deep nutrient-rich water across the shelf edge and in the Mackenzie Trough (Macdonald et al., 1998; Williams et al., 2006; Forest et al., 2007). The polynya appeared to have no detectable effect on potential production estimates.

9.2.3 Benthic community respiration

There exists a well-documented coupling between benthic metabolism and pelagic production throughout the circumpolar North. Descriptions of this coupling include Carey (1991), Grebmeier and Barry (1991), Ambrose and

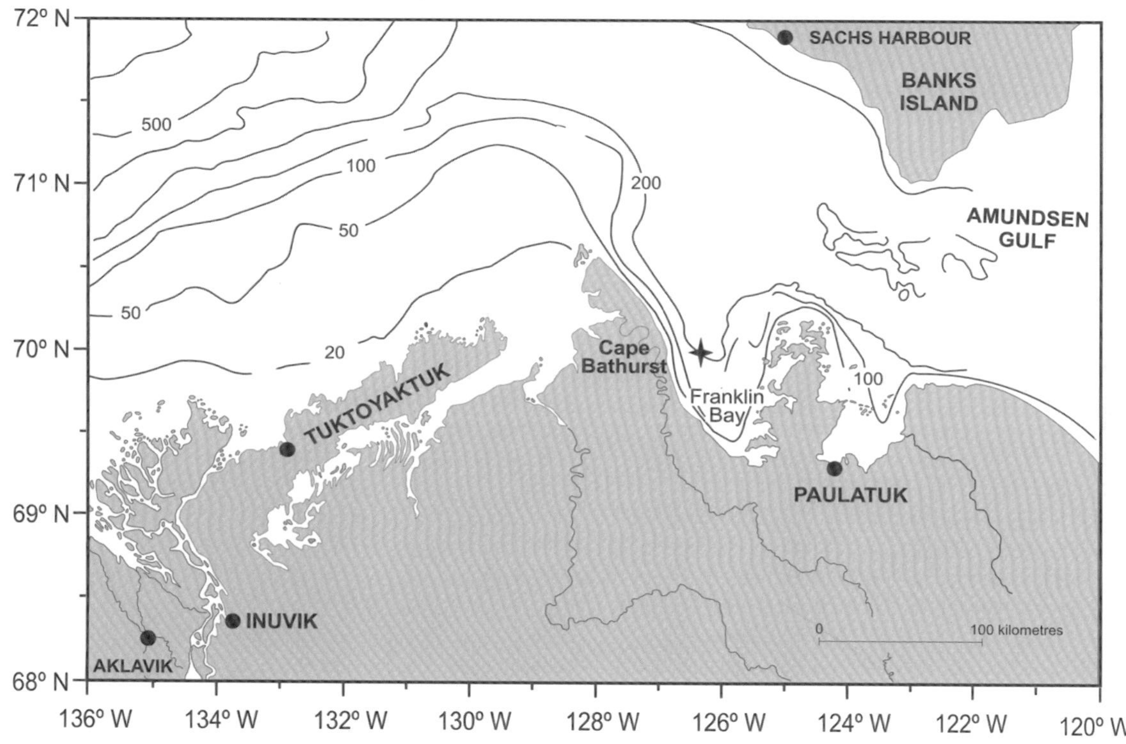

Figure 9.13

Map of the southeastern Beaufort Sea. The cross identifies the study site for the time-series in Franklin Bay. Depth is expressed in meters (after Renaud et al., 2006).

of the benthos to an ice algal bloom (Renaud et al., 2006). Data for this study was collected at a site within Franklin Bay (70°02'N, 126°18'W; Fig. 9.13) between December, 2003, and June, 2004.

The timing and magnitude of primary productivity in this region varies considerably (Arrigo and van Dijken, 2004) and is closely linked to ice dynamics. For our study, multiple ice cores were collected using a manual corer. Estimates of chlorophyll *a* biomass (chl *a*) on 10 cm sections over the total length of ice cores were performed on three occasions (25 March, 24 April, and 4 May, 2004). 95% of the total integrated chl *a* biomass in the ice cores was observed in the bottom <10 cm. Concentrations of chl *a* were determined fluorometrically and the concentration of chlorophyll *a* was calculated according to Holm-Hansen et al. (1965). Vertical pigment fluxes were measured from February 23 to June 20, 2004, using sequential deployments of particle interceptor traps. The particle interceptor traps were deployed for 8 days from March to mid-May, after which they were deployed for 6 days from mid-May to the end of June (in response to higher sinking fluxes). Water at the ice-water interface was collected using a hand pump with Tygon tubing held parallel to the under surface of the ice. Samples from particle interceptor traps and interfacial water were analyzed for chl *a* and total pigments.

Bottom sediment was sampled on 5 dates between 14 January and 7 May, 2004, and then again on 4 July, 2004. Replicate sub-cores (10 cm diameter x 20-25 cm deep with as much overlying water preserved as possible) were taken from the sediment cores to estimate the respiration of the entire infaunal sediment community. The protocol for whole core respiration determination was documented by Renaud et al. (2006). To estimate how much of the whole-core respiration was due to micro- and meio- fauna, additional minivial

Renaud (1995), Hobson et al. (1995), Piepenburg et al. (1997) and Grant et al. (2002). Most of these studies were conducted late in the open water season and excluded the potential effects of ice algae deposition on benthic-pelagic coupling. Ice algae typically exhibit maximum productivity in late spring, while phytoplankton exhibit maximum productivity during the summer. However, a temporal and/or spatial separation of the peak bloom of ice algae and phytoplankton does not necessarily mean that both organic matter sources are not important in fuelling the benthos at a given location. Here, we present initial efforts to quantify the response respiration determinations were performed following the methodology of Grant et al. (2002).

Small sub-cores used to analyze sediment pigments, organic carbon and nitrogen were obtained from the bottom sediment samples. Pigments were analyzed fluorometrically according to Holm-Hansen et al. (1965). Samples for organic carbon and total nitrogen content were dried at 60 °C for 24–48 hours and ground until homogeneous. Homogenized sediment was then run on a CHN analyzer with acetanilide as a standard.

Respiration, sediment pigment, carbon, and nitrogen data were analyzed by one-way analyses of variance (ANOVA) or Kruskal-Wallis (KW) tests (unequal variances) using the sampling date as a discriminating factor and JMP-IN (SAS Institute) software. Where ANOVA tests were significant, the Tukey HSD post-hoc test was performed to elucidate the differences. Bonferroni contrasts were performed following significant KW tests. Before performing the ANOVA tests, the homogeneity of variances was tested using Bartlett's test (Sokal and Rohlf, 1995).

There was a clear seasonal trend in the ice algae biomass in the bottom ice, as evidenced by monthly averaged chl *a* concentrations for high and low snow covers (Fig. 9.14a). Average bottom ice chl *a* concentrations were < 0.3 mg chl *a* m^{-2} during February-March. These concentrations increased by more than 10-fold in April (average = 5.6 mg chl *a* m^{-2}), and reached 15 mg chl *a* m^{-2} in May. A strong decrease in bottom ice chl *a* was observed during the month of June, with an average value of 2.5 mg chl *a* m^{-2}. April and May concentrations were significantly higher than those in February-March. There were no phaeopigments observed in the bottom ice throughout the ice season. Surface water (top 1 m under the ice) chl *a* was consistently low (< 0.60 mg chl *a* m^{-2}) until June, when average values increased to

4.0 mg chl *a* m^{-2} (Fig. 9.14b). Biomass values were significantly higher in June than during the other periods. Chl *a* made up more than 90% of surface water total pigments between February and June.

Seasonal trends in the sinking fluxes of chl *a* at 1 m under the ice were similar to those for ice algae (Fig. 9.14c). Low sinking fluxes were observed during February-March (average = 0.02 d^{-1}). These fluxes increased by 7-fold in April (average = 0.13 mg chl *a* m^{-2} d^{-1}), and continued to increase during May and June (the maximum sinking flux during this period was 0.38 mg chl *a* m^{-2} d^{-1}). Means from all months were significantly different, except for May and June when results were indistinguishable (p < 0.001; ANOVA). Seasonal trends in chl *a* sinking fluxes at 15 and 25 m closely followed those observed at 1 m beneath the sea ice. The pigment sinking fluxes at 25 m made up 69% of those observed at 1 m (% variance = 27% for the whole sampling season). The percent chl *a* in total pigments (chl *a* + phaeopigments) in the 1 m traps varied between 86 to 99% during the season, with no significant differences among sampling months (p > 0.11; Fig. 9.14c).

Sediment community oxygen demand varied by more than one order of magnitude (1.75 to 21.0 mmol O$_2$ m^{-2} d^{-1} between February 10 and April 6) over the sampling period (Fig. 9.15a). Respiration rates measured on April 6 and April 27 were significantly higher than the rates measured on January 14, February 10 and July 4 (p < 0.0001; ANOVA). The range in oxygen demand by micro- and meio- fauna in the minivial incubations was less than that for the entire community and only varyied by a factor of two (Fig. 9.15a). The mean respiration rate on April 27 was significantly higher than the rates measured in January, February, May, and July; the rate on 6 April was also significantly higher than on the last two sampling dates (p < 0.0001; ANOVA).

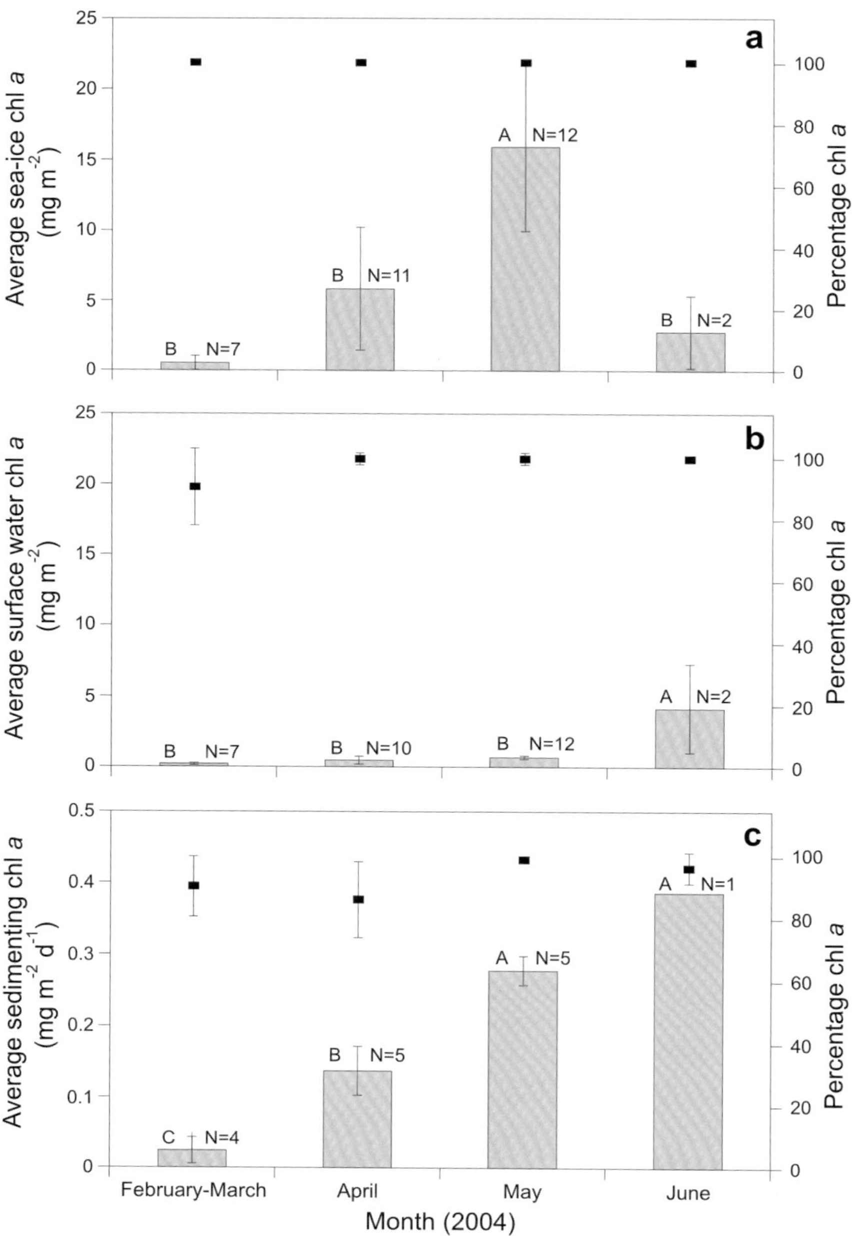

Phytopigment (chlorophyll *a* and phaeopigments) concentrations in the top 9 cm of the sediment at the over-winter station showed an opposite trend to that of ice algae and respiration. Phaeopigment concentration was highest from mid-January to mid-February. It decreased between February and early April, remained fairly stable until mid-May, and became slightly higher again in July (Fig. 9.15b). ANOVA failed to detect significant differences among sampling dates. However, a p-value of 0.053 suggested a trend toward lower values during April and May. Chl *a* values in the top 9 cm showed no significant differences among sampling dates ($p > 0.05$; ANOVA; Fig. 9.15b). Organic carbon and total nitrogen content in the top 2 cm of the sediments did not vary among sampling dates. Organic carbon values ranged from 1.28% (s.d. = 0.07) to 1.41% (0.14), and total nitrogen values varied between 0.13% (0.0005) and 0.16% (0.03). P-values for both ANOVAs were higher than 0.50.

Figure 9.14

Seasonal trends in average chlorophyll a concentration from (a) the bottom of the sea ice, (b) the top 1 m of seawater directly under the ice, and (c) particle interceptor traps deployed 1 m below the ice at a station in the southeastern Beaufort Sea. Data are averages of 1 or 2 months of samples in 2004. N = number of samples represented by the means. All error bars represent ±1 standard deviation. Black squares represent the average percentage of total pigments from each sampling period that was chlorophyll a (right axis). Date was determined to have a significant effect on chlorophyll a concentrations in all material (p < 0.001; Kruskal-Wallis test); bars with the same upper case letter are not significantly different from each other (Bonferroni test at alpha = 0.05). Percent chlorophyll a was not significantly different among time periods in any of the material sampled (p >> 0.05; after Renaud et al., 2006).

Ice algae have recently been considered an important food source for benthic communities particularly during the early season before the onset of pelagic production (Ambrose and Renaud, 1997; Hargrave et al., 2002; Carroll and Carroll, 2003). The CASES data presented here represents a first attempt to simultaneously track ice algae community development, its sinking from the ice, and associated benthic processes. We observed a sharp increase in both the sinking flux of chl *a* and benthic oxygen demand during the onset of an ice algal bloom between April and mid-May (Figs. 9.14a & b; Fig. 9.15a). Chl *a* concentrations measured immediately below the sea ice were more than an order of magnitude lower than those in the bottom ice during April and May (Figs. 9.14 a & b). The low concentrations of chl *a* within surface waters coupled with an increase in the chl *a* sinking flux at this time may have been linked to the release of ice algae from the bottom ice. The release of ice algae in early April could not be explained by ice melt (since it was a period of ice growth); instead, it is possible that ice algae were released from the bottom ice as a result of brine drainage (Melnikov, 1998).

Increased chl *a* sedimentation and benthic oxygen demand were not accompanied by an increase in sediment pigments, organic carbon, or total nitrogen content. Sediment phytopigment (chl *a* and phaeopigments) concentrations showed an opposite trend to that of ice algae and respiration. Phaeopigment concentrations were highest from mid-January to mid-February, decreased between February and early April, remained fairly stable until mid-May, and were slightly higher again during July (Fig. 9.15b). While the onset of ice algal growth and sedimentation coincided with increasing oxygen demand by the sediment community, it is unlikely that the input of ice algae alone was sufficient to explain the ten-fold increase in respiration rates from early April to mid-May. Using a 1:1 stoichiometric relationship between oxygen and carbon utilization, and a respiration coefficient of 0.85 (Smith, 1978), the benthic carbon demand during the period of high sediment oxygen demand was estimated to be around 210 mg m^{-2} d^{-1}. Carbon fluxes measured in under-ice traps did not approach this value until June (ranging between 20 and 28 mg m^{-2} d^{-1} from February to May; Renaud et al., 2006). While some benthic fauna derive significant portions of their early-season food from sedimenting ice algae (McMahon et al., 2006), it is unlikely that the measured pigment sedimentation rates were sufficient to account for the observed increase in benthic community respiration in April and May. It is possible, however, that the arrival of fresh phytodetritus to the sea floor served as a cue to benthic fauna to increase their activity. Infauna stimulated directly by sedimenting ice algae may have consumed all the new ice algal material that was deposited plus pigmented matter from the sediment inventory.

Epifaunal invertebrates (including echinoderms, amphipods and isopods) are abundant and important components of Arctic shelf communities (Welch et al., 1992; Piepenburg and von Juterzenka 1994; Piepenburg et al., 1995; Bluhm et al., 1998; Ambrose et al., 2001; Renaud et al., 2007). For example, the population density of ophiuroid echinoderms (brittlestars; 700 individuals m^{-2}) could account for up to 80% of total benthic-community metabolism in these areas (Piepenburg et al., 1995; Renaud et al., 2007). Bottom photographs and non-quantitative traps indicated that ophiuroids and large amphipods (like *Anonyx* spp.) were quite abundant at our sampling station in Franklin Bay. These organisms are known to consume ice algae (Ambrose et al., 2001; Hobson et al., 1995 & 2002). Through feeding, epibenthic organisms can bioturbate surface sediment, and alter its biological (Ambrose, 1993) and

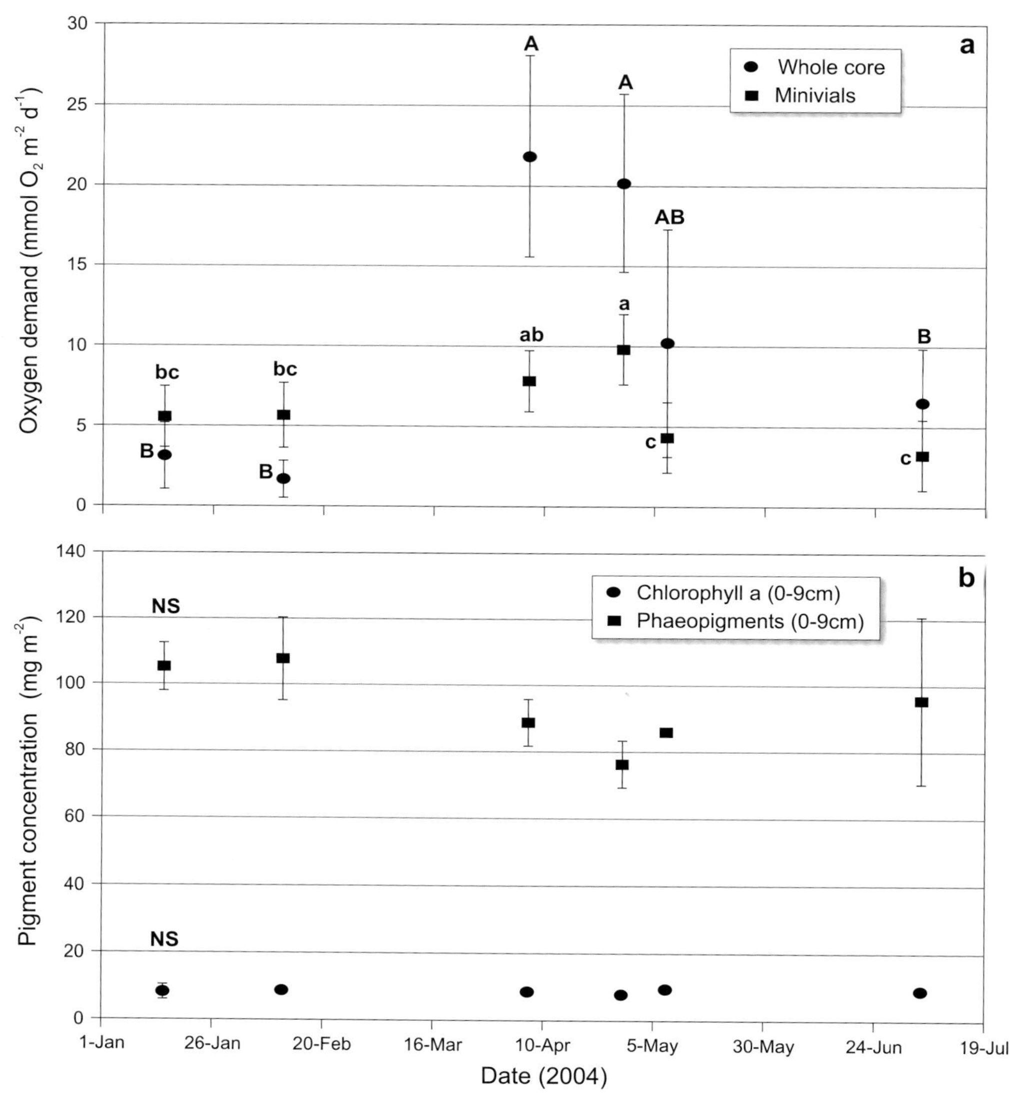

Figure 9.15

Time-series measurements of a) sediment oxygen demand from whole-core (circles) and minivial (squares) incubations, and b) concentration of sediment chlorophyll a (circles) and phaeopigments (squares). When analysis of variance indicated significant date effects (p < 0.05), letters appear beside or above symbols. Symbols marked with the same letter are not significantly different by Tukey's HSD test. NS indicates that the measurements did not vary significantly among dates. In (a), upper case letters refer to results of statistical tests for whole-core incubations; lower case letters reflect results from minivial incubations. All error bars represent ±1 standard deviation (after Renaud et al., 2006).

geochemical structure (Lohrer et al., 2004; Solan et al., 2004; Wenzhöfer and Glud, 2004). For instance, bioturbation has been found to enhance oxygen exchange in sediments by a factor of 1.5–3 (Glud et al., 2000). This kind of geochemical enhancement could possibly be expected at our study site given the large densities of epifauna we observed (4747 individuals m^{-2}). The benthic community was dominated by small deposit-feeding polychaetes (*Tharyx kirkegaarde/maroni*, *Lumbrineris impatiens*, *Prionospio cirrifera/steenstrupi*) and the larger, deep-burrowing *Maldane sarsi*. The direct enhancement of benthic respiration due to the deposition of ice algae may have been augmented by the increased feeding and burrowing activities of epifauna and infauna scavenging for newly deposited organic detritus.

9.3 Implications of this work

Carmack and McLaughlin (2001) and Piepenburg (2005) have predicted a number of feedback effects on the Arctic as a result of changes in pack ice which would likely impact the continental shelf benthos. These are: longer ice-free periods, particularly in the seasonal ice zone; increased wind-mixing, upwelling and wintertime brine rejection, which would increase the availability of nutrients to phytoplankton; increased availability of underwater light to phytoplankton (and benthic primary producers), especially in the seasonal ice zone; increased river flow and export of organic terrestrial material (POC, DOC) to the shelf (therefore increasing turbidity); a decrease in the extent of the ice algae community (and a corresponding decrease in its flux to the benthos); increased coastal erosion (resulting in increased turbidity); and shifting water mass fronts and currents, resulting in different migration pathways and the potential introduction of foreign species. The consequence of these changes

to the benthos would be a diminished carbon supply (if it is intercepted by zooplankton and the microbial loop), and changes in the quality, timing, and source of its carbon (Lovvorn et al., 2005). It would also likely favour changes in species composition (based on feeding and reproductive characteristics) and reduce or redistribute benthic biomass (due to a lower pelagic carbon supply). Lower benthic biomass would affect bottom predators, including mammals and sea birds, and favour smaller predators such as fish. Warmer water would also have physiological consequences for stenothermal, cold-adapted animals, particularly in terms of their limitation for oxygen uptake (Peck, 2002; Pörtner and Knust, 2007).

On the Canadian Beaufort Shelf, the physical and geochemical variability of Mackenzie River inflow already sets conditions for inshore fauna which are adaptable to changeable conditions. Climate warming in this area would likely first affect these fauna through greater river inflow and wave effects, thinner ice, the shorter duration of ice cover, more storms, greater mixing and supply of nutrients, enhanced sediment transport, and coastal erosion. The expected outcome would be an expansion of these stress-tolerant fauna populations over the shelf, and the immigration of other species with corresponding adaptations. We might also anticipate a greater import of Pacific origin species via increased buoyancy boundary flow from the Pacific (Carmack and McLaughlin, 2001).

9.4 Recommendations

The impact of climate change on carbon cycling within Arctic marine communities is difficult to predict (Wassmann et al., 2004). However, our study suggests that the factors which influence benthic community composition and ice algae production on Arctic shelves may also have significant impact on carbon processing

Coast guard helicopter in a Fjord. Photo: Hiroshi Hattori.

and storage within bottom sediments. It is becoming increasingly clear that both the structure and seasonal activity of benthic faunal communities as well as primary productivity determine carbon preservation and nutrient regeneration patterns on Arctic shelves. Given the logistical challenge of conducting scientific studies in the Canadian Arctic, most benthic sampling programs to date have been constrained to the summer and early fall when sea ice cover is minimal. Such programs have largely focused on describing spatial variations in bottom community structure (e.g. Dale et al., 1989; Aitken and Fournier, 1993; Aitken and Gilbert, 1996) and respiration (e.g. Renaud et al., 2007). Longitudinal studies of benthic community structure are virtually non-existent; however, there are a few studies pertaining to temporal changes in bottom community structure, including Conlan and Kvitek (2005; reporting the recolonization of the ice-scoured seafloor of Barrow Strait over several decades) and Cusson et al. (2007; reporting temporal variations in benthic community structure in Frobisher Bay). In the context of rapid climate warming, it has become imperative to document more temporal variations in bottom community structure and respiration to form a baseline from which future changes can be assessed. Specifically, future research must be directed towards: the availability of various forms of organic materials which provide nutrition to benthic organisms; the partioning of the food supply amongst potential production, respiration, and storage in bottom sediments; and the linkages between the benthos and their megafaunal predators, especially the marine mammals which are harvested to support indigenous subsistence or commercial industry.

References

Arctic Climate Impact Assessment, 2004. Impacts of a warming Arctic: Arctic climate impact assessment. Cambridge University Press. URL: www.acia.uaf.edu.

Aitken, A. E., and R. Gilbert, 1996. Marine Mollusca from Expedition Fiord, Western Axel Heiberg Island, Northwest Territories, Canada. Arctic. 29: 29-43.

Aitken, A.E., and J. Fournier, 1993. Macrobenthos communities in Cambridge, McBeth and Itirbilung Fiords, Baffin Island, Northwest Territories, Canada. Arctic. 46: 60-71.

Ambrose, W.G., Jr., 1993. Effects of predation and disturbance by ophiuroids on soft-bottom community structure in Oslofjord: results of a mesocosm study. Mar. Ecol. Prog. Ser. 97: 225-236.

Ambrose, W.G., Jr., L.M. Clough, P.R.Tilney, and L. Beer, 2001. Role of echinoderms in benthic remineralization in the Chukchi Sea. Mar. Biol. 139: 937-949.

Ambrose, W. G., Jr., and P.E. Renaud, 1995. Benthic response to water column productivity patterns: evidence for benthic-pelagic coupling in the Northeast Water Polynya. J. Geophys. Res. 100: 4411-4421.

Ambrose, W.G., Jr., and P.E. Renaud, 1997. Does a pulsed food supply to the benthos affect polychaete recruitment patterns in the Northeast Water Polynya? J. Mar. Syst. 10:483-495.

Arrigo, K.R., and G.L. van Dijken, 2004. Annual cycles of sea ice and phytoplankton in Cape Bathurst polynya, southeastern Beaufort Sea, Canadian Arctic. Geophys. Res. Lett. 31, L08304, 1-4.

Atkinson, E. G., and J.W. Wacasey, 1989. Benthic invertebrates collected from the western Canadian Arctic, 1951 to 1985. Can. Data Rep. Fish. Aquat. Sci. 745.132 pp.

Barry, J.P., J.M. Grebmeier, J. Smith, and R.B. Dunbar, 2003. Oceanographic versus seafloor-habitat control of benthic megafaunal communities in the S.W. Ross Sea, Antarctica. In: Di Tullio, G.R., Dunbar, R.B. (eds.), Biogeochemistry of the Ross Sea. Antarctic Research Series vol. 78, American Geophysical Union, Washington, D.C., pp. 327-354.

Berger, T. R., 1977. Northern Frontier, Northern Homeland. The report of the Mackenzie Valley Pipeline Inquiry: Volume One. Ministry of Supply and Services Canada. 213 pp.

Blazewicz-Paszkowycz, M., and J. Sekulska-Nalewajko, 2004. Tanaidacea (Crustacea, Malacostraca) of two polar fjords: Kongsfjorden (Arctic) and Admiralty Bay (Antarctic). Polar Biol. 27: 222-230.

Blazewicz-Paszkowycz, M., and R. Ligowski, 2002. Diatoms as a food source indicator for some Antarctic Cumacea and Tanaidacea (Crustacea). Antarctic Sci. 14 : 11-15.

Bluhm, B., D. Piepenburg, and K. von Juterzenka, 1998. Distribution, standing stock, growth, mortality, and production of *Strongylocentrotus pallidus* (Echinodermata: Echinoidea) in the northeastern Barents Sea. Polar Biol. 20: 325-334.

Brey, T., 2004. Empirical relations in aquatic populations. In: Population Dynamics in Benthic Invertebrates: A virtual handbook. http://www.awi-bremer-haven.de/ Benthic/Ecosystem/ FoodWeb/ Handbook/main.html. Alfred-Wegener Institute, Bremerhaven.

Callaway, R. 2006. Tubeworms promote community change. Mar. Ecol. Progr. Ser. 308: 49-60.

Carey, A.G., Jr., 1991. Ecology of North American Arctic continental shelf benthos: a review. Cont. Shelf Res. 11: 865-883.

Carmack, E.C., R.W. Macdonald, and S. Jasper, 2004. Phytoplankton productivity on the Canadian Shelf of the Beaufort Sea. Mar. Ecol. Progr. Ser 277: 37-50.

Carmack, E.C., and R.W. Macdonald, 2002. Oceanography of the Canadian Shelf of the Beaufort Sea: A setting for marine life. Arctic, 55, Supplement 1: 29-45.

Carmack, E., and F. McLaughlin, 2001. Arctic Ocean change and consequences to biodiversity: A perspective on linkage and scale. Memoirs of the National Institute of Polar Res. Spec. Issue 54: 365-375.

Carroll, M.L., and J. Carroll, 2003. The Arctic seas. In: K. D. Black, and G. B. Shimmield (eds.), Biogeochemistry of marine systems. Blackwell, p. 127-156.

Clarke, K. R., 1993. Non-parametric multivariate analyses of changes in community structure. Austr. J. Ecol. 18: 117-143.

Clarke, K. R., and R.H. Green, 1988. Statistical design and analysis for a 'biological effects' study. Mar. Ecol. Progr. Ser. 46: 213-226.

Clarke, K. R., and R. N. Gorley, 2006. PRIMER v6: User Manual/Tutorial. PRIMER-E: Plymouth.

Clarke, K. R., and R.M. Warwick, 2001. Change in Marine Communities: An Approach to Statistical Analysis and Interpretation, 2nd edition. PRIMER-E Ltd., Plymouth.

Clarke, K. R., and R.M. Warwick, 1998. A taxonomic distinctness index and its statistical properties. J. App. Ecol. 35: 523-531.

Clough, L.M., W.G., Jr., Ambrose, J.K. Cochran, C. Barnes, P.E. Renaud, and R.C. Aller, 1997. Infaunal density, biomass, and bioturbation in the sediments of the Arctic Ocean. Deep-Sea Res. II, 44: 1683-1704.

Clough, L.M., P.E. Renaud, and W.G., Jr., Ambrose, 2005. Sediment oxygen demand, infaunal biomass, and sediment pigment concentration in the western Arctic Ocean. Can. J. Fish. Aquat. Sci. 62: 1756-1765.

Conlan, K., and R. Kvitek, 2005. Recolonization of soft-sediment ice scours on an exposed Arctic coast. Mar. Ecol. Progr. Ser. 286: 21-42.

Conlan, K., A. Aitken, E. Hendrycks, C. McClelland, and H. Melling. Distribution patterns of Canadian Beaufort Shelf macrobenthos J. Mar. Syst. 74: 864-887.

Curtis, M. A., 1972. Depth distributions of benthic polychaetes in two fiords on Ellesmere Island, N.W.T. J. Fish. Res. Bd. Can. 29: 1319-1327.

Cusson, M, P. Archambault, and A. Aitken, 2007. Biodiversity of benthic assemblages on the Arctic continental shelf: historical data from Canada. Mar. Ecol. Progr. Ser. 331: 291-304.

Dale, J.E., A.E. Aitken, R. Gilbert, and M.J. Risk, 1989. Macrofauna of Canadian arctic fiords. Mar. Geol. 85: 331-358.

Dauwe, B., P. M. J. Herman, and C. H. R. Heip, 1998. Community structure and bioturbation potential of macrofauna at four North Sea stations with contrasting food supply. Mar. Ecol. Progr. Ser. 173: 67-83.

Dayton, P.K., and Oliver, J.S., 1977. Antarctic soft-bottom benthos in oligotrophic and eutrophic environments. Science, 197: 55-58.

De Grave, S., Casey, and D.A.Witaker, 2001. The accuracy of density standardization of infaunal benthos. J. Mar. Biol. Ass. U.K. 81: 541-542.

Denisenko, N.V., S.G. Denisenko, K.K. Lehtonen, A.-B. Andersin, and H.R. Sandler, 2007. Zoobenthos of the Cheshskaya Bay (southeastern Barents Sea): spatial distribution and community structure in relation to environmental factors Polar Biol. 30: 735-746 DOI 10.1007/s00300-006-0232-4.

Denisenko, S.G., 2001, Long-term changes of zoobenthos biomass in the Barents Sea. Proc. Zoo. Inst. Russ. Acad. Sci. 289: 39-46.

Dickson, D.L., and H.G. Gilchrist, 2002. Status of marine birds of the southeastern Beaufort Sea. Arctic. 55: 46-58.

Duchêne, J.-C., and R. Rosenberg, 2001. Marine benthic faunal activity patterns on a sediment surface assessed by video numerical tracking. Mar. Ecol. Progr. Ser. 223: 113-119.

Dunton, K.H., T. Weingartner, and E.C. Carmack, 2006. The nearshore western Beaufort Sea ecosystem: Circulation and the importance of terrestrial carbon in arctic coastal food webs. Progr. Oceanogr. 71: 362-378.

Fauchald, K., and P. A. Jumars, 1979. The diet of worms: a study of polychaete feeding guilds. Oceanogr. Mar. Biol. Ann. Rev. 17: 193-284.

Feder, H. M., S. C. Jewett, and A. L. Blanchard, 2007, Southeastern Chukchi Sea (Alaska) macrobenthos. Polar Biol. 30: 261-275.

Forest, A., M. Sampei, H. Hattori, R. Makabe, H. Sasaki, M. Fukuchi, P. Wassmann, and L. Fortier, 2007. Particulate organic carbon fluxes on the slope of the Mackenzie Shelf (Beaufort Sea): Physical and biological forcing of shelf-basin exchanges. J. Mar. Syst. doi:10.1016/j.jmarsys.2006.10.008

Frontier, S., and D. Pichot-Viale, 1998. Écosystèmes : Structure, Fonctionnement, Évolution. Dunod, Paris.

Frost, K.J., and L.F. Lowry, 1984. Trophic relationships of vertebrate consumers in the Alaskan Beaufort Sea. In: Barnes, P. W., Schell, D. M., Reimnitz, E. (Eds.), The Alaskan Beaufort Sea: Ecosystems and Environments. Academic Press, Orlando, p. 381-401.

Gilbert, G., and K. Pedersen, 1987. Ice scour data base for the Beaufort Sea. Environmental Studies Revolving Funds Report No. 055.

Glud, R.N., N. Risgaard-Petersen, B. Thamdrup, H. Fossing, and S. Rysgaard, 2000. Benthic carbon mineralization in a high-Arctic sound (Young Sound, NE Greenland). Mar. Ecol. Progr. Ser. 206: 59-71.

Golikov, A.N., and V.G. Averintzev, 1977. Distribution patterns of benthic and ice biocenoses in the high latitudes of the polar basin and their part in the biological structure of the world ocean. In: Dunbar, M. J. (ed.), Polar oceans. McGill University, Montreal, p. 331-364.

Grant, J., B. Hargrave, and P. Macpherson, 2002. Sediment properties and benthic-pelagic coupling in the North Water. Deep-Sea Res. II 49: 5259-5275.

Grebmeier, J.M., and J.P. Barry, 1991. The influence of oceanographic processes on pelagic-benthic coupling in Polar Regions: A benthic perspective. J. Mar. Syst. 2: 495-518.

Grebmeier J.M., and L.W. Cooper, 1995. Influence of the St. Lawrence Island polynya upon the Bering Sea benthos. J. Geophys. Res. 100: 4439-4460.

Grebmeier, J.M., H.M. Feder, and C.P. McRoy, 1989. Pelagic-benthic coupling on the shelf of the northern Bering and Chukchi Seas. II. Benthic community structure. Mar. Ecol. Progr. Ser. 51: 253-268.

Hargrave, B.T., I.D. Walsh, and D.W. Murray, 2002. Seasonal and spatial patterns in mass and organic matter sedimentation in the North Water. Deep-Sea Res II. 49: 5227-5244.

Harwood, L.A., and T.G. Smith, 2002. Whales of the Inuvialuit settlement region in Canada's Western Arctic: An overview and outlook. Arctic 55: Supp. 1: 77-93.

Harwood, L. A., and I. Stirling, 1992. Distribution of ringed seals in the southeastern Beaufort Sea during late summer. Can. J. Zoo. 70: 891-900.

Héquette, A., M. Desrosiers, and P. W. Barnes, 1995. Sea ice scouring on the inner shelf of the southeastern Canadian Beaufort Sea. Mar. Geol. 128: 201-219.

Highsmith, R.C., and K.O. Coyle, 1990. High productivity of northern Bering Sea benthic amphipods. Nature. 344: 862-863.

Hobson, K.A., W.G., Jr., Ambrose, and P.E. Renaud, 1995. Sources of primary production, benthic-pelagic coupling, and trophic relationships within the Northeast Water Polynya: Insights from δ^{13}C and δ^{15}N analysis. Mar. Ecol. Prog. Ser. 128: 1-10.

Hobson, K.A., A. Fisk, N. Karnovsky, M. Holst, J.-M. Gagnon, and M. Fortier, 2002. A stable isotope (δ^{13}C, δ^{15}N) model for the North Water food web: implications for evaluating trophodynamics and the flow of energy and contaminants. Deep-Sea Res. II. 49: 5131-5150.

Holm-Hansen, O., C.J. Lorenzen, R.W. Holms, and J.D. Strickland, 1965. Fluorometric determination of chlorophyll. J. Cons. Int. Explor. Mer. 30: 3-15.

Holte, B., 1998. The macrofauna and main functional interactions in the sill basin sediments of the pristine Holandsfjord, North Norway, with autecological reviews for some key-species. Sarsia 83: 55-68.

Holte, B., and B. Gullikesen, 1998. Common macrofaunal dominant species in the sediments of some north Norwegian and Svalbard glacial fjords. Polar Biol. 19: 375-382.

Intergovernmental Panel on Climate Change, 2001. Climate Change 2001: The Scientific Basis. In: Houghton, J.T., Ding, Y., Griggs, D.J., Noguer, M., van der Linden, P.J. Xiaosu, D. (eds.), Contributions of Working Group I to the Third Assessment Report of the Intergovernmental Panel on Climate Change (IPCC). Cambridge Univ. Press.

Johannessen, O.M., L. Bengtsson, M.W. Miles, S.I. Kuzmina, V.A. Semenov, G.V. Alekseev, A.P. Nagurnyi, V.F. Zakharov, L.P. Bobylev, L.H. Pettersson, K. Hasselmann, and H.P. Cattle, 2004. Arctic climate change: observed and modelled temperature and sea-ice variability. Tellus, 56A: 328-341.

Jangoux, M., and J. M. Lawrence, 1982. Echinoderm nutrition. A. A. Balkema, Rotterdam. 654 p.

Kamp, A., and U. Witte, 2005. Processing of ^{13}C-labelled phytoplankton in a fine-grained sandy-shelf sediment (North Sea): relative importance of different macrofauna species. Mar. Ecol. Progr. Ser. 297: 61-70.

Lee, S. H., D. M. Schell, T. L. McDonald, and W. J. Richardson, 2005. Regional and seasonal feeding by bowhead whales *Balaena mysticetus* as indicated by stable isotope ratios. Mar. Ecol. Progr. Ser. 285: 271-287.

Levin, L.A., D.F. Boesch, A. Covich, C. Dahm, C. Erséus, K.C. Ewel, R.T. Kneib, A. Moldenke, M.A. Palmer, P. Snelgrove, D. Strayer, and J.M.

Weslawski, 2001. The function of marine critical transition zones and the importance of sediment biodiversity. Ecosystems. 4: 430-451.

Lewis, C. F. M., and S.M. Blasco, 1990. Character and distribution of sea-ice and iceberg scours: keynote address. In: Clark, J. I. (Ed.). Workshop on ice scouring and design of offshore pipelines, Calgary, Alberta, April 18-19, 1990. Canada Oil and Gas Lands Administration, Energy, Mines and Resources Canada and Indian and Northern Affairs Canada.

Lohrer, A.M., S.F. Thrush, and M.M. Gibbs, 2004. Bioturbators enhance ecosystem function through complex biogeochemical interactions. Nature. 431: 1092-1095.

Lovvorn, J.R., L.W. Cooper, M.L. Brooks, C.C. De Ruyck, J.K. Bump, and J.M. Grebmeier, 2005. Organic matter pathways to zooplankton and benthos under pack ice in late winter and open water in late summer in the north-central Bering Sea. Mar. Ecol. Progr. Ser. 291: 135-150.

Macdonald, R.W., S.M. Solomon, R.E. Cranston, H.E. Welch, M.B. Yunker, and C. Gobeil, 1998. A sediment and organic carbon budget for the Canadian Beaufort Shelf. Mar. Geol. 144: 255-273.

McMahon, K.W., W.G. Jr. Ambrose, B.J. Johnson, M-Y. Sun, G.R. Lopez, L.M. Clough, and M.L. Carroll, 2006. Benthic community response to ice algae and phytoplankton in Ny Ålesund, Svalbard. Mar. Ecol. Progr. Ser. 310:1-14.

Melnikov, I.A., 1998. Winter production of sea ice algae in the western Weddell Sea. J. Mar. Syst. 17: 195-205.

Mills, E.L., 1967. The biology of an ampeliscid amphipod crustacean sibling species pair. J. Fish. Res. Bd. Can. 24: 305-355.

Montgomery, D.C. 1991. Design and Analysis of Experiments. John Wiley & Sons, Toronto.

Moritz, R.E., C.M. Bitz, and E.J. Steig, 2002. Dynamics of recent climate change in the Arctic. Science 297:1497-1502.

Myers, R., S. Blasco, G. Gilbert, and J. Shearer, 1996. 1990 Beaufort Sea ice scour repetitive mapping program. Volume 1. Environmental Studies Research Funds Report No. 129.

Norkko, A., J. E. Hewitt, S. F. Thrush, and G. A. Funnell, 2001. Benthic-pelagic coupling and suspension-feeding bivalves: Linking site-specific sediment flux and biodeposition to benthic community structure. Limnol Oceanogr. 46: 2067-2072.

Norling, K., R. Rosenberg, S. Hulth, A. Grémare, and E. Bonsdorff, 2007. Importance of functional bio-diversity and species-specific traits of benthic fauna for ecosystem functions in marine sediment. Mar. Ecol. Progr. Ser. 332: 11-23.

O'Brien, M.C., R.W. Macdonald, H. Melling, and K. Iseki, 2006. Particle fluxes and geochemistry on the Canadian Beaufort Shelf: implications for sediment transport and deposition. Cont. Shelf Res. 26: 41-81.

Oliver, J. S., M.A. Slattery, M.A. Silberstein, and E.F. O'Connor, 1983. A comparison of gray whale, *Eschrichtius robustus*, feeding in the Bering Sea and Baja California. U.S. Nat. Mar. Fish. Ser. Fish. Bull. 81: 513-522.

Papaspyrou, S., T. Gregersen, E. Kristensen, B. Christensen, and R.P. Cox, 2006. Microbial reaction rates and bacterial communities in sediment surrounding burrows of two nereid polychaetes (*Nereis diversicolor* and *N. virens*). Mar. Biol. 148: 541-550.

Peachey, R. L., and S.S. Bell, 1997. The effects of mucous tubes on the distribution, behaviour and recruitment of seagrass meiofauna. J. Exp. Mar. Biol. Ecol. 209: 279-291.

Peck, L. S., 2002. Ecophysiology of Antarctic marine ectotherms: limits to life. Polar Biol. 25: 31-40.

Peterson, G.H., and M.A. Curtis, 1980. Differences in energy flow through major components of subarctic, temperate and tropical marine shelf ecosystems. Dana, 1: 53-64.

Piepenburg, D., 1988. On the composition of the benthic-fauna of the Western Fram Strait. Ber. Polarforsch. 52: 1-4.

Piepenburg, D., 2005. Recent research on Arctic benthos: common notions need to be revised. Polar Biol. 28: 733-755.

Piepenburg, D., W.G., Jr. Ambrose, A. Brandt, P.E. Renaud, M.J. Ahrens, and P. Jensen, 1997. Benthic community patterns reflect water column process in the Northeast Water Polynya (Greenland). J. Mar. Syst. 10: 476-482.

Piepenburg, D., T.H. Blackburn, D.V. von Dorrien, J. Gutt, P.O. Hall, S. Hulth, M.A. Kendall, K.W. Opalinski, E. Rachor, and M.K. Schmid, 1995. Partitioning of benthic community respiration in the Arctic (northwestern Bering Sea). Mar. Ecol. Prog. Ser. 118: 199-213.

Peipenberg, D., and M.K. Schmid, 1996. Distribution, abundance, biomass, and mineralization potential of the epibenthic megafauna of the Northeast Greenland shelf. Mar.Biol.125: 321-332.

Piepenburg, D., and K. von Juterzenka, 1994. Abundance, biomass and spatial distribution patterns of brittle stars (Echinoderrmata: Ophiuroidea) on the Kolbeinsey Ridge north of Iceland. Polar Biol. 14: 185-194.

Pörtner, H. O., and R. Knust, 2007. Climate change affects marine fishes through the oxygen limitation of thermal tolerance. Science. 315: 95-97.

Renaud, P.E., N. Morata, W.C. Ambrose, Jr., J.J. Bowie, and A. Chiuchiolo, 2007. Carbon cycling by seafloor communities on the eastern Beaufort shelf. J. Exp. Mar. Biol. Ecol. 349: 248-260.

Renaud, P. E., A. Riedel, C. Michel, N. Morata, M.Gosselin, T. Juul-Pedersen, and A.Chiuchiolo, 2006. Seasonal variation in benthic community oxygen demand: A response to an ice algal bloom in the Beaufort Sea, Canadian Arctic? J. Mar. Syst. 67: 1-12.

Ricciardi, A., and E. Bourget, 1998. Weight-to-weight conversion factors for marine benthic macro-invertebrates. Mar. Ecol. Progr. Ser. 163: 245-251.

SAS, 1999. The SAS System for Windows. Release 8.02 Edition. SAS Institute Inc, Cary, NC.

Scherrer, B. 1984. Biostatistique. Gaëtan Morin, Montréal.

Schmid, M.K., Piepenberg, D., Golikov, A. A., von Juterzenka, K., Petryashov, V. V., Spindler, M., 2006. Trophic pathways and carbon flux patterns in the Laptev Sea. Progr. Oceanogr. 71: 314-330.

Seuthe, L., G. Darnis, C.W. Riser, P. Wassmann and L. Fortier, 2006. Winter-spring feeding and metabolism of Arctic copepods: insights from faecal pellet production and respiration measurements in the southeastern Beaufort Sea. Polar Biol. DOI 10.1007/s00300-006-0199-1.

Smith, K.L., 1978. Benthic community respiration in the N.W. Atlantic Ocean: *in situ* measurements from 40 and 5200 m. Mar. Biol. 47: 337-347.

Snelgrove, P.V.R., J.F. Grassle, J.P. Grassle, R.F. Petracca, and K.I. Stocks, 2001. The role of colonization in establishing patterns of community composition and diversity in shallow water sedimentary communities. J. Mar. Res. 59: 813-831.

Sokal, R.R: and Rohlf, F.J., 1995. Biometry, 3rd ed. W.H. Freeman and Co., New York.

Steffens, M., D. Piepenburg, and M. K., Schmid, 2006. Distribution and structure of macrobenthic fauna in the eastern Laptev Sea in relation to environmental factors. Polar Biol. 29 : 837-848.

Tremblay, J.-E., H. Hattori, C. Michel, M. Ringuette, Z.-P.Mei, C. Lovejoy, L. Fortier, K.A. Hobson, D. Amiel, and K.Cochran, 2006. Trophic structure and pathways of biogenic carbon flow in the eastern North Water Polynya. Progr. Oceanogr. 71: 402-425.

Vinnikov, K.Y., D.J. Cavalieri, and C.L. Parkinson, 2006. A model assessment of satellite observed trends in polar sea ice extents. Geophy. Res. Letters, 33: L05704, doi:10.1029/2005GL025282.

Vinnikov, K.Y., A. Robock, R.J. Stouffer, J.E. Walsh, C.L. Parkinson, D.L. Cavalieri, J.F.B. Mitchell, D.Garrett, and V.F. Zakharov, 1999. Global warming and northern hemisphere sea ice extent. Science 286: 1934-1937.

Wacasey, J.W., 1975. Biological productivity of the southern Beaufort Sea: zoobenthic studies. Department of the Environment, Victoria, B.C. Beaufort Sea Technical Report No. 12b, 39.

Wacasey, J.W., E.G. Atkinson, L. Derick, and A. Weinstein, 1977. Zoobenthos data from the southern Beaufort Sea, 1971-1975. Fisheries and Marine Service Data Report 41. 187 pp.

Walsh, J.J.,. D.A. Dieterle, W. Maslowski, and T.E. Whitledge, 2004. Decadal shifts in biophysical forcing of Arctic marine food webs: numerical consequences. J. Geophys. Res. 109:C05031, 10.1029/2003JC001945.

Warwick, R. M., and K. R. Clarke, 1995. New 'biodiversity' measures reveal a decrease in taxonomic distinctness with increasing stress. Mar. Ecol. Progr. Ser. 129 : 301-305.

Wassmann, P., E. Bauerfeind, M. Fortier, M. Fukuchi, B. Hargrave, S. Honjo, B. Moran, T. Noji, E. M. Nöthig, and R. Peinert, 2004. Particulate organic carbon flux to the seafloor of the Arctic Ocean: quantity, seasonality and processes. In: R. Stein, and R. W. MacDonald (eds.), The Organic Carbon Cycle in the Arctic Ocean. Springer–Verlag, p. 131-138

Welch, H.E., M.A. Bergman, T.D. Sifern, K.A. Martin, M.F. Curtis, R.E. Crawford, R.J. Conover, and H. Hop, 1992. Energy flow through the marine ecosystem of the Lancaster Sound region, Arctic Canada. Arctic. 45: 343-357.

Wenzhöfer, F., and R.N. Glud, 2004. Small-scale spatial and temporal variability in coastal benthic O_2 dynamics : effects of fauna activity. Limnol. Oceanogr. 49 : 1471-1481.

Ice floe. Photo: Trecia Schell.

Williams, W.J., E.C. Carmack, K Shimada, H. Melling, K. Aagaard, R.W.Macdonald, and R.G. Ingram, 2006. Joint effects of wind and ice motion in forcing upwelling in Mackenzie Trough, Beaufort Sea. Cont. Shelf Res. 26 : 2352-2366.

Wlodarska-Kowalczuk, and M., T. Pearson, 2004. Soft-bottom macrobenthic faunal associations and factors affecting species distributions in an Arctic glacial fjord (Kongsfjord, Spitsbergen). Polar Biol. 27: 155-167.

Zar, J.H. 1999. Biostatistical Analysis. Prentice-Hall, Inc., Englewood Cliffs, New Jersey

AMUNDSEN

CHAPTER 10

An Ocean of Data: The CASES Legacy

Josée Michaud[1,3]*, Leah Braithwaite[2], Marie-Ève Garneau[1], Christine Barnard[3], Warwick F. Vincent[1]

10.1 Introduction and Rationale

In the era of information technology, advances in data acquisition and storage have experienced rapid acceleration. The ability to integrate various fields of scientific research through vast datasets clearly acts as an accelerator to scientific progress—both by making efficient use of scarce research resources and by helping us better formulate our understanding of the surrounding world. Unfortunately, sometimes data can be lost or become irretrievable. This can be due to several factors (Strong and Leach, 2005), including:

- inappropriate storage, media degradation, or software/hardware obsolescence;
- inaccessible or inadequate archiving;
- insufficient descriptive information about the available datasets (i.e., metadata);
- inattention to long-term preservation beyond the use of individual researchers or organizations;
- inadequate planning or resources allocated to data management and archiving; and
- challenges in respecting the rights of the data originator(s) or in managing privacy considerations.

Preserving data such that it remains accessible to science in the long-term is a universal problem. Failing to address this issue can lead to the loss or under-use of information (which can severely setback future research) or even the inability to verify existing findings (Buneman et al., 2004).

Detection of environmental trends requires well-documented and consistent time-series information. With the increasing evidence of rapid environmental changes in Polar Regions, data *description* (i.e. metadata standardization), *archiving* and *open access* have become key issues in Arctic and Antarctic research programs. This interest has been further heightened by the launch of the fourth International Polar Year (IPY), from March 2007 to March 2009. Previous IPYs have not been entirely successful in the fields of data archiving and synthesis (Behr et al., 2007). This is unfortunate since environmental data from these earlier periods would have provided extremely valuable benchmarks for the analysis and interpretation of present-day climate conditions. There was no central data repository during the first IPY although rescued climate data from this period are available at: www.noaa.org.

The CCGS Amundsen at the foot of a glacier. Photo: Martin Fortier/ArcticNet.

LEFT: The CCGS Amundsen navigating in heavy ice conditions. Photo: Josée Michaud.

[1] Departement de Biologie, Université Laval, Québec, QC, G1V 0A6, Canada

[2] Canadian Ice Service, Environnent Canada, 373, promenade Sussex Ottawa, ON, K1A 0H3, Canada

[3] ArcticNet, Université Laval, Québec, QC, G1V 0A6, Canada

* Corresponding author: josee.michaud@arcticnet.ulaval.ca

Archiving efforts improved somewhat during the second IPY program (1932-1933) despite some loss of records during World War II. The third IPY (1957-1958) resulted in the establishment of World Data Centers (WDC; www.ngdc.noaa.gov/wdc), however not all the data were archived. In recognition of the vital need to improve data management, the fourth IPY has now identified the production of metadata catalogues (i.e., descriptions of data) and data archiving among its central objectives (www.ipy.org).

There are existing examples of successful data management, particularly within long-term environmental research programs. One outstanding example is the Long Term Ecological Research (LTER) program of the National Science Foundation. The LTER program supports multidecadal observations at 26 sites, including a polar marine site at Palmer Station (Antarctica). Its data catalogues and management systems are supported by a large allocation of resources and can be accessed through www.ilternet.edu. Fisheries and Oceans Canada (DFO) also maintains an extensive archive of oceanographic data, including the longest time-series available in the world from Ocean Weather Station Papa and Line-P, a series of oceanographic stations extending from the mouth of the Juan de Fuca Strait (south of Vancouver Island) to Station Papa at 50 °N 145 °W (in the Pacific Ocean). These datasets are some of the very few of sufficient quality and suitable length to be useful in examining the variability of the oceans, if only over a period of a few decades. The data are also archived in a consistent format, in order to provide optimal utility to the international scientific community and public at large (http://www.pac.dfo-mpo.gc.ca/sci/osap/projects/linepdata/default_e.htm).

In light of these programs and the increased international focus on scientific data management, the CASES Research Network has made metadata/data archiving one of its primary objectives. CASES represents one of the most ambitious multidisciplinary and international efforts to understand the biogeochemical and ecological impacts of the present decline in Arctic sea ice cover. Its scientific program encompassed all fields of oceanography and atmospheric sciences in an attempt to better understand the overall Arctic ecosystem. In keeping with IPY philosophy, multidisciplinary datasets resulting from this initiative will form the basis for several hundred publications and provides an important legacy to Arctic Ocean research. Therefore, data must be archived in a manner that will ensure long-term accessibility to the scientific community and to the public.

CASES datasets are varied and numerous, having been generated by eight subprojects within the program's mandate (Appendix 10.1). The CASES Research Network has involved over 50 principal investigators (PIs) and their research teams, representing a diversity of disciplines and institutions from Canada and abroad. The varied approaches to data management among individual laboratories, disciplines and institutions has represented a significant challenge in ensuring that CASES data is preserved adequately and is easily accessible, but while preserving the rights of the data originator(s) for publication. In keeping with this, the objectives of the CASES Data Management and Archive Strategy were to ensure that all data are:

- integrated into comprehensive databases;

- available to all participants of the CASES Research Network under conditions that respect the rights of the data originator(s);

- archived and ultimately accessible to the broader scientific community and the public; and

- available to researchers in other national and international polar research programs such as ArcticNet and IPY.

10.2 The Database

10.2.1 Strategy and implementation

To guarantee the longevity of CASES datasets, archiving is now underway through existing infrastructures and data services. Researchers are encouraged to use CASES databases appropriate to their discipline wherever possible, so long as the four objectives of the Data Management and Archive Strategy are met. In addition, CASES is using data management tools and systems which are available (and are continuing to be developed) within ArcticNet (a Network of Centres of Excellence program; www.arcticnet.ulaval.ca). This data management initiative is made in partnership with the Canadian Cryospheric Information Network (CCIN: www.ccin.ca) and the Integrated Science Data Management division of DFO (ISDM, formerly the Marine Environmental Data Service, MEDS: www.meds-sdmm.dfo-mpo.gc.ca). The ArcticNet Data Management Committee has adopted an eight-step implementation plan that also applies to CASES (Fig. 10.1). CCIN and ISDM house the CASES metadatabase (within the ArcticNet metadatabase, now called the Polar Data Catalogue) and also provide archiving facilities for current CASES datasets (with some holdings at other locations, depending on the subprogram). The datasets are being archived using national and international protocols and format requirements in order to allow compatibility with multiple platforms.

10.2.2 "Discovery" metadata database

The term *metadata* is simply defined as data about data. It is a description of the nature of a selection of data: where, when and by whom it was collected, as well as its current location. The purpose of metadata is to facilitate the understanding, use and management

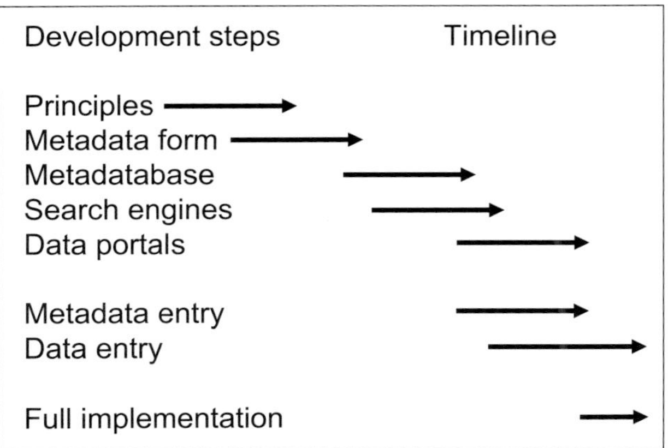

Figure 10.1

The data management implementation plan adopted by ArcticNet and CASES.

of data. It might effectively be considered the most crucial component of the data archiving process. The ArcticNet/CCIN Polar Data Catalogue has been designed to provide both structured "discovery" information (in order to facilitate a user's ability to find data) as well as descriptive information about the nature of the data and how it was collected. More detailed information about the data will be part of the actual dataset.

The infrastructure of the ArcticNet/CCIN metadata portal is shown schematically in Figure 10.2. The Polar Data Catalogue has been structured according to Federal Geographic Data Committee (FGDC) format, which means that it respects the ISO-standards for metadata and can be exported to other international databases. The input forms and procedures have been designed to streamline the process and keep user input to a minimum. The overall system includes an input and information tag module (including a keyword dictionary), a data registration module, a password-protected edit facility,

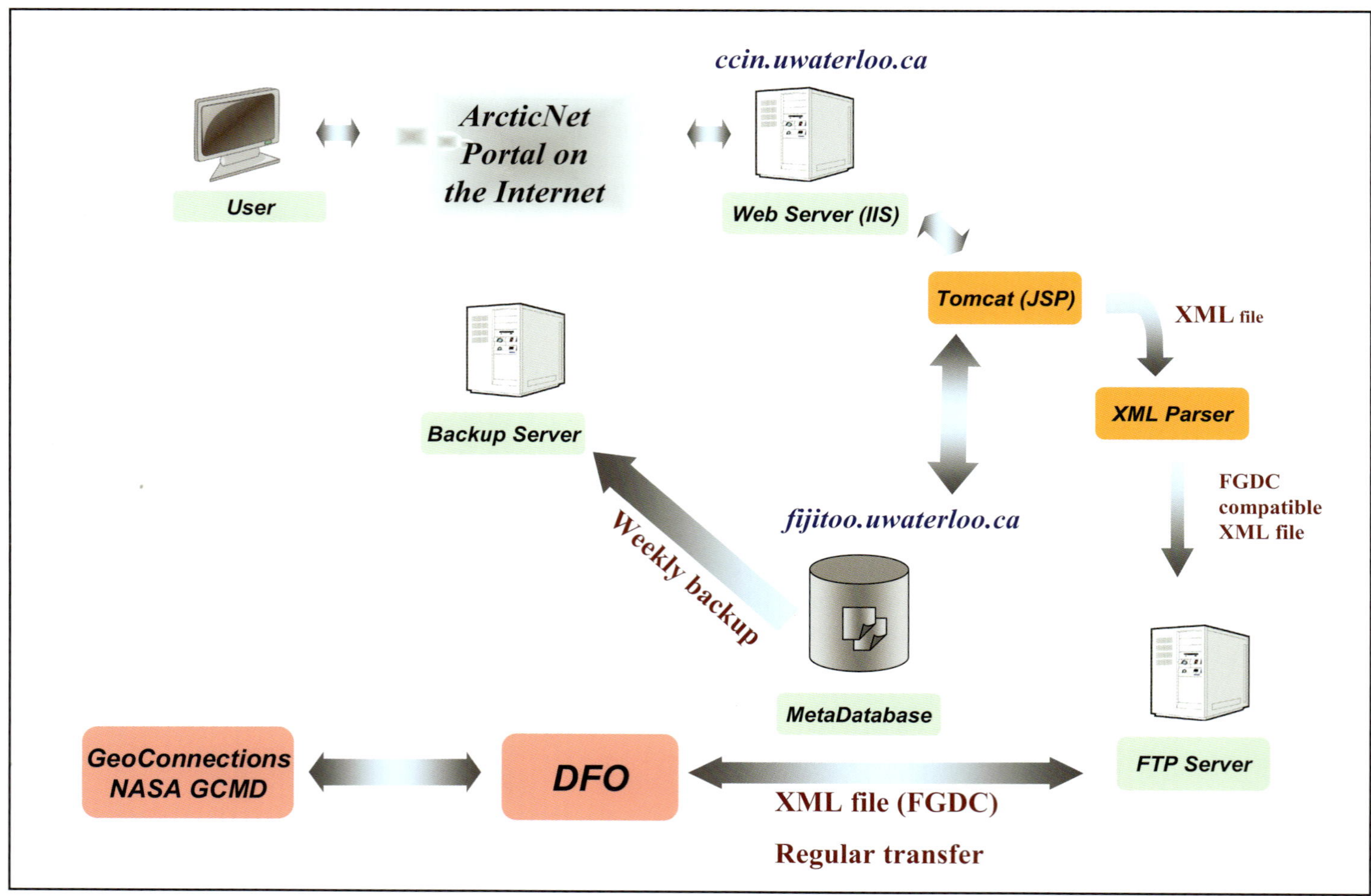

Figure 10.2

Infrastructure of the ArcticNet/CCIN metadata portal for the Polar Data Catalogue.

a keyword search facility, and a translation program for XML files being transferred to ISDM (and ultimately GeoConnections and the NASA Global Change Master Directory, GCMD: http://gcmd.nasa.gov). ISDM is the DFO branch which manages/archives marine data collected internally or acquired through national and international programmes conducted in ocean areas adjacent to Canada. It is therefore especially appropriate as a central holder of CASES metadata and data. In turn, the NASA GCMD database provides international access to a holding of more than 19,000 datasets pertaining to environmental or earth sciences. Back-up copies of the Polar Data Catalogue are created on a regular basis and stored at different locations.

The CASES/ArcticNet metadata input form is accessed through the intranet link of the ArcticNet portal (www.arcticnet.ulaval.ca) or directly at www.polardata.ca. The form is divided into sections, each containing mandatory fields. The first step in creating metadata is to register and create a user account; after which, the user has access to the metadata input form (Fig. 10.3). Required input fields include the study title, study site, how the data should be cited, the purpose of the data collection, an abstract, geographic coordinates, relevant time periods, research programs (which allows CASES to be explicitly identified as the source of the data), and keywords. Keywords are selected using a dropdown menu (Fig. 10.4) and are divided into seven categories: geographic locations, northern communities, natural sciences, health sciences and contaminants, social sciences, economics and policy, and Inuktitut keywords. New keywords can be registered by clicking on the link "Add a new keyword?" which sends an Email to the ArcticNet data manager who will assess the request and add the new word if it meets the ArcticNet keyword rules (minimum duplication, maximum clarity, and maximum utility). The metadata can only be modified or updated by the PI responsible for its original entry (or the delegated data entry person in his/her research team), and all entries are checked by the data manager to ensure that spurious input is avoided. Links to the actual data are also determined by the PI and can either be a website (where the data is stored) or an email address to the PI.

10.2.3 Searching the CASES metadatabase

As a first step towards implementing the ArcticNet/CCIN Polar Data Catalogue, a basic search facility has been programmed to allow the user to search listed keywords (Fig. 10.5). This search function provides a list of titles (or metadatabase records) which can be

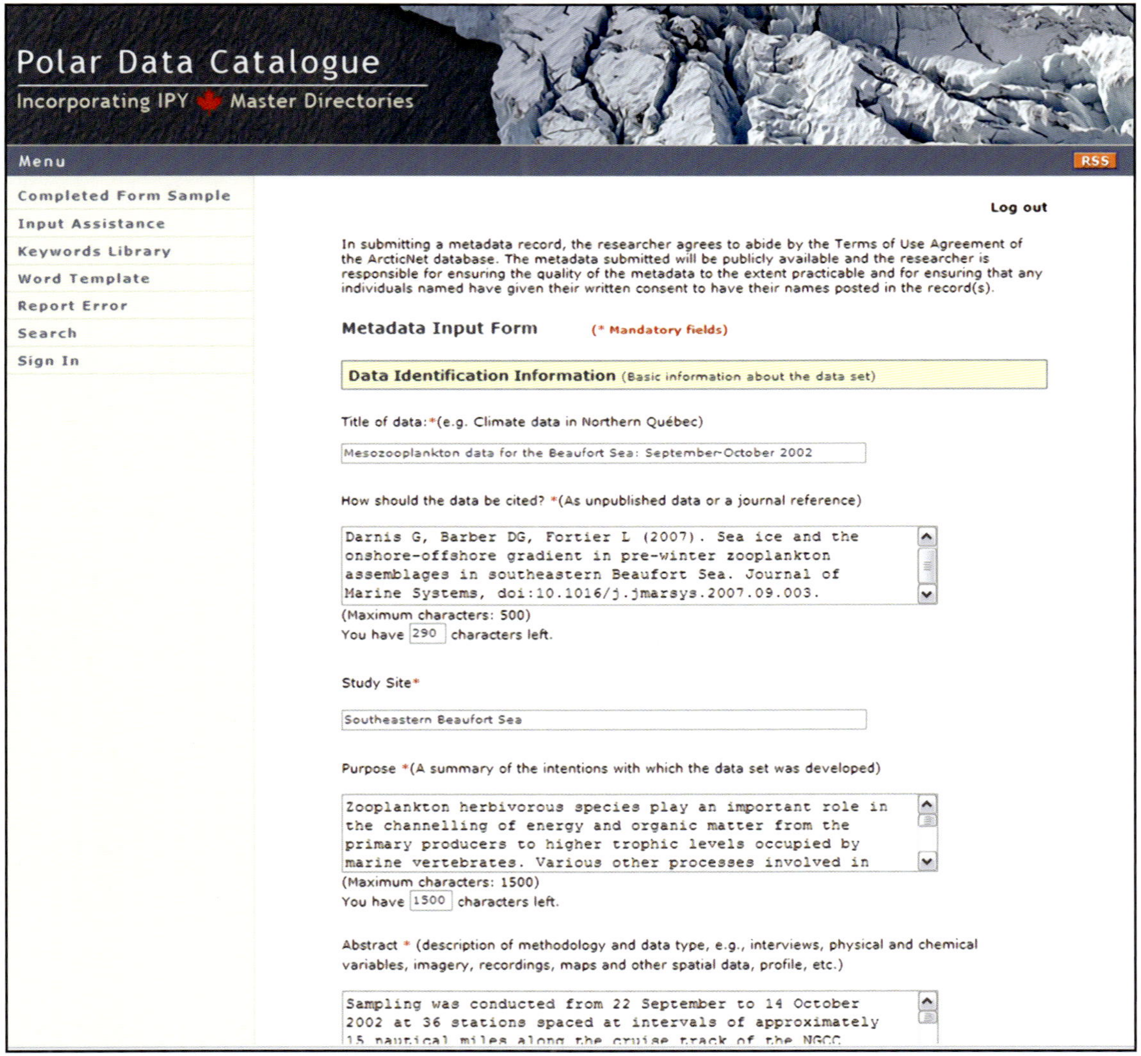

Figure 10.3

The ArcticNet metadata input form.

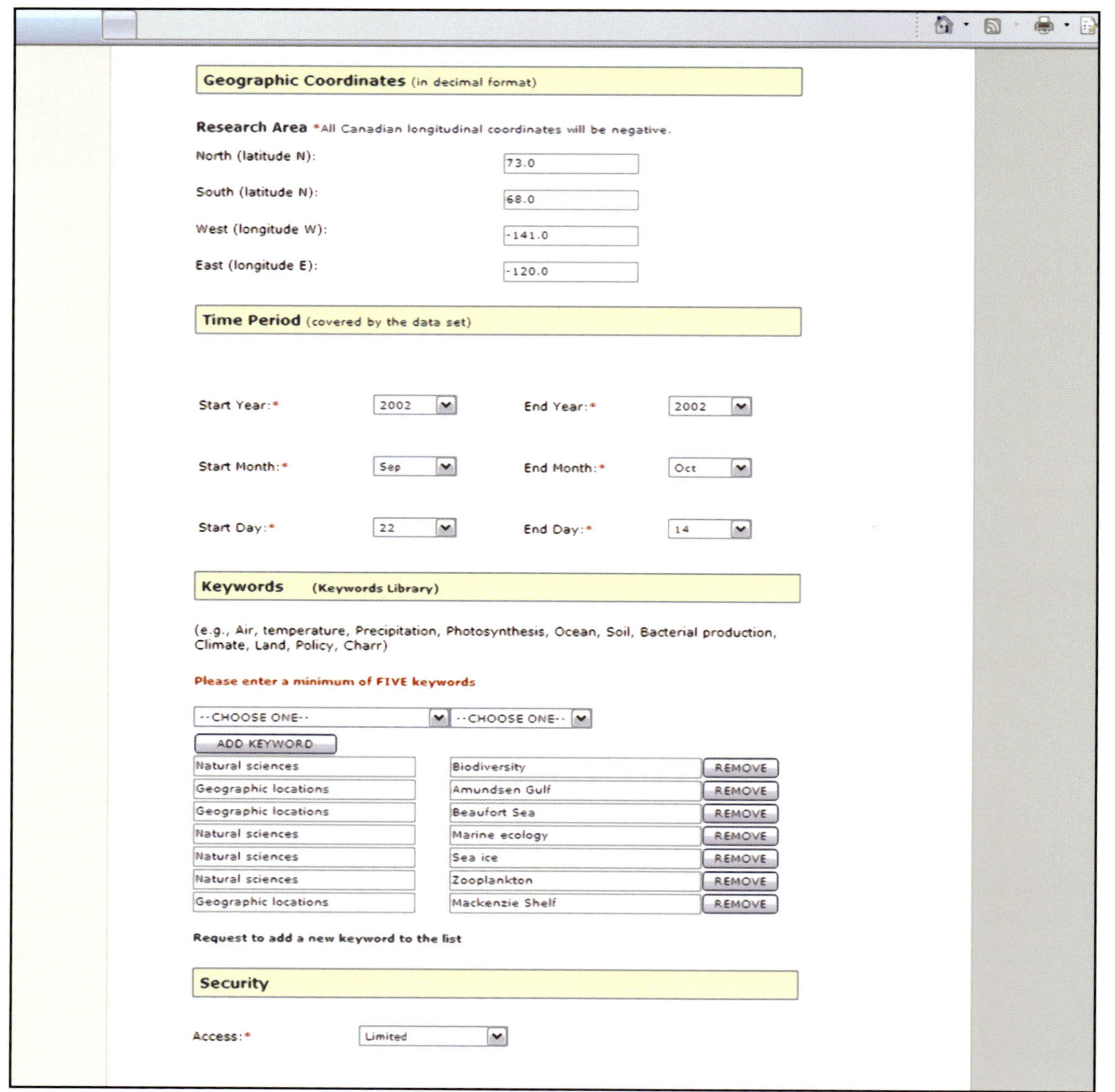

Figure 10. 4

The 'keywords' section of the metadata entry form.

accessed by all users, but only edited by the PI or his/her delegate by means of appropriate user name and password. More sophisticated searching facilities are currently under development by ArcticNet/CCIN. These include a map-based system produced by Noetix-CCIN-ArcticNet in conjunction with GeoConnections (Department of Natural Resources Canada).

10.2.4 Permanent archiving

Priniciple Investigators in the CASES Network are responsible for producing their datasets in a coherent format which is understandable to future data users. Each dataset will include much more detailed metadata than in the Polar Data Catalogue; for example, exact latitudes and longitudes, calibration data, validation information, cautionary notes, and the units and interpretation of each field of data. Accuracy and quality of the data are also the responsibility of each investigator, as well as its transfer to a publicly accessible data repository (such as CCIN and ISDM) within three years of the end of the program (i.e., by 18 June 2010). In keeping with IPY policy, the PI is responsible for ensuring that the initial metadata is catalogued soon as possible in the Polar Data Catalogue, and that this information becomes immediately open to public access.

10.3 Future Developments

The ArcticNet/CCIN Polar Data Catalogue and surrounding data management initiatives have generated considerable interest among researchers, northern communities and government agencies: demonstrated progress, for example, allowed DFO to commit resources toward the advancement of the metadatabase during the final year of CASES; ArcticNet has dedicated funding to a full time manager for the metadatabase; CCIN has worked closely with Noetix Research Inc.

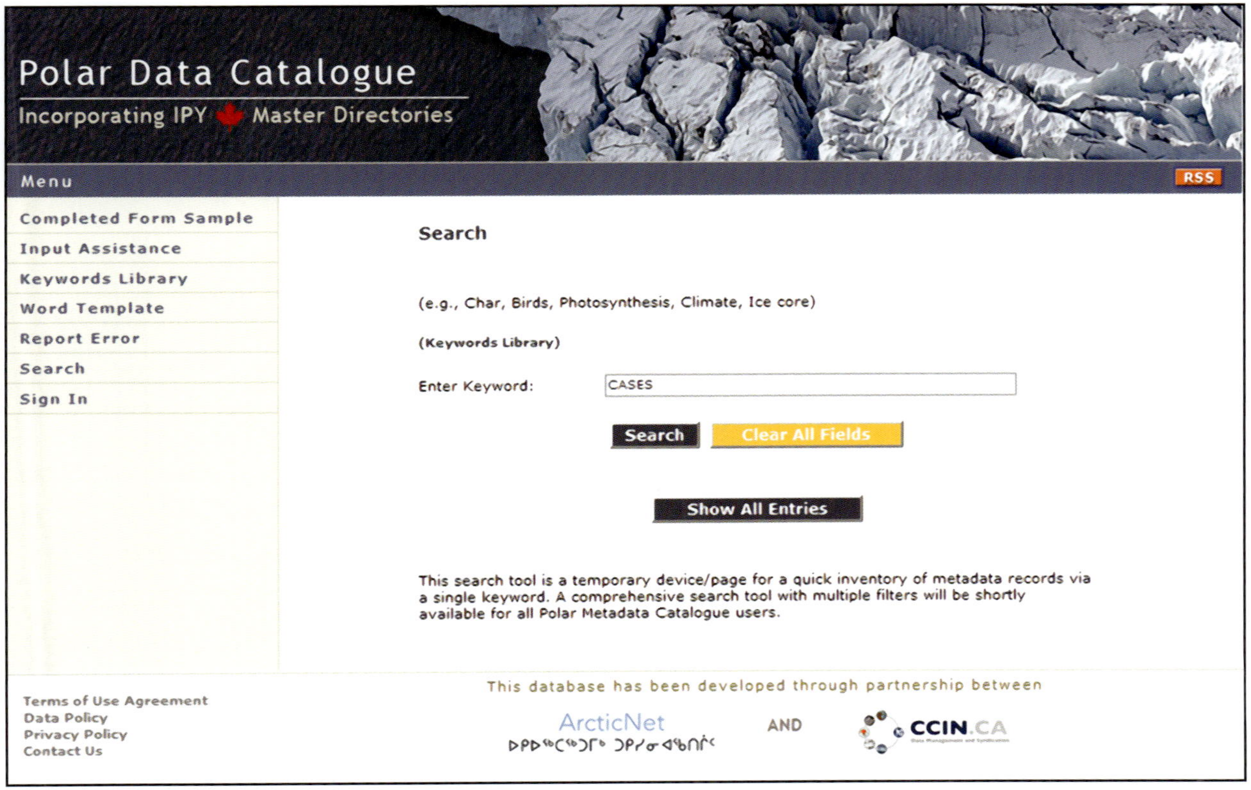

Figure 10.5

The ArcticNet metadatabase keyword search engine.

to obtain new support for geospatial tools that can be used to interrogate and display the metadata and data holdings; and close links have been established with northern communities (for example, via Inuit Tapiriit Kanatami and the northern science institutes) with the objective of developing data access tools of relevance to northern stakeholders. IPY Canada, through its NSERC and federal programs, adopted the metadata catalogue and management strategies for its projects, including the Canada Flaw Lead study (CFL) and the Microbial and Ecological Responses to the Global Environment (MERGE) study. Such interest, as well as any subsequent national/international collaborations in data sharing and preservation, will only increase the value of programs like CASES in the furthering of multidisciplinary scientific knowledge.

Acknowledgements

We would like to thank Ellsworth LeDrew, director of CCIN, for his enthusiasm, support and encouragement toward all CASES/ArcticNet data management initiatives; Peter Yoon, Science Manager at CCIN, and his staff, for advice and systems development; Bob Keeley, ISDM-DFO, for his expert guidance and ongoing support; members of the CASES Data Management Committee; members of the ArcticNet Data Management Committee; and NSERC for funding support.

References

Behr, S., R. Coen, W.K. Warnick, H. Wiggins, and A. York, 2007. IPY history reflects progress in science and society. Witness the Arctic. 12:1-4.

Buneman, P., K. Sanjeev, K. Tajima, and W.-C. Tan, 2004. Archiving Scientific Data. ACM Trans. Database Syst. 29 (1): 2–42.

Strong, D.F., and P.B. Leach, 2005. National Consultation on Access to Scientific Research Data. Final Report, January 31, 2005. Canada Institute for Scientific and Technical Information. National Research Council Canada, Ottawa, Canada.

Appendix 10.1

CASES data collected in the Mackenzie Shelf and Beaufort Sea (2002-2004)

DATA COLLECTED	Location
SUBGROUP-1 Atmospheric and sea ice forcing of coastal circulation	
CTD Water column pressure (Gratton et al.)	INRS
CTD Water column transmissivity (Gratton et al.)	INRS
CTD Water column salinity (Gratton et al.)	INRS
CTD Water column fluorescence (Gratton et al.)	INRS
CTD Water column salinity (Gratton et al.)	INRS
CTD Water column density sigma (S,T,P) (Gratton et al.)	INRS
CTD Water column density sigma-t (S,T,0) (Gratton et al.)	INRS
CTD Water column sigma theta (S, theta, 0) (Gratton et al.)	INRS
CTD Water column specific volume anomaly (Gratton et al.)	INRS
CTD Water column Brunt- Vaisala frequency (Gratton et al.)	INRS
CTD Water column potential temperature (Gratton et al.)	INRS
CTD Water column dissolved oxygen concentration (Gratton et al.)	INRS
CTD Water column Freezing temperature (Gratton et al.)	INRS
CTD Water column dissolved oxygen concentration (Gratton et al.)	INRS
CTD Water column pH (Gratton et al.)	INRS
CTD Water column Nitrates (Gratton et al.)	INRS
CTD Water column PAR pressure (Gratton et al.)	INRS
CTD Water column PAR (Gratton et al.)	INRS
CTD Water column surface PAR (Gratton et al.)	INRS
MPV (Horizontal)	
Moored current data	UBC
ADCP (150 kHz)	
Thermo-salinometer	
SUBGROUP-2 Ice-atmosphere interactions & biological	
Relative humidity	U. Manitoba
Air temperature	U. Manitoba
Wind speed and direction	U. Manitoba
Solar radiation	U. Manitoba
Long wave radiation	U. Manitoba

Cloud height	U. Manitoba
Occurrence of precipitation	U. Manitoba
Atm. Pressure	U. Manitoba
Atm. Temperature	U. Manitoba
At. Relative humidity	U. Manitoba
Snow and ice optical properties	U. Manitoba
Snow and ice microwave properties	U. Manitoba
Satellite imagery	U. Manitoba
Aerial survey	U. Manitoba
Surface based radiometry	U. Manitoba
Ice pCO2, DIC, and At (Papakyriakou and Miller)	IOS, U.
Air-surface CO2 fluxes (Papakyriakou)	U. Manitoba
SUBGROUP-3 Light, nutrients, primary and export production in ice free waters	
Ice Water nutrients (NO3, NO2, P, Si, NH4,DON) (Gosselin, Michel, Poulin)	UQAR
Ice Chla (> and < 5 um) (Gosselin, Michel, Poulin)	UQAR
Ice Bacterial abundance (Gosselin, Michel, Poulin)	UQAR
Ice Flagellates abundance (Gosselin, Michel, Poulin)	UQAR
Ice Exoplymeric substances (Gosselin, Michel, Poulin)	UQAR
Ice Taxonomy (Gosselin, Michel, Poulin)	UQAR
Ice POC-PON (Gosselin, Michel, Poulin)	UQAR
Ice PIC (Gosselin, Michel, Poulin)	UQAR
Ice BioSi (Gosselin, Michel, Poulin)	UQAR
Ice TOC-TN (Gosselin, Michel, Poulin)	UQAR
Ice DOC-DN (Gosselin, Michel, Poulin)	UQAR
Ice Nutrients (Gosselin, Michel, Poulin)	UQAR
Ice Salinity (Gosselin, Michel, Poulin)	UQAR
Ice PH (Gosselin, Michel, Poulin)	UQAR
Ice Cytometry (Gosselin, Michel, Poulin)	UQAR
Ice PAM (Gosselin, Michel, Poulin)	UQAR
Ice Virus (Gosselin, Michel, Poulin)	UQAR
Ice PI curves PP (Gosselin, Michel, Poulin)	UQAR
Ice a* (Gosselin, Michel, Poulin)	UQAR
Ice HPLC (Gosselin, Michel, Poulin)	UQAR

Ice CHN (Gosselin, Michel, Poulin)	UQAR
Trap Chla (< >5um) (Michel, Gosselin)	UQAR
Trap POC-PON (Michel, Gosselin)	DFO (FWI)
Trap PIC (Michel, Gosselin)	DFO (FWI)
Trap BioSi (Michel, Gosselin)	DFO (FWI)
Trap EPS (Michel, Gosselin)	DFO (FWI)
Trap Cell taxonomy (Michel, Gosselin)	DFO (FWI)
Trap Fecal pellet (Michel, Gosselin)	DFO (FWI)
Trap PAM & cytometry (Michel, Gosselin)	DFO (FWI)
Trap Enzymes (Michel, Gosselin)	DFO (FWI)
Trap Dry weight (Michel, Gosselin)	DFO (FWI)
Trap Total matter (Michel, Gosselin)	DFO (FWI)
Trap 210Pb (Japan)	Japan data
Trap PI curves (Japan)	Japan data
Trap a* (Japan)	Japan data
Trap HPLC (Japan)	Japan data
Boxcore Chla (Nozais)	UQAR
Boxcore Granulometry (Nozais)	UQAR
Boxcore Taxonomy (Nozais)	UQAR
Boxcore Meiofauna counts (Nozais)	UQAR
Boxcore PI curves (Nozais)	UQAR
Water column Chla (>< 5um and tot) (Demers)	UQAR
Water column HPLC(Demers)	UQAR
Water column BioSI (Demers)	UQAR
Water column cell Taxonomy (Demers)	UQAR
Water column cytometry phytoplankton (Demers)	UQAR
Water column cytometry Bacteria (Demers)	UQAR
Water column PAM (Demers)	UQAR
Water column PPP (Demers)	UQAR
Water column PI Curves (Demers)	UQAR
Water column nutrients (NO3,NO2,NH4,PO4(-3), Si(OH)4 (Price, Tremblay)	UQAM, Laval
Water column Urea (Price, Tremblay)	UQAM, Laval
Water column DON (Price, Tremblay)	UQAM, Laval
Water column NO3 uptake (Price, Tremblay)	UQAM, Laval

Water column NH4 uptake (Price, Tremblay)	UQAM, Laval
SUBGROUP-5 Pelagic food web: structure function and contaminants	
Zooplankton abundance (Fortier et al.)	U. Laval
Copepod egg production rate (Fortier et al.)	U. Laval
Larval, juvenile arctic cod abundance (Fortier et al.)	U. Laval
Arctic cod hatching date and age (Fortier et al.)	U. Laval
Arctic cod larvae stomach content (Fortier et al.)	U. Laval
Amphipods abundance, length, stage, wet weight, dry weight (Fortier et al.)	U. Laval
Ek60 Echosounder Arctic cod abundances (Simard, Fortier)	UQAR, Laval
Arctic cod length, weight , age, sex, stomach content (Gagné)	DFO (IML)
Bottom nepheloid layer SPOM POC (Deibel et al.)	Memorial U.
Bottom nepheloid layer SPOM PON (Deibel et al.)	Memorial U.
Bottom nepheloid layer SPOM particulate phosphorus (Deibel et al.)	Memorial U.
Bottom nepheloid layer SPOM C:N:P ratio (Deibel et al.)	Memorial U.
Bottom nepheloid layer SPOM del 13-C (Deibel et al.)	Memorial U.
Bottom nepheloid layer SPOM del 15-N (Deibel et al.)	Memorial U.
Bottom nepheloid layer SPOM fatty acid profiles (Deibel et al.)	Memorial U.
Benthic boundary layer zooplankton carbon content (Deibel et al.)	Memorial U.
Benthic boundary layer zooplankton nitrogen content (Deibel et al.)	Memorial U.
Benthic boundary layer zooplankton phosphorus content (Deibel et al.)	Memorial U.
Benthic boundary layer zooplankton del 13-C (Deibel et al.)	Memorial U.
Benthic boundary layer zooplankton del 15-N (Deibel et al.)	Memorial U.
Benthic boundary layer zooplankton fatty acid profiles (Deibel et al.)	Memorial U.
Benthic boundary layer zooplankton lipid classes (Deibel et al.)	Memorial U.
Benthic boundary layer zooplankton species composition (Deibel et al.)	Memorial U.
Benthic boundary layer zooplankton abundance (Deibel et al.)	Memorial U.
Autonomous video plankton recorder (Seascan) raw files (Deibel et al.)	Memorial U.
Autonomous video plankton recorder (Seascan) extracted images (Deibel et al.)	Memorial U.
In situ plankton video camera (Tiseliu) images of zooplankton and Arctic cod (Deibel	Memorial U.
Pressure, temperature and salinity from in situ plankton video camera casts (Deibel	Memorial U.
Water column zooplankton carbon content (Deibel et al.)	Memorial U.
Water column zooplankton nitrogen content (Deibel et al.)	Memorial U.
Water column zooplankton lipid classes (Deibel et al.)	Memorial U.
Water column zooplankton fatty acids (Deibel et al.)	Memorial U.

Box core samples (Scott et al.)	Dalhousie, GSC
Piston core samples and C-14 (Scott et al.)	Dalhousie, DFO
Gravity (cores) (Scott et al.)	Dalhousie, GSC
Sediment foraminferal data (Scott et al.)	Dalhousie, GSC
Dinoflagellate and pollen data (Rochon et al.)	UQAR
Water column Urea uptake (Price, Tremblay)	UQAM, Laval
Water column NAP absorp. (Larouche)	UQAR
Water column phyto absorp. (Larouche)	UQAR
Water column CDOM absorp. (Larouche)	UQAR
Water column AC-9, TSM, Minerals (Larouche)	UQAR
Water column HPLC (Larouche)	UQAR
Water column CHN(Larouche)	UQAR
Water column PAM (Japan)	Japan data
Water column PI curves (Japan)	Japan data
Water column PP (Japan)	Japan data
Water column a* (Japan)	Japan data
Water column HPLC (Japan)	Japan data
Water column 210Pb (Japan)	Japan data
Water column CHN (Japan)	Japan data
Water column delta 13C, 15 (Japan)N	Japan data
SUBGROUP-4 Microbial communities and heterotrophy	
Bacterial abundance (bact/ml, Deming et al.)	USA data
Dissolved and particulate EPS in melted ice core horizons (Deming et al.)	USA data
DNA Filters for Archaeal and bacterial diversity (Deming et al.)	USA data
FISH Filters for Archaeal and bacterial diversity (Deming et al.)	USA data
Extracellular enzyme kinetics on ice core horizons (Deming et al.)	USA data
Smoking Hills mineral samples for DNA analysis leg 5 (Deming et al.)	USA data
Bacterial production (thymidine and leucine rate measurements)	U. Laval
Virus variables	U. Laval
Picocyanobacteria	U. Laval
Picoeukaryotes	U. Laval
Bacterial heterotroph	U. Laval
Microzooplankton grazing (Vaqué, Spain)	U. Laval
CDOM	U. Laval

HPLC pigment analysis (Vincent et al.)	U. Laval
Baterial substrate analysis (Aliò et al.)	Spain
ARDEX data-optics and biological	U. Laval
ARDEX data-chemical	Simon Fraser U.
ARDEX data-photochemical (USA data)	Naval res. lab.
SUBGROUP-6 Organic and inorganic fluxes	
Trap Mass flux (mg m-2 d-1) (Fortier et al.)	U. Laval
Trap POC and PON flux (Fortier et al.)	U. Laval
Trap Fecal POC flux (Fortier et al.)	U. Laval
Trap Protistal POC flux (Fortier et al.)	U. Laval
Trap Carbon DW (Fortier et al.)	U. Laval
Trap TPOC, POCnio,. POCterr, POC (Fortier et al.)	U. Laval
Trap Estimated PPConsumption (Fortier et al.)	U. Laval
234Th and 210Pb fluxes (Cochran et al.)	SUNY
SPM concentration (Vincent, Larouche and Mucci)	Laval, UQAR, McGill
Floc properties	
Sediment geochemical properties (Sundby et al.)	UQAR, McGill
Water column DOC and TOC (Miller)	McGill, IOS
Water column Oxygen-18 (Macdonald)	IOS
Atmospheric carbon monoxide mixing ratios (Xie et al.)	UQAR
Surface water CO concentrations (Xie et al.)	UQAR
Depth profiles of CO in the water column and sea ice (Xie et al.)	UQAR
Biological CO consumption rates in surface water and sea ice (Xie et al.)	UQAR
Apparent quantum yieds of CO photoproduction for melt sea ice and wate and water column (Xie et al.)	UQAR
SUBGROUP-7 Benthic processes and carbon cycling	
Water depth (Aitken et al.)	CMN
Benthic fauna biomass (wet weight) (Aitken et al.)	CMN
Identification guides for benthic fauna (Aitken et al.)	CMN
Sediment texture (Aitken et al.)	CMN
Sediment organic matter (Aitken et al.)	CMN
Biomass and distribution of benthic organisms (Aitken et al.)	CMN
Bottom camera images (Renaud et al.)	Ak-niva

Benthic fauna distribution, abundance (Aitken et al.)	CMN
Sediment particle size distribution (Renaud et al.)	Ak-niva
Oxygen depletion data (experiments) (Renaud et al.)	Ak-niva
SUBGROUP-8 Decadal-Millenial variability in run-off, sea ice and carbon fluxes	
Multibeam bottom survey data (Hughes-Clarke et al., Blasco et al., Scott et al.)	Dalhousie, GSC, UNB

The Schools on Board students and mentors in front of the CCGS Amundsen during CASES. People in the photo are:

Katie Gurnsey, David Babb, MB Amy Kroeker, Clint Surrey, Lucette Barber, Grant Ingram, Marionne Lugay, Chetwynd Secondary School, Sheldon Clark, Candace Clouthier, Jack Miller,, Derrick Seabrook, Cassie Omoerah, and Angela Wolki.

RIGHT: Photo by Alexandre Forest.

The Schools on Board Program

Schools on Board was created and piloted as an outreach program to CASES in 2004. The aim of the program was to promote Arctic sciences to high schools across Canada by providing spaces onboard the *CCGS Amundsen*, for students and teachers to join the CASES expedition. The group boarded the ship in the early evening of February 23rd, and immediately became integrated in the activities of the science teams.

The program included science lectures, lab activities and fieldwork with scientists from across Canada and around the world. The successes of the 2004 program resulted in its inclusion in the internationally recognized ArcticNet and IPY-CFL research project, and recognition in the scientific and education communities, as a leading scientific outreach program that is effectively promoting Arctic sciences, increasing awareness of environmental issues related to the Arctic and climate change, and providing new and exciting learning opportunities to this country's next generation of science enthusiasts.

Other publications available from Aboriginal Issues Press

Aboriginal Health, Identity and Resources

Pushing the Margins: Native and Northern Studies

Working with Aboriginal Elders/Working with Indigenous Elders

Inuit Annuraangit: Our Clothes

Coats of Eider

Native Voices in Research

Aboriginal Cultural Landscapes

Gambling and Problem Gambling in First Nations Communities

Seeing the World with Aboriginal Eyes, a Four Directional Perspective on Human and non-Human Values, Cultures and Relationships on Turtle Island

Climate Change: Linking Traditional and Scientific Knowledge

Residents' Perspectives on the Churchill River

Aboriginal Connections to Race, Environment and Traditions

Facets of the Sacred

Sacred Landscapes

Beluga Co-Management Issues in Umiujaq and Kuujjuarapik, Nunavik

For more information on the following six pre-1999 publications, please contact Ms. Elaine Maloney at the Canadian Circumpolar Institute (780) 492-4512:

Human Ecology: Issues in the North, Volumes I, II, and III

Issues in the North, Volumes I, II, and III

Call for Papers

Aboriginal Issues Press produces one volume of refereed papers each year and welcomes scholarly papers relating to Aboriginal issues from all fields of study, including: traditional knowledge, social, physical and natural sciences, law, education, architecture, management, medicine, nursing, social work, physical education, engineering, environment, agriculture, art, music, drama, continuing education, and others. Aboriginal and non-Aboriginal authors from a wide range of backgrounds and from all geographic locations are welcome, including: scientists, poets, educators, elders, chiefs, students, and government personnel. All papers (ranging from scientific papers to poetry) are reviewed by scholars working in related fields.

Papers must be submitted in hard copy and on disk (Word) following the APA Writer's Style Manual 5th Edition. Maximum paper length is 10 pages, double spaced, 2.5 cm margins, 12 pt font, Times New Roman. Any photographs, charts, or graphs must be provided in digital format and are included in the 10 page count. Sources are included directly in the text; provide the author and date for paraphrased information, for example (Flett 1937, Graham 2001) and the author, date and page number for direct quotations, for example (Boas 1964, 33). Include full references to all sources in a section titled References at the end of the paper; for economy of space footnotes or endnotes will not be included. Capitalize Aboriginal Peoples, Native Peoples and First Nations. Convert all English measurements to metric. Include a one-sentence biography, a 50 to 75 word abstract, and your return address with your submission. For further information please contact aboriginal-issues-press@umanitoba.ca

Books

Aboriginal Issues Press also publishes sole-authored books. Scholars interested in submitting a book manuscript for review must provide the following information to aboriginal-issues-press@umanitoba.ca

1. Table of contents;

2. Two sample chapters representative of the writing style;

3. A curriculum vitae of the author(s);

4. Samples of illustrations, photos, graphs, etc. for the publication;

5. A prospectus which includes the purpose, objectives, and how the author(s) acquired the knowledge shared in the manuscript;

6. A description of the audience and market; and

7. An explanation of why you chose Aboriginal Issues Press as a possible publisher.